AF356531

Non-Newtonian Microfluidics

Non-Newtonian Microfluidics

Editors

Lanju Mei
Shizhi Qian

MDPI • Basel • Beijing • Wuhan • Barcelona • Belgrade • Manchester • Tokyo • Cluj • Tianjin

Editors
Lanju Mei
Department of Engineering
and Aviation Sciences
University of Maryland
Eastern Shore
Princess Anne
United States

Shizhi Qian
Department of Mechanical
and Aerospace Engineering
Old Dominion University
Norfolk
United States

Editorial Office
MDPI
St. Alban-Anlage 66
4052 Basel, Switzerland

This is a reprint of articles from the Special Issue published online in the open access journal *Micromachines* (ISSN 2072-666X) (available at: www.mdpi.com/journal/micromachines/special_issues/nonNewtonian_microfluidics).

For citation purposes, cite each article independently as indicated on the article page online and as indicated below:

LastName, A.A.; LastName, B.B.; LastName, C.C. Article Title. *Journal Name* **Year**, *Volume Number*, Page Range.

ISBN 978-3-0365-4642-1 (Hbk)
ISBN 978-3-0365-4641-4 (PDF)

Contents

micromachines

MDPI

Editorial

Editorial for the Special Issue on Micromachines for Non-Newtonian Microfluidics

Lanju Mei [1,*] **and Shizhi Qian** [2]

1   Department of Engineering and Aviation Sciences, University of Maryland Eastern Shore, Princess Anne, MD 21853, USA
2   Department of Mechanical and Aerospace Engineering, Old Dominion University, Norfolk, VA 23529, USA; sqian@odu.edu
*   Correspondence: lmei@umes.edu

Citation: Mei, L.; Qian, S. Editorial for the Special Issue on Micromachines for Non-Newtonian Microfluidics. *Micromachines* **2022**, *13*, 906. https://doi.org/10.3390/mi13060906

Received: 3 June 2022
Accepted: 6 June 2022
Published: 8 June 2022

Publisher's Note: MDPI stays neutral with regard to jurisdictional claims in published maps and institutional affiliations.

Microfluidics has seen a remarkable growth over the past few decades, with its extensive applications in engineering, medicine, biology, chemistry, etc. Many of these real applications of microfluidics involve the handling of complex fluids such as whole blood, protein solutions, and polymeric solutions which exhibit non-Newtonian characteristics—specifically, viscoelasticity. The elasticity of the non-Newtonian fluids induces intriguing phenomena such as elastic instability and turbulence even at extremely low Reynolds numbers. This is the consequence of the nonlinear nature of the rheological constitutive equations. The nonlinear characteristics of non-Newtonian fluids can dramatically change the flow dynamics, and is useful to enhance mixing at the microscale. Electrokinetics in the context of non-Newtonian fluids are also of significant importance, with their potential applications in micromixing enhancement and bio-particles manipulation and separation. This Special Issue comprises 14 original contributions.

Regarding the instability of non-Newtonian fluids, Zhang et al. [1] demonstrate the viscoelastic fluid instability in an asymmetric nozzle–square microchannel. It is found that the critical Weissenberg number is different for the forward-directed flow and the backward-directed flow in the same microchannel. Ji et al. [2] numerically study the electroosmotic flow (EOF) of Oldroyd-B viscoelastic fluid through a 10:1 constriction microfluidic channel. Compared to EOF of Newtonian fluid, EOF of viscoelastic fluid becomes unstable when the PAA concentration and electric field exceed some critical values.

Joo's group numerically and experimentally conducts novel studies on viscoelastic droplet motion. Ren et al. [3] numerically study migration of Oldroyd-B viscoelastic droplets on rigid surfaces with wettability gradients. The effects of parameters including droplet size, relaxation time, solvent viscosity, and polymer viscosity of the liquid on the migration speed and distance are investigated. Bai et al. [4] also investigate Janus droplet formation in a double Y-type microfluidic device filled with a power law shear-thinning fluid. Compared with Newtonian fluid, the Janus droplet is more readily generated in shear-thinning fluid. In the experimental study on the dynamics of liquid droplets under an electric field, Wei et al. [5] observe that viscoelastic droplets differ from Newtonian droplets, and further discuss the effects of viscoelasticity, the wettability, and the droplet size.

In terms of heat transfer for non-Newtonian fluids, Deng et al. [6] present transient hydrodynamical features and corresponding heat transfer characteristics in two-layer electroosmotic flow of power-law nanofluids in a microchannel. The results show that increase in nanoparticle volume fraction promotes heat transfer performance, and shear thickening feature of conducting nanofluid tends to suppress the effects of viscous dissipation and electrokinetic width on heat transfer. Hou et al. [7] numerically study heat and mass transport in a pseudo-plastic fluid past over a stretched porous surface in the presence of the Soret and Dufour effects. The effects of the tri-hybrid nanoparticles, the Dufour number, the heat generation parameter, the Eckert number, the buoyancy force parameters, and the porosity parameters on fluid temperature are reported. Hou et al. [8] also investigate the thermal transport of hydro-magnetized Carreau–Yasuda liquid passing over a permeable stretched surface, considering several important effects, including Joule heating, viscous dissipation, and heat generation/absorption. Sohail et al. [9] further report heat and mass transfer in three-dimensional second grade non-Newtonian fluid in the presence of a variable magnetic field. The maximum heat energy and improvement in motion of fluid particles are achieved using higher values of second grade fluid number. Parveen et al. [10] conduct heat transfer and entropy generation analysis in MHD axisymmetric flow of hybrid nanofluid. Detailed rheological impacts of involved parameters on flow variables and entropy generation number are demonstrated. Zhou et al. [11] provide a mathematical model for non-Newtonian Maxwell nanofluid flow with heat transmission over a porous spinning disc. The effects of some mathematical parameters on velocity, energy, concentration, and magnetic power are discussed. Rojas-Altamirano et al. [12] calculate the effective thermal conductivity of human skin using the Fractal Monte Carlo Method. In the study, tissue is described as a porous medium, and blood is considered a Newtonian and non-Newtonian fluid for comparative and analytical purposes.

Particle focusing in non-Newtonian fluids is also of great interest. Feng et al. [13] investigate particle focusing and separation in viscoelastic flow in a spiral channel. They explain the particle focusing position by the effects of inertial flow, viscoelastic flow, and Dean flow, and show that particle separation resolution can be improved in viscoelastic flow.

For the mixing in non-Newtonian fluids, Mei et al. [14] investigate electroosmotic micromixing of power law non-Newtonian fluid in a microchannel with wall-mounted obstacles and surface potential heterogeneity. Significant improvement in the mixing efficiency is achieved by increasing the obstacle surface zeta potential, the flow behavior index, the obstacle height, and the EDL thickness.

We would like to thank all the authors for submitting their papers to this Special Issue. We also thank all the reviewers for dedicating their time and helping to improve the quality of the submitted papers.

Conflicts of Interest: The authors declare no conflict of interest.

References

1. Zhang, W.; Wang, Z.; Zhang, M.; Lin, J.; Chen, W.; Hu, Y.; Li, S. Flow Direction-Dependent Elastic Instability in a Symmetry-Breaking Microchannel. *Micromachines* **2021**, *12*, 1139. [CrossRef] [PubMed]
2. Ji, J.; Qian, S.; Liu, Z. Electroosmotic flow of viscoelastic fluid through a constriction microchannel. *Micromachines* **2021**, *12*, 417. [CrossRef]
3. Ren, Y.J.; Joo, S.W. The Effects of Viscoelasticity on Droplet Migration on Surfaces with Wettability Gradients. *Micromachines* **2022**, *13*, 729. [CrossRef] [PubMed]
4. Bai, F.; Zhang, H.; Li, X.; Li, F.; Joo, S.W. Generation and Dynamics of Janus Droplets in Shear-Thinning Fluid Flow in a Double Y-Type Microchannel. *Micromachines* **2021**, *12*, 149. [CrossRef]
5. Wei, B.S.; Joo, S.W. The Effect of Surface Wettability on Viscoelastic Droplet Dynamics under Electric Fields. *Micromachines* **2022**, *13*, 580. [CrossRef]
6. Deng, S.; Xiao, T. Transient Two-Layer Electroosmotic Flow and Heat Transfer of Power-Law Nanofluids in a Microchannel. *Micromachines* **2022**, *13*, 405. [CrossRef]
7. Hou, E.; Wang, F.; Nazir, U.; Sohail, M.; Jabbar, N.; Thounthong, P. Dynamics of tri-hybrid nanoparticles in the rheology of pseudo-plastic liquid with dufour and soret effects. *Micromachines* **2022**, *13*, 201. [CrossRef] [PubMed]
8. Hou, E.; Wang, F.; El-Zahar, E.R.; Nazir, U.; Sohail, M. Computational Assessment of Thermal and Solute Mechanisms in Carreau–Yasuda Hybrid Nanoparticles Involving Soret and Dufour Effects over Porous Surface. *Micromachines* **2021**, *12*, 1302. [CrossRef]
9. Sohail, M.; Nazir, U.; Bazighifan, O.; El-Nabulsi, R.A.; Selim, M.M.; Alrabaiah, H.; Thounthong, P. Significant involvement of double diffusion theories on viscoelastic fluid comprising variable thermophysical properties. *Micromachines* **2021**, *12*, 951. [CrossRef]
10. Parveen, N.; Awais, M.; Awan, S.E.; Khan, W.U.; He, Y.; Malik, M.Y. Entropy Generation Analysis and Radiated Heat Transfer in MHD (Al2O3-Cu/Water) Hybrid Nanofluid Flow. *Micromachines* **2021**, *12*, 887. [CrossRef] [PubMed]
11. Zhou, S.-S.; Bilal, M.; Khan, M.A.; Muhammad, T. Numerical analysis of thermal radiative maxwell nanofluid flow over-stretching porous rotating disk. *Micromachines* **2021**, *12*, 540. [CrossRef]

12. Rojas-Altamirano, G.; Vargas, R.O.; Escandón, J.P.; Mil-Martínez, R.; Rojas-Montero, A. Calculation of Effective Thermal Conductivity for Human Skin Using the Fractal Monte Carlo Method. *Micromachines* **2022**, *13*, 424. [CrossRef] [PubMed]
13. Feng, H.; Jafek, A.R.; Wang, B.; Brady, H.; Magda, J.J.; Gale, B.K. Viscoelastic particle focusing and separation in a spiral channel. *Micromachines* **2022**, *13*, 361. [CrossRef] [PubMed]
14. Mei, L.; Cui, D.; Shen, J.; Dutta, D.; Brown, W.; Zhang, L.; Dabipi, I.K. Electroosmotic mixing of non-newtonian fluid in a microchannel with obstacles and zeta potential heterogeneity. *Micromachines* **2021**, *12*, 431. [CrossRef] [PubMed]

Article

Flow Direction-Dependent Elastic Instability in a Symmetry-Breaking Microchannel

Wu Zhang [1,*], Zihuang Wang [1], Meng Zhang [2], Jiahan Lin [1], Weiqian Chen [1], Yuhong Hu [1] and Shuzhou Li [3]

[1] College of Physical and Material Engineering, Guangzhou University, Guangzhou 510006, China; 1919400068@e.gzhu.edu.cn (Z.W.); 1619200027@e.gzhu.edu.cn (J.L.); wqchen@e.gzhu.edu.cn (W.C.); 1919400035@e.gzhu.edu.cn (Y.H.)

[2] Precision Medicine Institute, the First Affiliated Hospital of Sun Yat-Sen University, Sun Yat-Sen University, Guangzhou 510080, China; meng.zhang_china@outlook.com

[3] School of Materials Science and Engineering, Nanyang Technological University, Singapore 639798, Singapore; lisz@ntu.edu.sg

* Correspondence: zhangwu@gzhu.edu.cn

Abstract: This paper reports flow direction-dependent elastic instability in a symmetry-breaking microchannel. The microchannel consisted of a square chamber and a nozzle structure. A viscoelastic polyacrylamide solution was used for the instability demonstration. The instability was realized as the viscoelastic flow became asymmetric and unsteady in the microchannel when the flow exceeded a critical Weissenberg number. The critical Weissenberg number was found to be different for the forward-directed flow and the backward-directed flow in the microchannel.

Keywords: viscoelastic fluid; elastic instability; microfluid; direction-dependent

Citation: Zhang, W.; Wang, Z.; Zhang, M.; Lin, J.; Chen, W.; Hu, Y.; Li, S. Flow Direction-Dependent Elastic Instability in a Symmetry-Breaking Microchannel. *Micromachines* **2021**, *12*, 1139. https://doi.org/10.3390/mi12101139

Academic Editors: Lanju Mei and Shizhi Qian

Received: 19 August 2021
Accepted: 15 September 2021
Published: 23 September 2021

Publisher's Note: MDPI stays neutral with regard to jurisdictional claims in published maps and institutional affiliations.

1. Introduction

In Newtonian fluid, flow complexity originates mostly from the nontrivial inertial effect, which is mainly induced in the macroscale condition [1]. In microfluidics, the Reynolds (Re) number of Newtonian fluid is usually a very small value, and the inertial effect is negligible. Therefore, only creeping flow is induced in the microscale Newtonian fluid [2]. For viscoelastic fluid, on the other hand, complex flow behavior can be stimulated from the polymer molecules or surfactant dispersed in the fluid [3–5]. In certain designed microchannel structures, the polymer molecules or surfactant can be compressed or stretched to induce significant normal stress in the fluid. As a result, various complex flows such as turbulent flow, Couette flow, or swirling flow can be obtained in viscoelastic fluid even at the micrometer scale [6–9]. Analytically, complex flow can be understood from the Navier–Stokes equation, whereby the nonlinearity of Newtonian fluid depends on the advective term in the equation, while viscoelastic fluid nonlinearity is also contributed by the rheological effects from the normal stress in the fluid [10], which can be applied for mixing [11,12] or sorting [13] in microfluidics.

In particular, the elastic instability of viscoelastic fluid in specially designed microfluidic channels has been intensively studied [14,15]. One classical geometry of the microfluidic channel is a cross-slot, which consists of two perpendicular intersecting straight channels [16–18]. In the cross-slot, a stagnation point can be found at the cross center when fluid is injected into both ends of one straight channel and flows out from both ends of the other straight channel. Due to the structure symmetry, the flow velocity is 0 at the stagnation point while the velocity gradient is finite. This introduces an extensional flow in the microchannel, which can stretch or compress the polymer molecule in the fluid. The normal stress is, thus, induced and leads to elastic instability of the flow when the Weissenberg (Wi) number of the fluid exceeds a critical value. Below the critical Wi number, the viscoelastic flow is steady and symmetric, behaving as a Newtonian fluid.

The flow pattern becomes asymmetric as the flow rate and *Wi* number increase beyond a critical vale. As the *Wi* number further increases, the flow pattern fluctuates and becomes time-dependent. Numerically, elastic instability was studied by investigating the flow of an upper-convicted Maxwell (UCM) fluid in a cross-slot channel, and different rheological conditions were analyzed for different instability types [19]. The Oldroyd-B model and a simplified Phan–Thien–Tanner model were also established for the analysis [20–22]. In addition, elastic instability has been demonstrated experimentally using different types of viscoelastic fluids such as polymer solution or micellar solution [23,24]. Instability was also studied by analyzing the flow under different aspect ratios of the cross-slot geometry [19]. The cross-slot was also used as a flow-focusing device to induce purely elastic instability [25]. Other geometries, such as a T-structured channel, were proposed, and a direct transition from symmetric flow to time-dependent flow was observed in the channel [26]. Contraction–expansion microchannels were also proposed, in which the extension and relaxation of polymer molecules were observed to study the elastic instability [27].

Previous work has mainly focused on the effect of microstructure geometry and fluid properties on the flow elastic instability. In these studies, the instability was fully characterized by the Weissenberg number. Here, we experimentally demonstrate a flow direction-sensitive elastic instability even at the same Weissenberg number. This can be used to stabilize or induce elastic instability by simply altering the flow direction without changing the flow rate. Unlike previous designs, the microchannel was a symmetry-breaking geometry, consisting of a square chamber and a nozzle structure. Therefore, the flow path in one direction was not the same as that in the opposite direction. This microchannel with carefully designed asymmetry has been intensively reported to realize a rectifying property, which induces different flow pressure depending on the flow direction under the same flow rate [28,29]. The microfluidic diode and memory were then developed on the basis of this rectifying function [30,31]. Here, we observed flow direction-dependent elastic instability, with symmetric steady flow evolving into an asymmetric unsteady flow. In addition, turbulence was also observed in certain *Wi* number conditions in the unsteady flow.

2. Methods

The microchannel applied for obtaining non-Newtonian fluid instability was fabricated using a standard soft lithography process. A SU8-3000 photoresist (MicroChem) layer with a thickness of 50 μm was first spin-coated on a Si wafer, followed by an optical lithography process to develop the microchannel pattern. Liquid PDMS with a base and curing agent mixing ratio of 10:1 was then spin-coated on the patterned SU8 layer and heated for 4 h at 65 °C, yielding a solid layer of 50 μm thickness. The solid PDMS layer was peeled off from the Si wafer and bonded with a flat PDMS layer. The fabricated microfluidic channel is illustrated in Figure 1. The channel was composed of a square chamber and a symmetric nozzle structure. The square chamber in the microfluidic channel induces expansion and contraction of the flow, which introduces normal stress into the viscoelastic fluid. The nozzle structure is located on one side of the square chamber, which increases the localized flow resistance and confers the whole microfluidic channel with a symmetric-breaking structure. As shown in Figure 1, the width and height of the microchannel were $w_0 = 100$ μm and $h = 50$ μm, respectively. The width and length of the square chamber were $w_1 = 300$ μm and $w_2 = 300$ μm, and the nozzle had a width of 50 μm. The channel between the left-side inlet and the nozzle structure, and the channel between the right-side inlet and the square chamber were both designed with a length of 4 mm, which was long enough for the full development of the flow. To clarify the two flow directions in the microchannel, the forward direction was defined as the direction with flow passing through the nozzle first, while the backward direction was defined as the direction with flow passing through the square chamber first, as indicated by the blue arrow and red arrow in Figure 1, respectively.

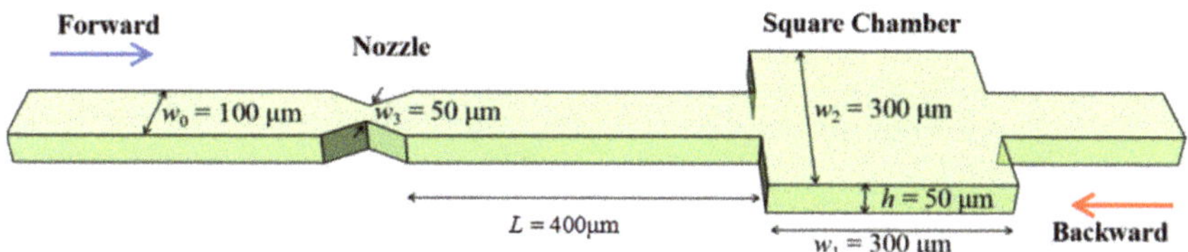

Figure 1. Schematic of microchannel consisting of a square chamber and nozzle structure.

The streamline of the flow was recorded under a microscope by adding polystyrene tracking particles with a diameter of 1 μm in a concentration of 0.6 μL/mL. The flow pattern was studied at different flow rates and, therefore, different *Wi* numbers for the viscoelastic fluid. The Weissenberg number, defined as $Wi = \lambda U/L$, describes the elasticity of the flow and qualifies the nonlinearity of the fluid, where L is the characteristic length of the channel, U is the flow rate in the channel, and λ is the relaxation time of the fluid, referring to the characteristic stretch–relax time of the polymer, which was measured as ~0.1 s for the PAM of 500 ppm. The Reynolds number is used to characterize the relationship between the inertial and viscous forces in the Newtonian fluid, expressed as $Re = \rho U L/\eta$, where ρ is the density of the fluid.

3. Results and Discussion

We used polyacrylamide (PAM) with a molecular weight of 18 million for the viscoelastic fluid instability demonstration. PAM of 100, 200, 500, and 1000 ppm was measured using a rotational rheometer (Malvern, Discovery HR1) with cone-plate geometry. The cone had a diameter of 60 mm and angle of 2.006°. The complex shear modulus of the PAM solutions are shown in Figure 2. The storage modulus G′ was larger than the loss modulus G″ at the lower frequency band, confirming the elastic property of the fluid. The two curves of G′ and G″ crossed at 2, 4, 5, and 10 Hz for PAM of 100, 200, 500, and 1000 ppm. An increased crossing frequency indicates a larger elasticity of the solution with a higher ppm value. The viscosity of the fluid at different shear rates is shown in Figure 3a. The viscosity continuously decreased from ~10 Pa·s to ~0.01 Pa·s in the shear rate range from $0.1\ s^{-1}$ to $100\ s^{-1}$. For comparison, the viscosity of two typical Newtonian fluids, glycerol solution and DI water, was also measured, as shown in Figure 3b. The viscosity remained almost constant at 0.1595 and 0.0054 Pa·s for the 80% and 50% volume concentrations of glycerol, respectively, and at 0.0008 Pa·s for the DI water. For the subsequent fluid instability investigation, we chose PAM of 500 ppm with moderate viscosity.

The flow of the PAM solution in the nozzle–square microchannel was first investigated in low *Wi* number conditions. The flow rate was controlled with a syringe pump (NE-300 Just Infusion™). It was first set to $Q = 100\ \mu L/h$, corresponding to $Wi = 5.56$. Figure 4a,b present the streamline patterns of the forward flow and backward flow in the square chamber, respectively. The streamline patterns were both symmetric and steady in the low Weissenberg condition for the forward and backward flow, indicating that no elastic instability was induced. The flow patterns for both directions were almost the same; therefore, the nozzle structure has little impact on the flow in low *Wi* number conditions. For both directions, the streamline expanded to the square chamber first when entering the contraction–expansion structure, and then gradually concentrated to the center with a small streamline curvature. On the other hand, at the two corners of the chamber where the fluid flowed out, the tracking particles remained static and the fluid formed a stationary regime.

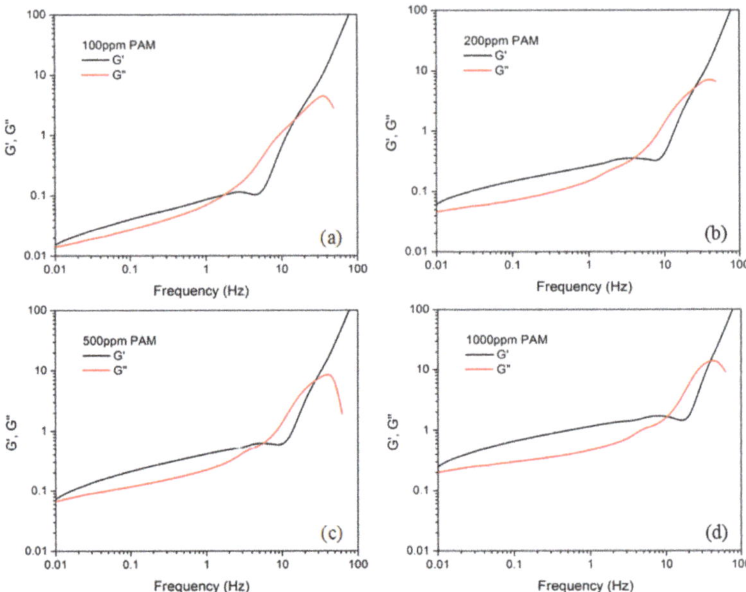

Figure 2. Complex shear modulus of the PAM solutions of (**a**) 100, (**b**) 200, (**c**) 500, and (**d**) 1000 ppm.

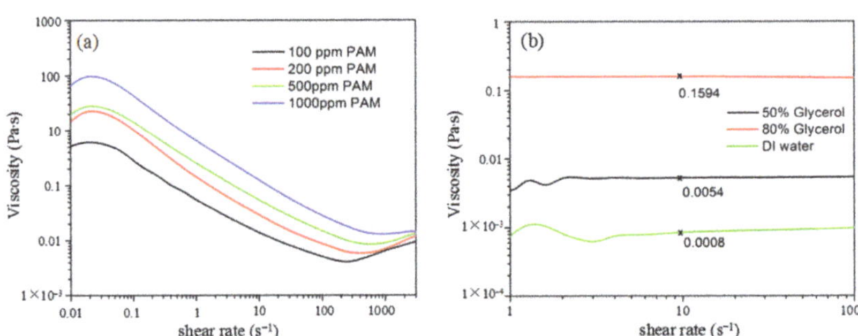

Figure 3. (**a**) Viscosity of non-Newtonian PAM solution and (**b**) Newtonian solutions.

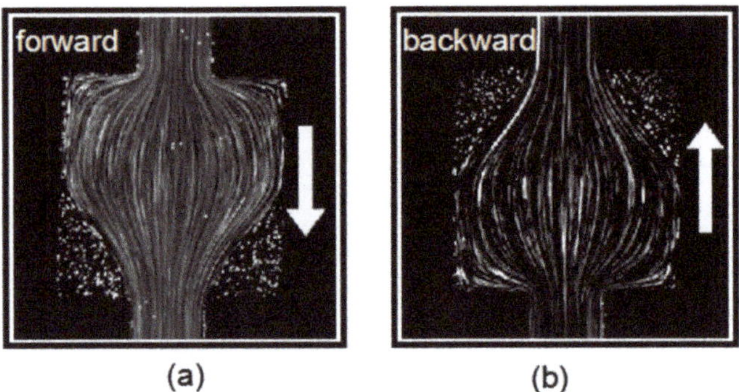

Figure 4. Viscoelastic flow pattern in the square chamber for flow along (**a**) the forward direction and (**b**) the backward direction at $Wi = 5.56$.

Then, we increased the flow rate of PAM solution to $Q = 300$ μL/h, and the *Wi* number was increased to 16.67. The patterns recorded at three different instants are shown in Figure 5a–c. Similarly, to Figure 4a, the streamline expanded to the square chamber first when entering the contraction–expansion structure, and then gradually concentrated to the center. However, the curvatures of the streamline on the left side and the right side of the flow were no longer always the same. In addition, the stationary triangle regime (highlighted by the red dashed line) increased in size compared to Figure 4a. It can be clearly seen that the forward flow pattern became asymmetric and time-dependent. As the inertial effect is negligible at this flow rate, the instability should be purely elastic, thus stemming from the normal stress in the viscoelastic fluid. On the other hand, the backward flow at the same flow rate and *Wi* number, as shown in Figure 5d, remained steady and symmetric, and the flow pattern was almost the same as that in Figure 4b. The significant difference between the forward and backward flow patterns indicates that the flow elastic instability was direction-dependent in the asymmetric microchannel structure. For forward flow, it became extensional at the nozzle structure, and the normal stress increased before the fluid entered the square chamber. As a result, it was more likely to induce elastic instability. On the other hand, for backward flow, the nozzle structure increased the flow resistance of the fluid passing through the square chamber, which stabilized the flow.

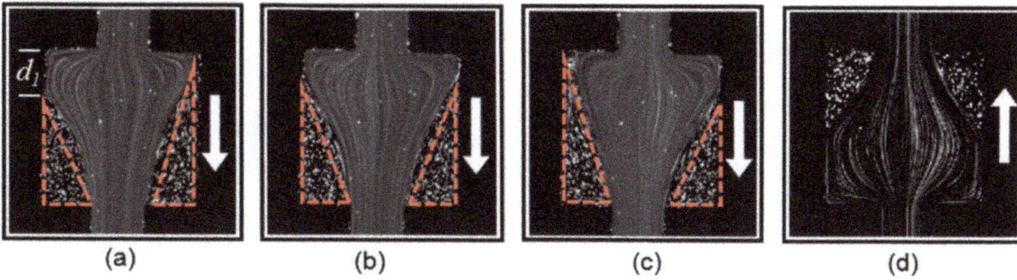

Figure 5. Viscoelastic flow pattern in the square chamber for flows along the forward direction at (**a**–**c**) three different instants, and (**d**) backward direction at *Wi* = 16.67.

As the flow rate of the PAM solution was increased to 1000 μL/h, the *Wi* number increased to 55.56. The instability of the forward flow became more significant, as shown in Figure 6a–c, which represent the flow pattern at different instants. In Figure 6a, the streamline was biased to the left side, and a large stationary regime formed on the right side of the square chamber. In particular, due to the large normal stress, turbulence occurred in the left-side triangle regime. In Figure 6b, the flow became symmetric and the turbulence disappeared. When the streamline become right-side-biased, as shown in Figure 6c, turbulence formed again in the right-side triangle regime. The streamlines in Figure 6a,c are almost mirror images. On the other hand, for backward flow, as shown in Figure 6d, the flow was still stabilized by the nozzle structure and remained symmetric and steady.

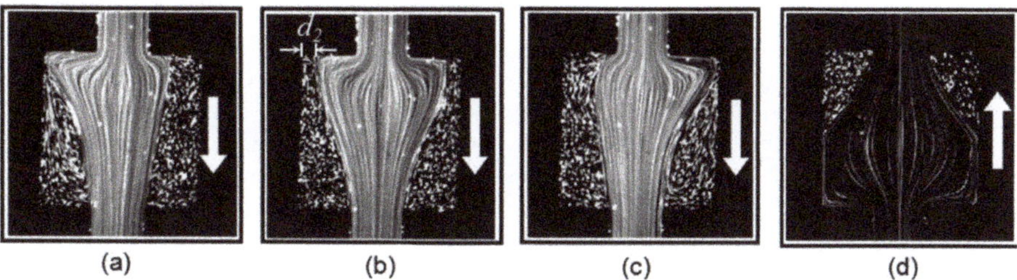

Figure 6. Viscoelastic flow pattern in the square chamber for flows along the forward direction at (**a**–**c**) three different instants, and (**d**) backward direction at *Wi* = 55.56.

As the flow rate of the PAM solution was increased to 2000 μL/h, the *Wi* number increased to 111.11. The forward flow was still asymmetric and unsteady with turbulence on the streamline biased side, as shown in Figure 7a–c. On the other hand, as shown in Figure 7d–f, the backward flow started to become asymmetric and unsteady in this high *Wi* number condition. The increase in elastic instability in the backward flow was due to the normal stress now being large enough to exceed the stabilizing effect by the nozzle structure.

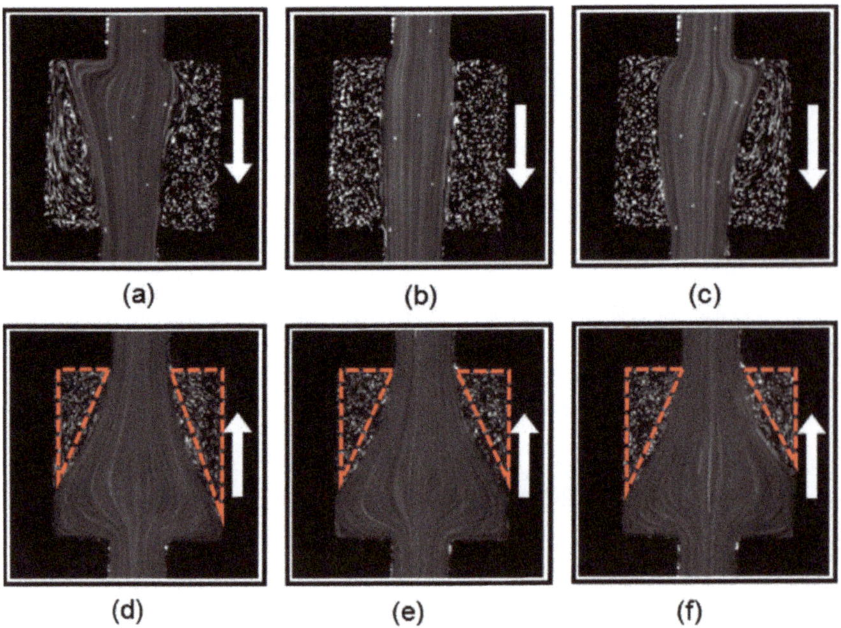

Figure 7. Viscoelastic flow pattern in the square chamber for (**a**–**c**) flows along the forward direction at three different instants, and (**d**–**f**) flows along the backward direction at three different instants at *Wi* = 111.11.

To quantitatively analyze the instability behavior, we measured the length between the left upper corner of the square chamber and the left edge of the forward-flowing streamline. We noted that length $d = d_1$ (as in Figure 5a) when the streamline edge was on the left side of the chamber, and $d = -d_2$ (as in Figure 6b) when the streamline edge was on the upper side of the chamber. The real-time change of d from 1 s to 30 s is plotted in Figure 8 for different *Wi* numbers. We can see obvious resonances of d for the cases *Wi* = 16.67 and 55.56, indicating a significant instability of the streamline. The streamline continuously oscillated with a period of ~10 s and ~20 s for *Wi* = 16.67 and 55.56, respectively. In addition, d was mostly above 0 for *Wi* = 16.67, indicating that the edge of the streamline was on the left-side edge of the square chamber. On the other hand, d fluctuated between positive and negative values for *Wi* = 55.56, indicating that the edge of the streamline was on the left-side edge and upper-side edge of the square chamber, respectively. When *Wi* = 111.11, d was a negative value because the edge of the streamline remained on the upper-side edge of the square chamber at all measured timepoints. Thus, the oscillation became less obvious.

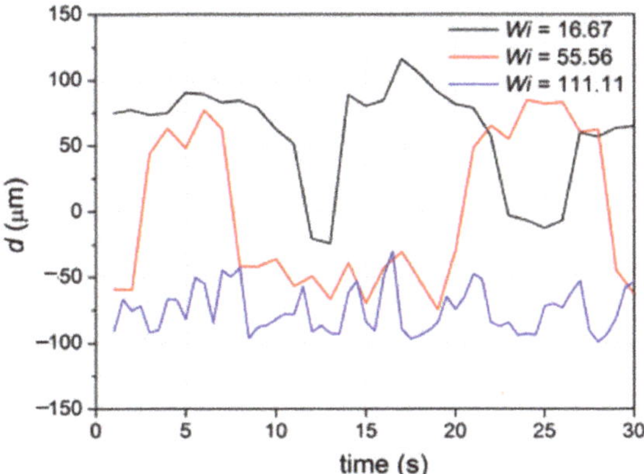

Figure 8. The edge length of the forward-flowing streamline on the left side of the flow.

To verify whether the above instability was elastic-related, we injected Newtonian fluid into the nozzle–square microchannel for comparison. Here, a glycerol solution of 80% volume concentration is used. The flow rate was set to $Q = 2000$ µL/h, corresponding to an Re number of 0.07. As shown in Figure 9a,b, the forward flow and backward flow of the glycerol solution were both symmetric and steady, and both flow patterns were almost the same. This verifies that the asymmetric flow pattern, the unsteady streamline, and the flow direction-dependent instability of the PAM solution were due to the fluid elastic property.

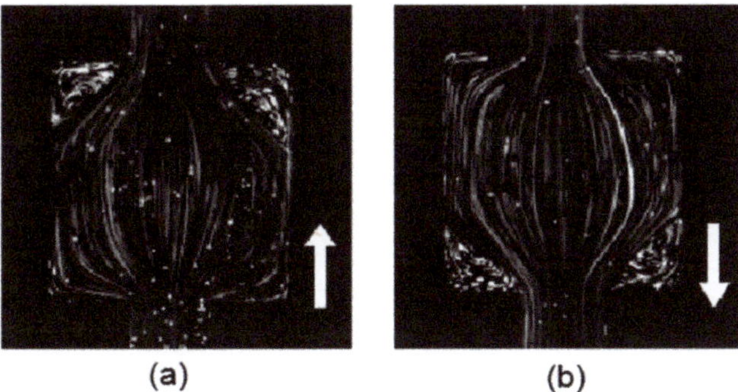

Figure 9. Flow pattern of the glycerol solution in (**a**) forward direction and (**b**) backward direction at $Q = 2000$ µL/h.

To further explain the effect of the nozzle structure, we prepared a microchannel without a nozzle structure using only the same square chamber. The microchannel structure was symmetric; thus, we only investigated the PAM flow along one direction. Figure 10a–c show the flow pattern of the PAM solution at different instants at $Q = 300$ µL/h and $Wi = 16.67$. The flow remained in a symmetric static state, and there was no flow instability. Figure 10d–f show the flow pattern of the PAM solution at different instants at $Q = 1000$ µL/h and $Wi = 55.56$. The flow became asymmetric and unsteady due to the increased normal stress when the flow entered the expansion square chamber and formed an extensional flow. As the flow rate Q increased to 2000 µL/h and $Wi = 111.11$, the asymmetry and instability of the flow pattern became more obvious, as presented by the three different patterns shown in Figure 10g–i.

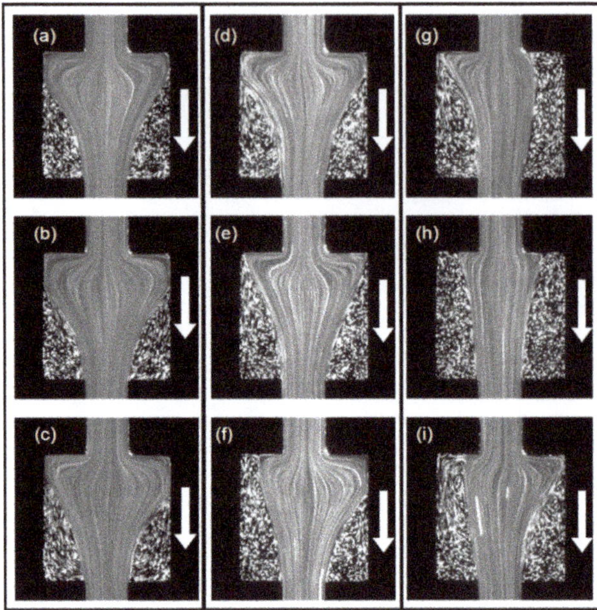

Figure 10. The PAM flow pattern in the symmetric square-only microchannel at three different instants (**a–c**) when $Wi = 16.67$, (**d–f**) when $Wi = 55.56$, and (**g–i**) when $Wi = 111.11$.

On the basis of the above analysis, the instability of the viscoelastic flow in the asymmetric nozzle–square microchannel and in the symmetric square microchannel is summarized in Figure 11. The critical Wi number for the flow instability was not only different for the square–nozzle microchannel and the square-only microchannel, but also different for the forward-directed flow and the backward-directed flow in the same nozzle–square microchannel. The critical Wi number for the forward-directed flow in the nozzle–square microchannel was the smallest, indicating that it is easiest to induce elastic instability in such a flow condition. This is because the polymers in the viscoelastic fluid were stretched by the nozzle structure before entering the square structure, facilitating the induction of normal stress in the subsequent extensional flow. On the other hand, the critical Wi number for the backward-directed flow in the nozzle–square microchannel was the largest, indicating that it is hardest to induce elastic instability in such a flow condition. This is because the nozzle structure increased the flow resistance before the flow entered the square structure, which stabilized the flow.

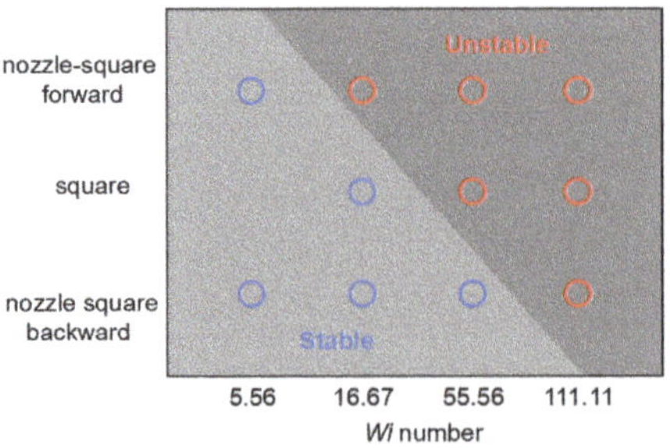

Figure 11. Critical Weissenberg number phase diagram for the two flow directions in the nozzle–square microchannel and the flow in the square-only microchannel.

4. Conclusions

In summary, we studied the viscoelastic fluid instability in an asymmetric nozzle–square microchannel. The instability was demonstrated to be purely elastic and dependent on the flow Weissenberg number. Beyond a certain *Wi* number, the flow pattern was converted from a symmetric steady state to an asymmetric unsteady state. The critical *Wi* number was demonstrated to be different for the two flow directions in the same nozzle–square microchannel. In other words, the flow instability was flow direction-dependent even at the same *Wi* number and in the same microchannel structure. This flow direction-dependent instability can not only be applied for fluidic rectifier applications, but also be used to stabilize viscoelastic flow in high-flow-rate conditions for mass transportation.

Author Contributions: Conceptualization, W.Z. and M.Z.; methodology, Z.W. and J.L.; validation, W.C. and Y.H.; formal analysis, W.Z. and M.Z.; investigation, W.Z. and S.L.; resources, W.Z.; data curation, M.Z. and Z.W.; writing—original draft preparation, W.Z.; writing—review and editing, W.Z.; visualization, M.Z.; supervision, W.Z.; funding acquisition, W.Z. All authors have read and agreed to the published version of the manuscript.

Funding: The research was funded by the National Natural Science Foundation of China (Grant No. 61905046), the China Postdoctoral Science Foundation (Grant No. 2020M683044), the Provincial Key Laboratory of Information Photonics Technology (Guangdong University of Technology, Grant No. GKPT20-08), the Science and Technology Projects in Guangzhou (Grant No. 202102010446), and the National College Student Innovation and Entrepreneurship Training Program of Guangzhou University (Grant No. 202111078049).

Data Availability Statement: Raw data presented in this study are available on request from the corresponding author.

Conflicts of Interest: The authors declare no conflict of interest.

References

1. Sahu, K.C.; Valluri, P.; Spelt, P.D.; Matar, O.K. Linear instability of pressure-driven channel flow of a Newtonian and a Herschel-Bulkley fluid. *Phys. Fluids* **2007**, *19*, 122101. [CrossRef]
2. Tian, F.; Zhang, W.; Cai, L.; Li, S.; Hu, G.; Cong, Y.; Liu, C.; Li, T.; Sun, J. Microfluidic co-flow of Newtonian and viscoelastic fluids for high-resolution separation of microparticles. *Lab Chip* **2017**, *17*, 3078–3085. [CrossRef] [PubMed]
3. Miller, E.; Cooper-White, J. The effects of chain conformation in the microfluidic entry flow of polymer-surfactant systems. *J. Non-Newton. Fluid Mech.* **2009**, *160*, 22–30. [CrossRef]
4. Li, X.B.; Li, F.C.; Cai, W.H.; Zhang, H.N.; Yang, J.C. Very-low-Re chaotic motions of viscoelastic fluid and its unique applications in microfluidic devices: A review. *Exp. Therm. Fluid Sci.* **2012**, *39*, 1–16. [CrossRef]
5. Zhang, M.; Zhang, W.; Wu, Z.; Shen, Y.; Wu, H.; Cheng, J.; Zhang, H.; Li, F.; Cai, W. Modulation of viscoelastic fluid response to external body force. *Sci. Rep.* **2019**, *9*, 9402. [CrossRef] [PubMed]
6. Sousa, P.C.; Pinho, F.T.; Alves, M.A. Purely-elastic flow instabilities and elastic turbulence in microfluidic cross-slot devices. *Soft Matter* **2018**, *14*, 1344–1354. [CrossRef]
7. Groisman, A.; Steinberg, V. Elastic turbulence in curvilinear flows of polymer solutions. *New J. Phys.* **2004**, *6*, 29. [CrossRef]
8. Groisman, A.; Steinberg, V. Mechanism of elastic instability in Couette flow of polymer solutions: Experiment. *Phys. Fluids* **1998**, *10*, 2451–2463. [CrossRef]
9. Larson, R.G.; Shaqfeh, E.S.; Muller, S.J. A purely elastic instability in Taylor–Couette flow. *J. Fluid Mech.* **2006**, *218*, 573–600. [CrossRef]
10. Yuan, C.; Zhang, H.N.; Li, Y.K.; Li, X.B.; Wu, J.; Li, F.C. Nonlinear effects of viscoelastic fluid flows and applications in microfluidics: A review. *Proc. Inst. Mech. Eng. Part C-J. Mech. Eng. Sci.* **2020**, *234*, 4390–4414. [CrossRef]
11. Zhang, M.; Zhang, W.; Wu, Z.; Shen, Y.; Chen, Y.; Lan, C.; Li, F.; Cai, W. Comparison of Micro-Mixing in Time Pulsed Newtonian Fluid and Viscoelastic Fluid. *Micromachines* **2019**, *10*, 262. [CrossRef]
12. Zhang, M.; Cui, Y.; Cai, W.; Wu, Z.; Li, Y.; Li, F.; Zhang, W. High Mixing Efficiency by Modulating Inlet Frequency of Viscoelastic Fluid in Simplified Pore Structure. *Processes* **2018**, *6*, 210. [CrossRef]
13. Nam, J.; Lim, H.; Kim, D.; Jung, H.; Shin, S. Continuous separation of microparticles in a microfluidic channel via the elasto-inertial effect of non-Newtonian fluid. *Lab Chip* **2012**, *12*, 1347–1354. [CrossRef] [PubMed]
14. Galindo-Rosales, F.J.; Campo-Deano, L.; Sousa, P.C.; Ribeiro, V.M.; Oliveira, M.S.; Alves, M.A.; Pinho, F.T. Viscoelastic instabilities in micro-scale flows. *Exp. Therm. Fluid Sci.* **2014**, *59*, 128–139. [CrossRef]
15. Pan, L.; Morozov, A.; Wagner, C.; Arratia, P.E. Nonlinear Elastic Instability in Channel Flows at Low Reynolds Numbers. *Phys. Rev. Lett.* **2013**, *110*, 174502. [CrossRef]

16. Davoodi, M.; Houston, G.; Downie, J.; Oliveira, M.S.; Poole, R.J. Stabilization of purely elastic instabilities in cross-slot geometries. *J. Fluid Mech.* **2021**, *922*, A12. [CrossRef]

17. Sousa, P.C.; Pinho, F.T.; Oliveira, M.S.; Alves, M.A. Purely elastic flow instabilities in microscale cross-slot devices. *Soft Matter* **2015**, *11*, 8856–8862. [CrossRef] [PubMed]

18. Poole, R.J.; Alves, M.A.; Oliveira, P.J. Purely elastic flow asymmetries. *Phys. Rev. Lett.* **2007**, *99*, 164503. [CrossRef]

19. Cruz, F.A.; Poole, R.J.; Afonso, A.M.; Pinho, F.T.; Oliveira, P.J.; Alves, M.A. Influence of channel aspect ratio on the onset of purely-elastic flow instabilities in three-dimensional planar cross-slots. *J. Non-Newton. Fluid Mech.* **2016**, *227*, 65–79. [CrossRef]

20. Canossi, D.O.; Mompean, G.; Berti, S. Elastic turbulence in two-dimensional cross-slot viscoelastic flows. *Epl* **2020**, *129*, 7. [CrossRef]

21. Zografos, K.; Burshtein, N.; Shen, A.Q.; Haward, S.J.; Poole, R.J. Elastic modifications of an inertial instability in a 3D cross-slot. *J. Non-Newton. Fluid Mech.* **2018**, *262*, 12–24. [CrossRef]

22. Afonso, A.M.; Pinho, F.T.; Alves, M.A. Electro-osmosis of viscoelastic fluids and prediction of electro-elastic flow instabilities in a cross slot using a finite-volume method. *J. Non-Newton. Fluid Mech.* **2012**, *179*, 55–68. [CrossRef]

23. Zhao, Y.; Shen, A.Q.; Haward, S.J. Flow of wormlike micellar solutions around confined microfluidic cylinders. *Soft Matter* **2016**, *12*, 8666–8681. [CrossRef] [PubMed]

24. Haward, S.J.; McKinley, G.H.; Shen, A.Q. Elastic instabilities in planar elongational flow of monodisperse polymer solutions. *Sci. Rep.* **2016**, *6*, 33029. [CrossRef] [PubMed]

25. Ballesta, P.; Alves, M.A. Purely elastic instabilities in a microfluidic flow focusing device. *Phys. Rev. Fluids* **2017**, *2*, 053301. [CrossRef]

26. Soulages, J.; Oliveira, M.S.; Sousa, P.C.; Alves, M.A.; McKinley, G.H. Investigating the stability of viscoelastic stagnation flows in T-shaped microchannels. *J. Non-Newton. Fluid Mech.* **2009**, *163*, 9–24. [CrossRef]

27. Tai, J.; Lim, C.P.; Lam, Y.C. Visualization of polymer relaxation in viscoelastic turbulent microchannel flow. *Sci. Rep.* **2015**, *5*, 16633. [CrossRef]

28. Sousa, P.C.; Pinho, F.T.; Oliveira, M.S.; Alves, M.A. Efficient microfluidic rectifiers for viscoelastic fluid flow. *J. Non-Newton. Fluid Mech.* **2010**, *165*, 652–671. [CrossRef]

29. Sousa, P.C.; Pinho, F.T.; Oliveira, M.S.; Alves, M.A. High performance microfluidic rectifiers for viscoelastic fluid flow. *Rsc Adv.* **2012**, *2*, 920–929. [CrossRef]

30. Adams, M.L.; Johnston, M.L.; Scherer, A.; Quake, S.R. Polydimethylsiloxane based microfluidic diode. *J. Micromech. Microeng.* **2005**, *15*, 1517–1521. [CrossRef]

31. Groisman, A.; Enzelberger, M.; Quake, S.R. Microfluidic Memory and Control Devices. *Science* **2003**, *300*, 955. [CrossRef] [PubMed]

Article

Electroosmotic Flow of Viscoelastic Fluid through a Constriction Microchannel

Jianyu Ji [1], Shizhi Qian [1,*] and Zhaohui Liu [2]

1 Department of Mechanical and Aerospace Engineering, Old Dominion University, Norfolk, VA 23529, USA; jji016@odu.edu
2 State Key Laboratory of Coal Combustion, School of Energy and Power Engineering, Huazhong University of Science and Technology, Wuhan 430074, China; zliu@mail.hust.edu.cn
* Correspondence: sqian@odu.edu; Tel.: +1-757-683-3304

Abstract: Electroosmotic flow (EOF) has been widely used in various biochemical microfluidic applications, many of which use viscoelastic non-Newtonian fluid. This study numerically investigates the EOF of viscoelastic fluid through a 10:1 constriction microfluidic channel connecting two reservoirs on either side. The flow is modelled by the Oldroyd-B (OB) model coupled with the Poisson–Boltzmann model. EOF of polyacrylamide (PAA) solution is studied as a function of the PAA concentration and the applied electric field. In contrast to steady EOF of Newtonian fluid, the EOF of PAA solution becomes unstable when the applied electric field (PAA concentration) exceeds a critical value for a fixed PAA concentration (electric field), and vortices form at the upstream of the constriction. EOF velocity of viscoelastic fluid becomes spatially and temporally dependent, and the velocity at the exit of the constriction microchannel is much higher than that at its entrance, which is in qualitative agreement with experimental observation from the literature. Under the same apparent viscosity, the time-averaged velocity of the viscoelastic fluid is lower than that of the Newtonian fluid.

Keywords: electroosmosis; microfluidics; elastic instability; non-Newtonian fluid; Oldroyd-B model

Citation: Ji, J.; Qian, S.; Liu, Z. Electroosmotic Flow of Viscoelastic Fluid through a Constriction Microchannel. *Micromachines* **2021**, *12*, 417. https://doi.org/10.3390/mi 12040417

Academic Editor: Stéphane Colin

Received: 6 February 2021
Accepted: 7 April 2021
Published: 9 April 2021

Publisher's Note: MDPI stays neutral with regard to jurisdictional claims in published maps and institutional affiliations.

1. Introduction

Electroosmotic flow (EOF) uses electric field to control fluid motion, and has been widely used in various microfluidic and nanofluidic applications such as fluid pump [1], mixing [2], and polymer translocation in biosensing [3]. The existing studies of EOF have been mainly focusing on Newtonian fluids [4,5]. However, in reality, EOF has been widely used to control and manipulate biological fluids (e.g., blood, saliva, lymph, protein, and DNA solutions) [6–8] and polymeric solutions [9], which exhibit non-Newtonian characteristics. Therefore, investigating EOF of viscoelastic fluids is of practical importance.

Bello et al. [10] conducted the pioneering study on EOF of non-Newtonian fluid, and measured EOF velocity of methyl cellulose solution in a capillary. Their results show that EOF velocity of such polymer solutions is much higher than that predicted with the classic Helmholtz-Smoluchowski velocity. Chang and Tsao [11] conducted similar experiments and found the effective viscosity decreased because of the sheared polymeric molecules inside the electrical double layer (EDL). Theoretically, non-Newtonian effects can be characterized by proper constitutive models relating the dynamic viscosity and the rate of shear. Such constitutive models include power-law model [12], Carreau model [13], WhiteMetzner model [14], Bingham model [15], Oldroyd-B (OB) model [16], PTT model [17], Moldflow second-order model [18], Giesekus model [19], etc. Das [20] developed an approximate solution for EOF velocity of power-law fluid between two parallel plates. Zhao et al. [21,22] derived a generalized Helmholtz–Smoluchowski velocity for EOF of power-law fluid in a slit microchannel. Later, Zhao and Yang [23,24] extended the study to a cylindrical microcapillary. Olivares et al. [25] experimentally investigated

EOF of a non-Newtonian polymeric solution and verified the generalized Helmholtz–Smoluchowski velocity. Tang et al. [26] numerically investigated EOF of power-law fluid using Lattice–Boltzmann method. Zimmerman et al. [13] carried out numerical simulation of EOF of Carreau fluid in a T-junction microchannel, and found that the flow field significantly depended on the non-Newtonian characteristics of the fluid. The aforementioned studies on EOF of non-Newtonian fluid are limited to inelastic constitutive models (i.e., power-law and Carreau models). However, some fluids show both viscous and elastic behaviors, which can be presented by viscoelastic constitutive models. There are existing literatures investigating the characteristics of EOF of viscoelastic fluids [27–31], showing that the viscoelasticity of the fluid affects the flow pattern and flow rate. Note that in the aforementioned studies, the EOF of non-Newtonian fluid was assumed in a steady state.

Recently, EOFs of non-Newtonian fluids have been reported to be time-dependent and show instabilities even at low Reynolds number (*Re*). Such EOFs are time-dependent because of the nonlinear viscosity and elasticity of non-Newtonian fluids. Bryce and Freeman [32] first reported the electro-elastic instability in EOF of PAA solutions through a 2:1:2 micro-scale contraction/expansion when the applied electric field exceeded a threshold value. Later, Bryce and Freeman [33] reported that such instabilities insignificantly enhanced the mixing in micro flows. Pimenta and Alves [34,35] later experimentally and numerically studied the electro-elastic instabilities of PAA solutions in both cross-slot and flow-focusing micro devices, and found that mixing efficiency was not improved significantly. Song et al. [36] experimentally and numerically studied the elastic instability in EOF of viscoelastic polyethylene oxide (PEO) solutions through T-shaped microchannels, and results demonstrated that the threshold electric field for onset of instability highly depended on the PEO concentration. Song et al. [37] later extended the work by experimentally investigating the fluid rheological effects on the elastic instability in EOF of six types of phosphate buffer-based aqueous solutions through T-shaped microchannels. They found that shear thinning effect of the fluid might account for the electro-elastic instabilities, however, the fluid with high elasticity alone did not have instability, which is inconsistent with the results of Pimenta [35]. The authors attribute the inconsistency to the neglect of microstructural effects (e.g., polymer-wall interaction and electric effect on molecular structure of polymer, etc.) of shear-thinning polymer solutions. However, this experimental result shows similarity to the work of Ko et al. [38], in which weakly shear-thinning, viscoelastic polyvinylpyrrolidone (PVP), and PEO solutions exhibited Newtonian-like EOF patterns, while shear-thinning and weakly elastic xanthan gum (XG) solution exhibited disturbance and vortices, suggesting that fluid elasticity alone has an insignificant impact on the steady-state EOF pattern. More recently, Sadek [39] experimentally investigated EOF of viscoelastic fluids through different microchannel configurations, including hyperbolic-shaped contractions followed by an abrupt expansion, and abrupt contractions followed by a hyperbolic-shaped expansion, and EOF showed instabilities of elastic origin at very low Weissenberg numbers (*Wi*) (i.e., *Wi* < 0.01).

There is only limited literature on numerical studies of electro-elastic instabilities. Afonso et al. [40] numerically investigated the elastic instability of EOF through a cross-slot geometry using the upper-converted Maxwell and the simplified Phan-Thien-Tanner models, and a direct flow transition from steady symmetric state to unsteady flow without crossing the steady asymmetric state at a critical *Wi* was observed. Pimenta and Alves [35] numerically investigated the electro-elastic instabilities in cross-slot and flow-focusing micro devices using OB model and Poisson-Boltzmann (PB) model. They found that strong shear-dominated flow within the EDL at the corners had a more significant contribution to the elastic instabilities than the extensionally dominated bulk flow. Song et al. [36] numerically investigated EOF of PEO solution through a T-shaped microchannel. Their model considered only the influence of PEO solution on the fluid viscosity, conductivity, and zeta potential. Due to the neglect of fluid elasticity effect in the mathematical model, only the electrokinetic flow phenomena of dilute PEO solution (i.e., $\leq$750 ppm) were captured.

Both experimental and numerical investigations in the EOF instabilities of viscoelastic fluid are limited, and the conditions proposed by various researchers for triggering the instabilities in the EOF of viscoelastic fluids show inconsistency and remain unclear. Inspired by the existing literature, in this work we numerically study EOF of viscoelastic fluids through a 10:1:10 contraction microchannel. The geometry, which consists of a constriction microchannel connecting two relatively big reservoirs on either end, is close to actual microfluidic device. The time-dependent OB model and PB model are adopted to describe the constitutive characteristics and the electrokinetic phenomenon, respectively. EOFs of PAA solutions with various weight concentrations under different applied electric fields are investigated. The effects of polymer concentration and applied electric field on the elastic instability are studied, and a map in polymer concentration-electric field space for predicting the onset of upstream vortices is formed.

2. Mathematical Model

We consider incompressible monovalent binary electrolyte solution such as KCl with bulk concentration c_0 mixed with PAA polymer solution of concentration c_p, which fills a microchannel of height H_c, length L_c, and width W connecting two identical reservoirs of height H_r and length L_r on either side. The solid walls of the constriction microchannel and the reservoirs are assumed to carry a constant negative zeta potential, ζ_0. When dealing with non-Newtonian fluids, a constant zeta potential has been widely accepted [41]. Huang et al. [41] compared theoretical and experimental results of PEO solutions, and a constant zeta potential was proven for various PEO concentrations. Therefore, in the current study, we neglect the effect of the polymer concentration on the wall zeta potential. Two electrodes are placed at both ends of the reservoirs, and an external potential bias U_0 is applied between the inlet (Anode) and outlet (Cathode). Through the interaction between the externally applied electric field and net charges accumulated within the EDL in the vicinity of the charged walls, EOF flowing from the anode reservoir through the constriction microchannel towards the cathode reservoir is generated. The apparent electric field between the inlet and outlet is defined as $E_{app} = U_0/(2L_r + L_c)$. In some applications, there are slit microchannels with width much larger than height [42,43]. For example, two-phase flow patterns were studied in a microchannel with 10-mm width and 50-μm height [42]. For microchannels with such geometries, the flow can be simplified to a 2D problem [44]. Therefore, in the current study we assume that the channel width is much larger than the channel height, and the flow can be simplified to a 2D problem as schematically shown in Figure 1. A Cartesian coordinate system with origin fixed at the center of the microchannel is adopted with the x-axis along the length direction and the y-axis in the height direction.

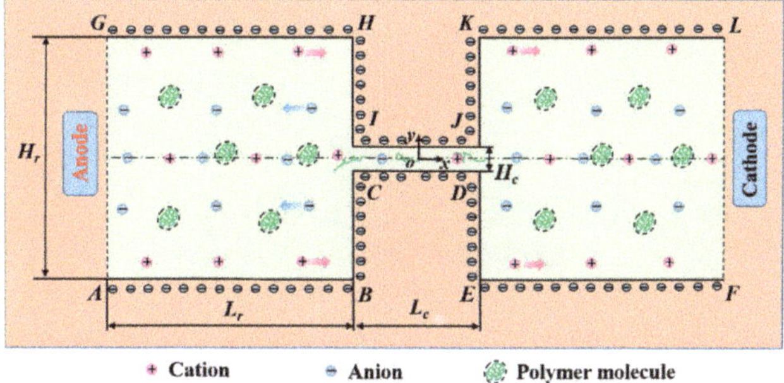

Figure 1. Schematic diagram of a constriction microchannel connecting two reservoirs at both ends. The solid walls of reservoirs and the constriction channel are negatively charged, and an electric field is imposed by applying a potential difference between anode and cathode positioned in two fluid reservoirs.

The induced EOF of viscoelastic fluid is governed by the continuity and Navier–Stokes equations:

$$\nabla \cdot u = 0, \tag{1}$$

$$\rho\left(\frac{\partial u}{\partial t} + u \cdot \nabla u\right) = -\nabla p + \eta_s \nabla^2 u + \nabla \cdot \tau - \rho_E \nabla \phi_{Ext}, \tag{2}$$

where u is the velocity; p is the pressure; ρ is the fluid density; η_s denotes the solvent dynamic viscosity; τ is the polymeric stress tensor accounting for the memory of the viscoelastic fluid; t represents time; ρ_E and ϕ_{Ext} represent, respectively, the volume charge density within the electrolyte solution and the externally applied electric potential. For different types of viscoelastic fluids, various constitutive models have been developed to relate the polymeric stress tensor τ and the deformation rate of the fluid, including WhiteMetzner model [14], which is commonly used for shear-thinning fluid; PTT model [17], which has good performance for prediction of viscosity at low shear rates; Giesekus model [19], which is suitable for concentrated polymer solutions; and OB model [16], which is suitable for dilute polymer solutions. Since OB model can properly fit the rheological behavior of aqueous PAA solutions [45], OB model is adopted in this study. In the OB model, the polymeric stress tensor, τ, is described as [16],

$$\tau = \frac{\eta_P}{\lambda}(c - I), \tag{3}$$

where η_p is the polymer dynamic viscosity; λ is the relaxation time of the polymer, which refers to the time it takes for polymer chains to return to equilibrium after being disturbed; c is the symmetric conformation tensor of the polymer molecules; and I is the identity matrix.

For the OB model, the conformation tensor c is governed by [16],

$$\frac{\partial c}{\partial t} + u \cdot \nabla c = c \cdot \nabla u + (\nabla u)^T \cdot c - \frac{1}{\lambda}(c - I). \tag{4}$$

Typically, numerical simulation of viscoelastic flow is difficult to converge for high Weissenberg number problem [46,47]. Computations were found to break down at frustratingly low values of Weissenberg number (usually around $Wi = 1$; precise critical value also depends on the flow geometry) [48]. Therefore, the log-conformation tensor approach [47] is adopted. In the log-conformation tensor method, a new tensor (Θ) is defined as the natural logarithm of the conformation tensor,

$$\Theta = \ln(c) = \mathbf{R}\ln(\Lambda)\mathbf{R}, \tag{5}$$

where Λ is a diagonal matrix whose diagonal elements are the eigenvalues of c; and $\mathbf{R}$ is an orthogonal matrix with its columns being the eigenvalues of c. Equation (4) for the conformation tensor written in terms of Θ then becomes [46],

$$\frac{\partial \Theta}{\partial t} + u \cdot \nabla \Theta = \Omega\Theta - \Theta\Omega + 2\mathbf{B} + \frac{1}{\lambda}\left(e^{\Theta} - I\right). \tag{6}$$

in the above, Ω and $\mathbf{B}$ are, respectively, the anti-symmetric matrix and the symmetric traceless matrix of the decomposition of the velocity gradient tensor ∇u [46].

Then, the conformation tensor c is recovered from Θ,

$$c = \exp(\Theta). \tag{7}$$

The total electric potential, Ψ, is decomposed in two variables, $\Psi = \phi_{Ext} + \psi$ [35], with ϕ_{Ext} representing the potential originated from the externally applied electric potential while ψ being the potential arising from the charge of channel walls. In this study, the EDL thickness is on the order of nanometers (the calculation of EDL thickness will be shown in the next paragraph), while the microchannel height is on the order of micrometers. Therefore, the Poisson–Boltzmann equation [49] is used to describe the potential, ψ:

$$\nabla \cdot (\varepsilon \nabla \psi) = \rho_E = Fc_0 \left(\exp\left(\frac{e\psi}{kT}\right) - \exp\left(-\frac{e\psi}{kT}\right) \right). \tag{8}$$

In the above, F is the Faraday's constant (i.e., 96,485.33289 C·mol^{-1}); e is the elementary charge (i.e., $1.6021766341 \times 10^{-19}$ C); k is Boltzmann's constant (i.e., 1.380649×10^{-23} J·K^{-1}); T is the absolute temperature of the fluid (i.e., 295 K); and ε represents the permittivity of the solution (i.e., 6.906266×10^{-10} F·m^{-1}). In this study, the bulk concentration c_0 is 0.01 mM; $z_1 = 1$ and $z_2 = -1$. In a biocompatible solution with pH of 7.4, the concentration of H$^+$ is $10^{-7.4}$ mol/L, and the concentration of OH$^-$ is $10^{-6.6}$ mol/L. The concentration of weak electrolyte is relatively low comparing with the background salt. Therefore, the weak electrolyte is not considered in the current study. The EDL thickness can be calculated by $\lambda_D = \sqrt{\frac{\varepsilon kT}{eF\left(z_1^2 c_0 + z_2^2 c_0\right)}}$, which is 95 nm. The potential ϕ_{Ext} is governed by the following Laplace equation [50],

$$\nabla^2 \phi_{Ext} = 0. \tag{9}$$

The boundary conditions are given as follows (Figure A4):

(1) At the Anode (edge AG in Figure A4): $\mathbf{n} \cdot \nabla \mathbf{u} = \mathbf{0}$; $p = 0$; $\boldsymbol{\tau} = \mathbf{0}$; $\phi_{Ext} = U_0$; $\mathbf{n} \cdot \nabla \psi = 0$; $\Theta = 0$; where $\mathbf{n}$ denotes the normal unit vector on the surface.

(2) At the Cathode (edge FL in Figure A4): $\mathbf{n} \cdot \nabla \mathbf{u} = \mathbf{0}$; $p = 0$; $\mathbf{n} \cdot \nabla \boldsymbol{\tau} = \mathbf{0}$; $\phi_{Ext} = 0$; $\mathbf{n} \cdot \nabla \psi = 0$; $\mathbf{n} \cdot \nabla \Theta = 0$.

(3) On the reservoir walls (edges ABC, DEF, GHI, and JKL in Figure A4) and the microchannel walls (edges CD and IJ in Figure A4): $\mathbf{u} = \mathbf{0}$; $\mathbf{n} \cdot \nabla \phi_{Ext} = 0$; $\psi = \zeta_0$; $\mathbf{n} \cdot \nabla \Theta = 0$; $\mathbf{n} \cdot \nabla p$ is obtained from the momentum equation; the components of $\boldsymbol{\tau}$ are linearly extrapolated.

The following initial conditions are specified within the domain: $\mathbf{u} = \mathbf{0}$; $p = 0$; $\boldsymbol{\tau} = \mathbf{0}$; $\phi_{Ext} = 0$; $\psi = \mathbf{0}$; $\Theta = 0$.

Note that the electric potentials and the flow are only one-way coupling. The electric potentials ϕ_{Ext} and ψ are in a steady state, and they are independent on the flow. However, the electric potentials affect the flow through the electrostatic force, which is the last term in Equation (2). For Newtonian fluid, the third term, $\nabla \cdot \boldsymbol{\tau}$, in the right-hand-side of Equation (2) is dropped, and the model includes Equations (1), (2), (8), and (9).

3. Numerical Method and Code Validation

The governing equations are numerically solved using the finite volume method by RheoTool (version 4.1, https://github.com/fppimenta/rheoTool, accessed on 1 June 2020), an open-source viscoelastic EOF solver [35] implemented in the open-source OpenFOAM platform. The details of the solver can be found from the work of Pimenta and Alves [34,50]. To numerically solve the coupled Equations (1), (2), (6), (8), and (9) along with the boundary and initial conditions, CUBISTA scheme [51] is used to discretize the convective terms in Equations (2) and (6). Central differences are used for the discretization of Laplacian and gradient terms. The time derivatives are discretized with three-time level explicit difference scheme [52], which is of the second order of accuracy. The exponential source term in Equation (8) is linearized using Taylor expansion up to the second term [53]. All of the terms in the momentum equation (i.e., Equation (2)), except the pressure gradient and the electric contribution, are discretized implicitly. A small time-step, $\Delta t = \lambda / 10^5$, is used to ensure the accuracy. The well-known SIMPLEC (Semi-Implicit Method for Pressure-Linked Equations-Consistent) algorithm [54] is used to resolve the velocity-pressure coupling. An inner-iteration loop is used to reduce the explicitness of the method and increase its accuracy and stability. The pressure field is computed by PCG (Preconditioned Conjugate Gradient) solver, of which the tolerance and maximum iteration are set to be 1×10^{-8} and 800, respectively. The velocity field is computed by PBiCG (Preconditioned Biconjugate

Gradient) solver, of which the tolerance and the maximum iteration are set to be 1×10^{-10} and 1000, respectively. The computational steps of the solver are as follows [34]:

Step 1. Initialize the fields $\{u, p, \tau, \phi_{\text{Ext}}, \psi, \Theta\}_0$ and time $(t = 0)$.

Step 1.1. Compute steady state ϕ_{Ext} from Equation (9) and ψ from Equation (8).

Step 2. Enter the time loop $(t = \Delta t)$.

Step 2.1. Enter the inner iteration loop $(i = 0)$.

Step 2.1.1. Compute Θ_i and τ_i by log-conformation method.
Step 2.1.2. Compute estimated velocity field u_i^* by solving the momentum equation.
Step 2.1.3. ompute pressure field p_i by enforcing the continuity equation.
Step 2.1.4. Correct the previously estimated velocity field using the correct pressure field.
Step 2.1.5. Increase the inner iteration index $(i = i + 1)$ and repeat the computation from Step 2.1.1, until the inner iteration criteria (i.e., maximum tolerance) is satisfied.
Step 2.1.6. Set $\{u, p, \tau, \phi_{\text{Ext}}, \psi, \Theta\}_t = \{u_i, p_i, \tau_i, \phi_{\text{Ext}i}, \psi_i, \Theta_i\}$.

Step 2.2. Increase time, $t = t + \Delta t$, and return to Step 2.1 until the simulation time is reached.

Step 3. Stop the simulation and exit.

Structural mesh is adopted to discretize the computational domain. 90° corners of the contraction channel (points I, J, C, and D in Figure 1) are smoothed by a fillet of 1 μm in radius to avoid sharp turns. The 90° corners of the reservoirs (points H, K, B, and E in Figure 1) are smoothed by a fillet of 2 μm in radius. To capture the EDL in the vicinity of the charged walls, a finer mesh is distributed near the charged reservoir and channel walls as shown in Figure 2. To reduce the number of mesh, we use a relatively low bulk concentration $c_0 = 0.01$ mM, and the EDL thickness is 95 nm in this study. In order to capture the details in the EDL and to guarantee the accuracy, the mesh size near the charged wall is set to be 10 nm so that there are 10 meshes within the EDL. There are 77,192 meshes in the whole geometry. A mesh independence study, described in the Appendix A, is performed to ensure the accuracy of the simulation.

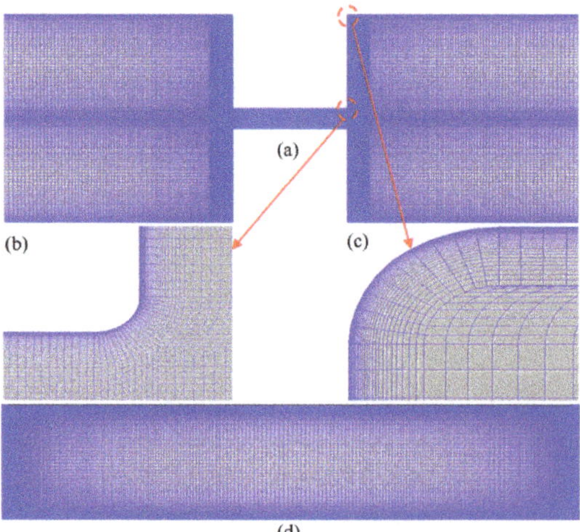

Figure 2. Computational mesh used in the numerical simulations. Mesh of the whole geometry (**a**) and detailed view of the mesh at channel corner (**b**), at reservoir corner (**c**), and in the constriction microchannel (**d**).

In this work, η_p and λ for 100 ppm, 250 ppm, and 1000 ppm PAA-water solutions [55] are adopted to accomplish curve fitting as shown in Figure 3. The values of η_p and λ were experimentally measured [56], and the slow retraction method was used to measure the relaxation time. The polymer dynamic viscosity can be expressed as $\eta_p = 2.22 \times 10^{-5} \cdot c_p$, and the relaxation time can be expressed as $\lambda = 3.69 \times 10^{-3} + 3.94222 \times 10^{-5} \cdot c_p + 9.68889 \times 10^{-5} \cdot c_p^2$, where c_p represents the weight concentration of PAA solution with the unit of ppm. The η_p and λ for other c_p studied in this work are estimated by the curve-fitting expressions.

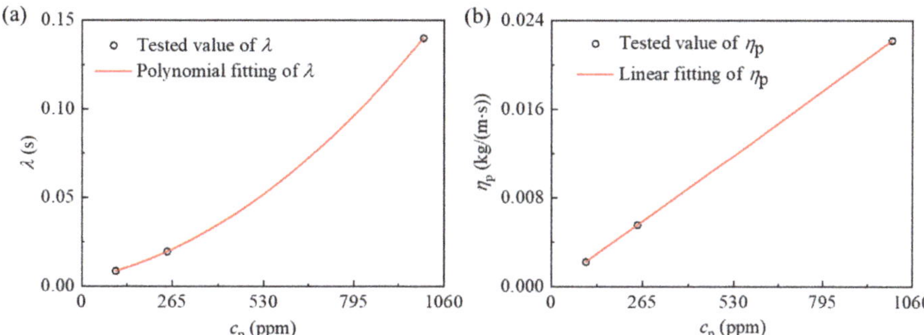

Figure 3. (**a**) Polymer dynamic viscosity η_p and (**b**) relaxation time λ as a function of the polyacrylamide (PAA) concentration, c_p.

In a microfluidic channel with EDL thickness much smaller than the channel height, the EOF velocity of a Newtonian fluid can be approximated by the Helmholtz–Smoluchowski velocity formula [57],

$$u_0 = -\frac{\varepsilon \zeta_0 E_x}{\eta_0}, \tag{10}$$

where E_x is the actual local electric field in the main stream direction, and η_0 is the total viscosity of the fluid. To check the accuracy of our code, we simulate EOFs of both Newtonian and viscoelastic fluids in the same geometry with $H_c = 40$ μm, $L_c = 200$ μm, $H_r = 400$ μm, and $L_r = 400$ μm. Other parameters are set as $U_0 = 60$ V, $\zeta_0 = -0.11$ V [58], and $\varepsilon = 6.906266 \times 10^{-10}$ F·m^{-1}. For Newtonian fluid, the total viscosity is set as $\eta_0 = \eta_s = 0.00322$ kg/(m·s). When the concentration of PAA solution is less than 2 ppm, the relaxation time is less than 0.1 ms [56], and the fluid can be approximately treated as Newtonian fluid. Therefore, for the OB model, parameters are set as $\eta_s = 0.00317$ kg/(m·s), $\eta_p = 0.00005$ kg/(m·s), $\eta_0 = \eta_s + \eta_p = 0.00322$ kg/(m·s), and $\lambda = 0.1$ ms. Figure 4a depicts electric potential $\phi_{\text{Ext}}(x, 0)$ along the x-axis when $E_{\text{app}} = 600$ V/cm. The electric field in the x-direction, $-\frac{\partial \phi_{\text{Ext}}}{\partial x}$, in the constriction microchannel is 1820 V/cm, which is about 10 times of the electric field in the reservoirs. This is because of the 10:1:10 contraction geometry and current conservation. With the same electric conductivity, the electric field is inversely proportional to the cross-sectional area of the geometry. Note that the actual electric field within the constriction microchannel is about three times of the apparent electric field, E_{app}, which does not consider the cross-sectional variation of the geometry. EOFs of both Newtonian fluid and viscoelastic fluid reach a steady state. Figure 4b shows the x-component velocity profiles, $u(0, y)$, of the Newtonian fluid (solid line) and the OB model (circles). The velocity first rises rapidly within the thickness of EDL, then reaches a plateau in the cross section of the channel. When $E_x = 1820$ V/cm, the calculated Helmholtz–Smoluchowski velocity is 4.29 mm/s, and the velocity at the center of the channel is 4.27 mm/s for both Newtonian and OB models. The relative difference between the approximated velocity and the simulated velocity is less than 0.5%. In addition, the result for OB model matches that of Newtonian fluid. Such consistency between Newtonian model and OB model is because the polymer dynamic viscosity η_p is much smaller than the solvent dynamic viscosity η_s, and the relaxation time of the polymer λ is also tiny. Under the considered condition, the elastic effect of the fluid is negligible and the OB fluid is almost the same as Newtonian fluid with the same total viscosity.

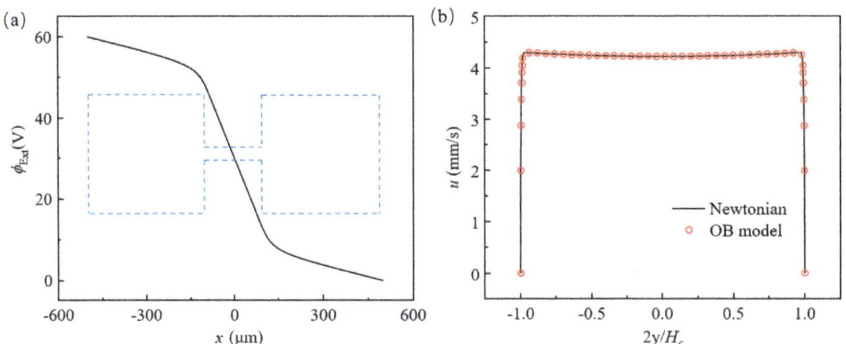

Figure 4. (**a**) Electric potential distribution (blue dash line shows the relative position of the geometry) along the x-axis; (**b**) the x-component velocity at the center of the constriction microchannel, $u(0,y)$, for Newtonian model (solid line) and OB model (symbol).

Afonso et al. [28] derived an analytical solution of viscoelastic EOF between two parallel plates based on the Debey–Hückel approximation, which is valid under the condition of low zeta potential (i.e., $\zeta_0 < 25$ mV). To further validate our code for OB model, EOF of viscoelastic fluid with $\eta_s = 0.001$ kg/(m·s), $\eta_p = 0.00222$ kg/(m·s), $\eta_0 = \eta_s + \eta_p = 0.00322$ kg/(m·s), and $\lambda = 8.6$ ms in a straight 2D channel (with height of 40 μm) is studied. These rheology parameters are corresponding to those of 100 ppm PAA solution. U_0 is set as 10 V, while ζ_0 is chosen as -10 mV and -110 mV, respectively. Under the considered conditions, the flows are steady state due to relatively low electric field strength. Figure 5 depicts the x-component velocity profile at the center of the channel, and our numerical results (triangles) are in excellent agreement with the analytical result (line). Although the analytical solution is based on the Debey–Hückel approximation, we find that the numerical result also agrees well with the analytical solution when the zeta potential ζ_0 is -110 mV. Therefore, the agreement of results attained from the OB model and Newtonian model, which are also validated by the Helmholtz–Smoluchowski approximation, as well as the agreement of analytical solution of OB model and numerical results for EOF of viscoelastic fluid in a straight channel, validate our code.

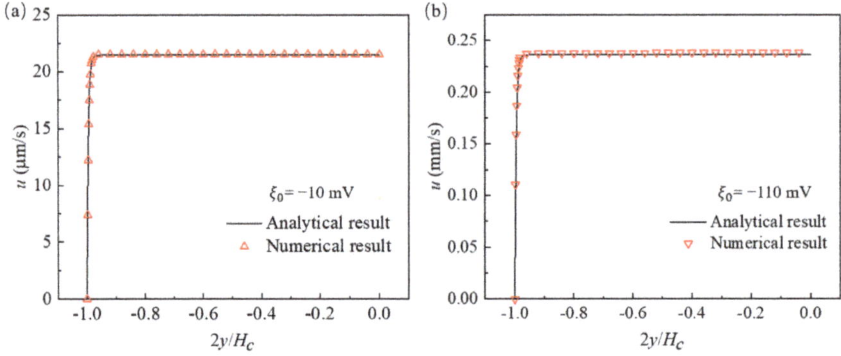

Figure 5. The x-component velocity profile of viscoelastic electroosmotic flow (EOF) between two parallel plates: (**a**) Zeta potential is -10 mV; (**b**) zeta potential is -110 mV. Analytical result of Afonso et al. [28] (solid line) and current numerical result (symbol). The analytical solution is described in Appendix A.

4. Results and Discussion

Newtonian fluid is investigated to provide the reference flow characteristic for the contraction geometry. For the PAA solution with different concentration c_p, the applied apparent electric field E_app is varied from low values to high (i.e., 100–600 V/cm). In this section, first, we describe the flow pattern of Newtonian fluid and the time-dependent flow patterns of PAA solutions. Then, the instabilities of PAA solutions with various E_app and c_p are discussed and a flow map is formed based on the investigated values of E_app and c_p. Finally, statistical results of cross-sectional average velocity are presented.

4.1. Instability of PAA Solutions

For Newtonian fluids with various total viscosities, the EOF reaches a steady state under all conditions of the applied electric field strengths. There is no vortex occurring in the reservoirs and the constriction microchannel. The streamlines of Newtonian fluid show excellent symmetry about the x-axis. Additionally, the magnitude of the velocity, $U(x, y)$, is symmetric about the y-axis, $U(x, y) = U(-x, y)$. For EOF of PAA solutions, when E_app and c_p are relatively low, the flow pattern is similar to that of Newtonian fluid, and the flow reaches a steady state without vortex. With increasing E_app and c_p, however, the viscoelastic flow becomes time dependent and significant instabilities are observed. Figure 6 depicts the streamlines at different times when $E_\mathrm{app} = 100$ V/cm and $c_\mathrm{p} = 500$ ppm. Figure 7 depicts the streamlines at different times when $E_\mathrm{app} = 600$ V/cm and $c_\mathrm{p} = 150$ ppm. Figure 8 shows the velocity magnitudes as a function of time at three different locations, namely, upstream of the constriction microchannel $(-3H_\mathrm{c}, 0)$, center of the constriction microchannel $(0, 0)$, and downstream of the constriction microchannel $(3H_\mathrm{c}, 0)$. For the EOF of both $c_\mathrm{p} = 150$ ppm and $c_\mathrm{p} = 500$ ppm, we observe strong instabilities and upstream vortices.

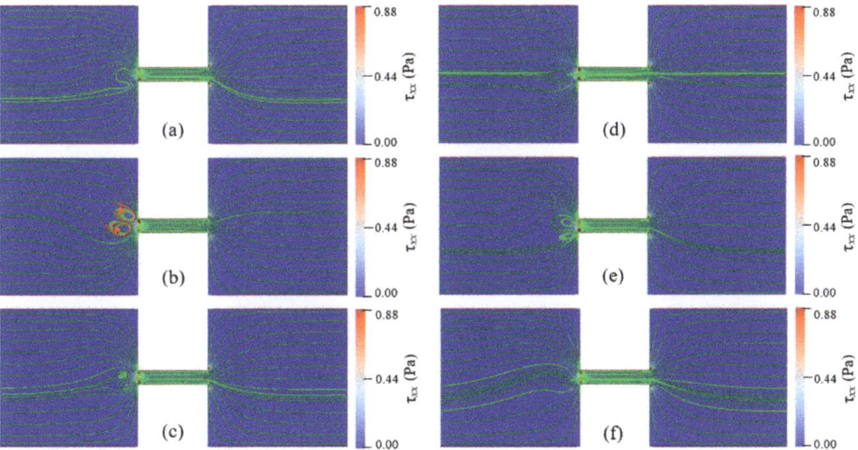

Figure 6. Instability of EOF with $c_\mathrm{p} = 500$ ppm and $E_\mathrm{app} = 100$ V/cm. Streamlines at different times: (**a**) 1.71 s, (**b**) 1.75 s, (**c**) 1.79 s, (**d**) 1.83 s, (**e**) 1.87 s, and (**f**) 1.91 s. The color bar represents the elastic normal stress τ_{xx}. (Supplementary Materials Video S1).

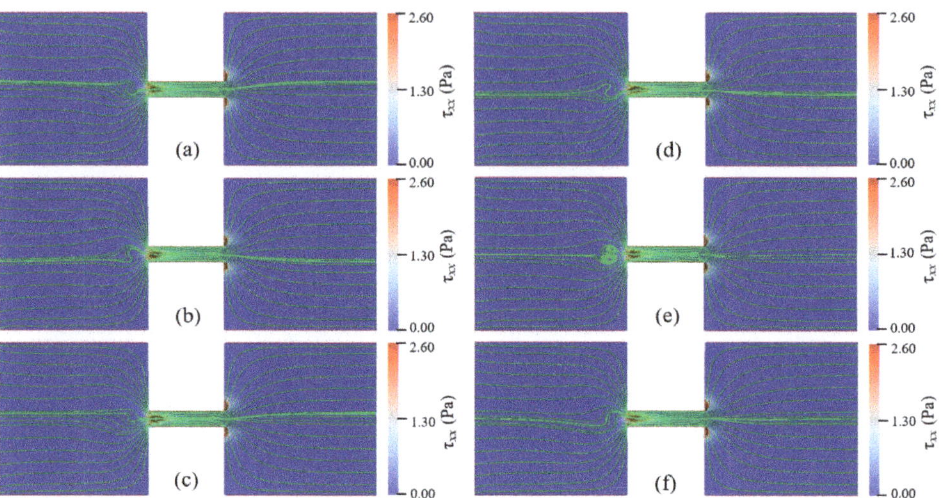

Figure 7. Instability of EOF with $c_p = 150$ ppm and $E_{app} = 600$ V/cm. Streamlines at different times: (**a**) 1.70 s, (**b**) 1.72 s, (**c**) 1.74 s, (**d**) 1.76 s, (**e**) 1.78 s, and (**f**) 1.80 s. The color bar represents the elastic normal stress τ_{xx}. (Supplementary Materials Video S2).

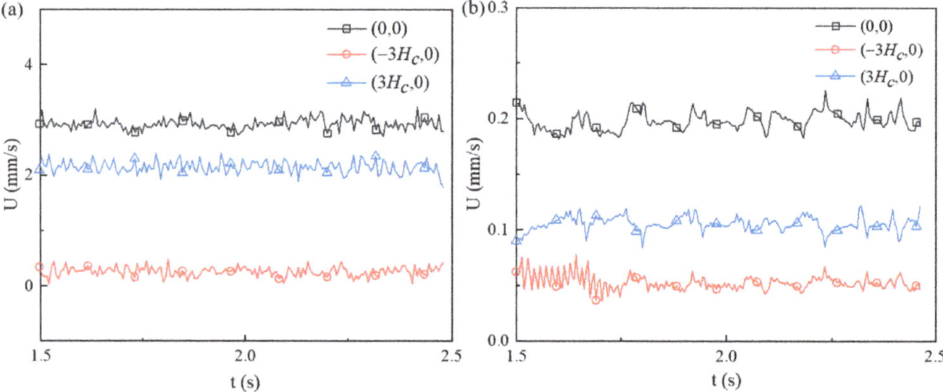

Figure 8. Velocity magnitudes at three different locations $((-3H_c, 0), (0, 0), (3H_c, 0))$. (**a**) $c_p = 150$ ppm and $E_{app} = 600$ V/cm, (**b**) $c_p = 500$ ppm and $E_{app} = 100$ V/cm.

Figures 6 and 7 show that the viscoelastic EOF is time-dependent. The streamlines in the left inlet reservoir far away from the solid walls (AB and GH in Figure 1) and near the entrance of the constriction microchannel show significant fluctuation and also become asymmetric about the *x*-axis. However, the streamlines near the solid walls of both reservoirs and in the outlet reservoir show insignificant change with time. Within 0.1 s, vortices continuously form and disappear within the inlet reservoir right before the entrance of the constriction microchannel. In Figure 6a, we observe significant curvature at the streamlines of the EOF upstream of the constriction microchannel. Then, the curvature of the streamlines further develops into a pair of vortices as shown in Figure 6b. Such vortices are also time-dependent. The size and shape of the vortices show notable differences at different times. After growing to the maximum size, the vortices start to shrink until the vortices break and disappear, as shown in Figure 6,d. Next, the vortices in the EOF keep forming and breaking repeatedly as shown in Figure 6e,f. Comparing Figure 6b,c, the central locations of the vortices are both spatially and temporally dependent. In Figure 6b, the direction of the circulation is marked by red curved lines. The pair of vortices are in opposite directions and form a stagnant region right before the entrance of the constriction microchannel. Therefore, we call the induced vortices as entrance-centerline vortices.

For $c_p = 150$ ppm and $E_{app} = 600$ V/cm, the width and length of the vortices are nearly the same as the height of the constriction microchannel (H_c), while for $c_p = 500$ ppm and $E_{app} = 100$ V/cm, the width and length of the vortices are about $2H_c$. In the similar geometry, however, Ko [38] did not observe vortices in their experiments with 200 ppm PAA solution under E_{app} ranging from 75 V/cm to 200 V/cm. The elastic instability increases with increasing polymer concentration and the applied electric field. Our later results discussed in the following section show that for $c_p = 200$ ppm, the vortices occur when the applied electric field exceeds the threshold value of 300 V/cm. Therefore, our numerical results qualitatively agree with the experimental observation of Ko [38] under their experimental condition. Table A1 in the Appendix A summarizes the EOF instabilities from the literature. In Sadek's [39] experimental study, small vortices at the entrance and large upstream circulation flows were observed. For highly concentrated polymer solution, downstream circulation flows were observed at a critical voltage. The upstream vortices found in this study are distinct from the small vortices and large circulation flows found in Sadek's study [39] in terms of location. Note that the geometry in our study differs significantly from the experimental study of Sadek [39], and we do not observe large circulating flows near the reservoir corners and channel lips.

For Newtonian fluid, the EOF reaches a steady state, and the velocity magnitudes are symmetric about the y-axis. Therefore, for Newtonian fluid, we have $U(-3H_c, 0) = U(3H_c, 0)$. However, as shown in Figure 8, the velocity magnitudes at three points $(-3H_c, 0)$, $(0, 0)$, and $(3H_c, 0)$ fluctuate around certain values and velocity magnitudes do not show symmetry about the y-axis. For $c_p = 500$ ppm and $E_{app} = 100$ V/cm, as shown in Figure 8b, the time-averaged velocity at the channel center is 0.201 mm/s with a standard deviation of 0.013 mm/s. The time-averaged velocity at downstream of the constriction microchannel is 0.108 mm/s, which is about 2 times of that at upstream of the constriction microchannel (i.e., 0.047 mm/s). For $c_p = 150$ ppm and $E_{app} = 600$ V/cm, the time-averaged velocities at the upstream, center, and downstream of the constriction microchannel are, respectively, 0.19 mm/s, 2.82 mm/s, and 2.05 mm/s. The ratio of the downstream velocity to upstream velocity is about 10 times. In contrast to Newtonian EOF, the flow velocity of viscoelastic fluid at the downstream is significantly higher than that at the upstream, which has also been experimentally observed in Ko's [38] experiments, where a fluid jet after the constriction microchannel was observed and the ratio of the velocity at the downstream centerline to that at upstream of the constriction microchannel varies between 1 and 2 under E_{app} ranging from 75 V/cm to 200 V/cm and $c_p = 200$ ppm.

Figure 8 also shows that EOF of $c_p = 500$ ppm and $E_{app} = 100$ V/cm presents stronger instabilities than that of $c_p = 150$ ppm and $E_{app} = 600$ V/cm. Comparing Figures 6 and 7, the streamlines show stronger fluctuation and larger upstream vortices for solution with relatively high polymer concentration. Such trend suggests that although the increase of both E_{app} and c_p can enhance the instabilities of the viscoelastic EOF, the polymer concentration, c_p, affects the instabilities of the EOF more significantly, which will be further discussed in next section.

Figures 6 and 7 also show the spatial distribution of elastic normal stress τ_{xx} with the color bar representing its magnitude. To clearly reveal it, Figure 9 depicts the spatial distribution of τ_{xx} in the whole geometry for $c_p = 150$ ppm and $E_{app} = 600$ V/cm at $t = 1.78$ s. Within the two reservoirs, the elastic normal stress is nearly zero at location far away from the constriction microchannel. However, significant elastic normal stress is induced near the entrance of the constriction microchannel and near the downstream lips. Due to the contraction geometry, the electric field within the constriction microchannel is about 10 times of that within the inlet reservoir as shown in Figure 4a, and the flow velocity in the microchannel is significantly higher than that in the reservoir. For example, Figure 8 shows that the ratio of the time-averaged velocity within the microchannel to that in the inlet reservoir, $U(0,0)/U(-3H_c,0)$, is 4.28 for $c_p = 500$ ppm and $E_{app} = 100$ V/cm and 10.79 for $c_p = 150$ ppm and $E_{app} = 600$ V/cm. Near the entrance of the microchannel, the high velocity gradient results in a strong extension of polymer molecules, and consequently induces significant elastic normal stress. Therefore, τ_{xx} experiences a rapid increase near the entrance of the constriction microchannel. At the exit of the constriction microchannel, similarly, a significant increase of τ_{xx} is induced at the exit lips. Figure 10a,b depict the streamlines in the constriction microchannel and the color bar represents the velocity magnitude, U, for $c_p = 150$ ppm and $E_{app} = 600$ V/cm and Newtonian fluid with the same total viscosity and E_{app} at $t = 1.78$ s, respectively. For the viscoelastic fluid, at both the entrance and exit of the constriction microchannel, as shown by the dashed circles in Figure 10a, velocity becomes spatially dependent along the y-axis. Velocity near the walls of the constriction microchannel is significantly higher than that at the centerline of the microchannel, and a local maximum occurs near the inlet/outlet corners of the constriction microchannel. However, in the EOF of Newtonian fluid, as shown in Figure 10b, at both the entrance and exit of the constriction microchannel, the velocity magnitude is more evenly distributed in the cross section of the constriction microchannel. Figure 10c depicts the velocity magnitude profile at the entrance ($2x/H_c = -5$) and exit ($2x/H_c = 5$) of the constriction microchannel. For Newtonian fluid, the velocity magnitude profile is identical at $2x/H_c = \pm 5$ and is symmetric about the x-axis. The ratio of the maximum velocity magnitude near the channel walls to that at the centerline is 1.6. However, for PAA solution, due to the elastic instability, the velocity magnitude profile is asymmetric about the x-axis at $2x/H_c = \pm 5$. In addition, the ratios of the maximum velocity magnitude near the channel walls to that at the centerline are 9.7 and 4 at $2x/H_c = -5$ and $2x/H_c = 5$, respectively, which are much higher than that of the Newtonian fluid. For Newtonian fluid, the velocity profile is symmetric about the centerline of the microchannel (i.e., $y = 0$). However, Figure 10c shows that the local maximum velocity near the top channel wall differs from that near the bottom channel wall, and the velocity profile is asymmetric about $y = 0$.

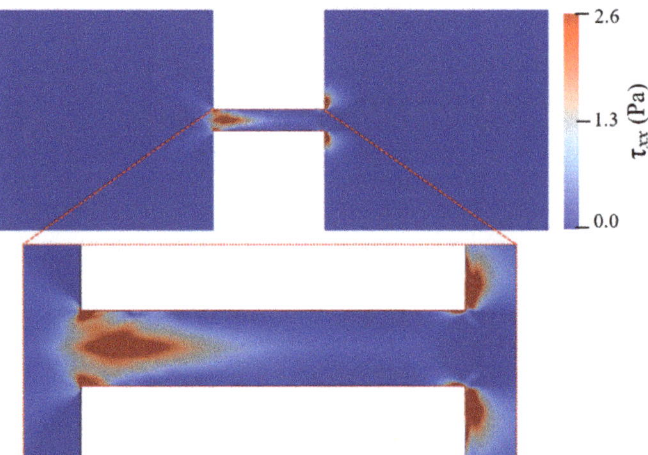

Figure 9. Spatial distribution of the elastic normal stress τ_{xx} for $c_p = 150$ ppm and $E_{app} = 600$ V/cm at $t = 1.78$ s.

As PAA solution flows from the microchannel into the outlet reservoir, fluid velocity first decreases when fluid exits the microchannel and then increases in the outlet reservoir, as shown by the region marked with a circle in Figure 10a and by the velocity magnitude as a function of x at $y = 0$ in Figure 10d. The two dashed lines in Figure 10d represent the entrance and exit of the constriction microchannel. EOF of Newtonian fluid within the constriction is a plateau, and its velocity magnitude within the constriction is much higher than those at both reservoirs, and this is because the electric field within the constriction microchannel is significantly higher than that in the reservoirs. However, the velocity of PAA solution becomes spatially dependent within the constriction, and a local maximum occurs before the exit and a local minimum occurs at the exit of the constriction microchannel. In addition, a local maximum occurs at the downstream outlet reservoir. Figure 10d also clearly shows that the velocity in the downstream outlet reservoir is significantly higher than that at the upstream inlet reservoir. For example, $U(2x/H_c = 10.0)/U(2x/H_c = -10.0) = 3.37$. The unexpected velocity decrease at the microchannel exit and velocity increase at the downstream outlet reservoir do not occur in Newtonian fluid as shown in Figure 10b,d. Such a phenomenon is probably because of the extrudate swell effect of polymers [59]. At the exit of the constriction microchannel, curved streamlines tilting toward the walls of the constriction microchannel are observed in viscoelastic fluid, suggesting that fluid tends to flow toward the charged walls of the microchannel. Such lateral velocity component results in the velocity's increase near the microchannel walls and velocity's decrease near the centerline at the exit of the constriction microchannel. In addition, the significant increase of τ_{xx} near the downstream lips observed in Figure 9 can also be attributed to the extrudate swell effect of polymers when polymer exits from the constriction microchannel to larger outlet reservoir.

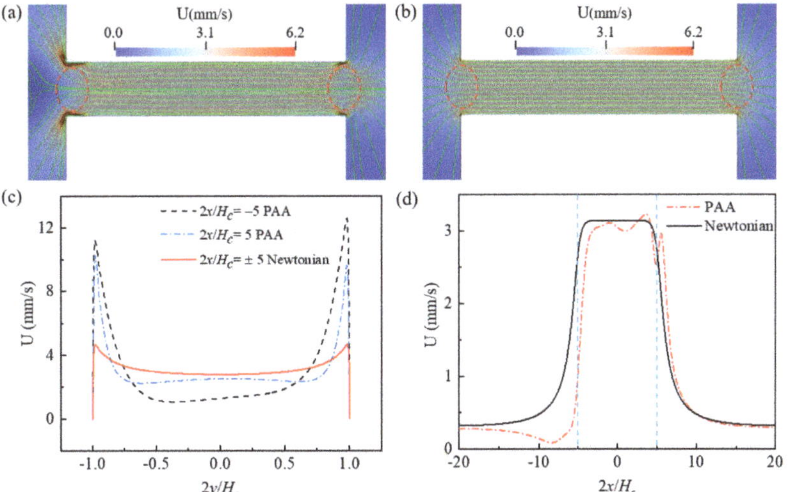

Figure 10. Streamlines and velocity magnitude for $c_p = 150$ ppm and $E_{app} = 600$ V/cm and Newtonian fluid at $t = 1.78$ s: (**a**) 150 ppm PAA solution, (**b**) Newtonian fluid with same total viscosity as 150 ppm PAA solution, (**c**) velocity magnitude profiles at $2x/H_c = \pm 5$, (**d**) velocity magnitudes profiles at $y = 0$ (The blue dash lines show the position of the contraction microchannel). The color bar represents the velocity magnitude U.

The generated elastic normal stress τ_{xx} was typically used to explain the formation of vortices in pressure-driven viscoelastic flows within curved geometries [60–62]. It has been reported that the development of the polymeric elastic stresses is caused by the flow-induced changes of the polymer conformation in the solution. Such changes of the polymer conformation are strain-dependent, anisotropic, and dependent on the flow. The extra elastic stresses are nonlinear under shear and can alter the flow behavior. At low Reynolds numbers where inertia is negligible, when the elastic normal stress exceeds by a certain amount the local shear stress, the flow transits from stable to unstable, and the vortices form at upstream of the constriction microchannel. Such elastic instabilities are often observed in flows with sufficient curvature [63–65], and some argue that curvature is necessary for infinitesimal perturbations to be amplified by the normal stress imbalances in the viscoelastic flows [66]. However, other theoretical studies reported that viscoelastic flows also showed a nonlinear instability in parallel shear flows, such as in viscoelastic flows within straight pipes at low Reynolds numbers [67]. Although the formation of the upstream vortices in the viscoelastic EOF shares the same mechanism as the pressure-driven flow, the locations of the vortices found in this study are distinct from the typical lip and corner vortices occurring in pressure-driven viscoelastic flows. This is probably because of the different velocity profiles in the pressure-driven flow and the EOF. In pressure-driven flow, the velocity is zero at solid walls and increases to a maximum at the centerline of the geometry. However, the EOF velocity profile is nearly a plug flow as shown in Figure 4b. The velocity increases from zero to a plateau within the EDL thickness, which is only on the order of a few nanometers. For the pressure-driven flow, the highest velocity is at the centerline of the geometry and the velocity near the wall is relatively low, resulting in the stagnant region near the solid boundaries (lips and corners). However, EOF velocity in the vicinity of the charged wall is almost the same as that in the channel centerline. For the extensional flow of viscoelastic fluids, the stretched polymer molecules lead to large elastic stresses, which significantly depend on the geometry and velocity profile. The induced elastic stresses render the primary flow unstable and cause an irregular secondary flow. The flow subsequently acts back on the polymer molecules and stretches them further, causing a strong disturbance of the EOF and yielding a time-dependent EOF.

4.2. Elastic Instabilities under Various E_{app} and c_p

In order to study the effects of E_{app} and c_p on the instabilities of viscoelastic EOF, c_p is varied from 100 ppm to 500 ppm and E_{app} is varied from 100 V/cm to 600 V/cm. Flow patterns under different conditions of E_{app} and c_p are shown in Figures 11–14. At certain E_{app} (c_p), EOF becomes more unstable with the increase of c_p (E_{app}). Figure 11 shows the streamlines for different PAA concentrations under $E_{app} = 600$ V/cm, and Figure 12 shows the streamlines within the constriction microchannel with the color bar representing pressure for Newtonian fluid and τ_{xx} for PAA solutions. As shown in Figures 11 and 12, when c_p increases, the polymeric stress τ_{xx} at the entrance of the constriction microchannel increases rapidly, resulting in the fluctuation of the streamlines at upstream of the microchannel. When c_p is relatively low (100 ppm), EOF of viscoelastic fluid is similar to that of Newtonian fluid, and the flow is in a steady state. With an increase in the PAA concentration up to 150 ppm, significant curvature of the centerline streamlines is observed, and the streamlines become asymmetric about the x-axis and y-axis. As c_p continuously increases up to 200 ppm, a pair of upstream vortices in opposite flow directions are induced at upstream of the constriction microchannel, forming a stagnant region as shown in Figure 11d. The width and length of the pair of vortices are about 1.6 time of the constriction microchannel height (i.e., $1.6H_c$). Such vortices are found to grow significantly in size with increasing c_p, which is in qualitative agreement with the experimental observations of Ko [38]. Within the constriction microchannel, as shown in Figure 12d, nearly 1/4 of the microchannel length (i.e., $\frac{1}{4}L_c$) from the entrance shows a significant increase of τ_{xx}. Near the downstream lips of the microchannel, a local maximum of the polymeric stress τ_{xx} is observed. When c_p increases to 250 ppm, the fluctuation of the streamlines and the size of the vortices grow dramatically as shown in Figure 11e, in which the width and length of the vortices are about 2.9 times of the microchannel height (i.e., $2.9H_c$). The region with significant value of τ_{xx} is near 1/3 of the microchannel length (i.e., $\frac{1}{3}L_c$), as shown in Figure 12e. When c_p further increases to 500 ppm, as shown in Figure 11f, the vortices grow into 4.4 times of the constriction microchannel height (i.e., $4.4H_c$). More than half of the microchannel length shows a significant increase in τ_{xx}. Furthermore, a small vortex is induced near the downstream lip of the microchannel, which is also reported in experimental studies of Ko [38].

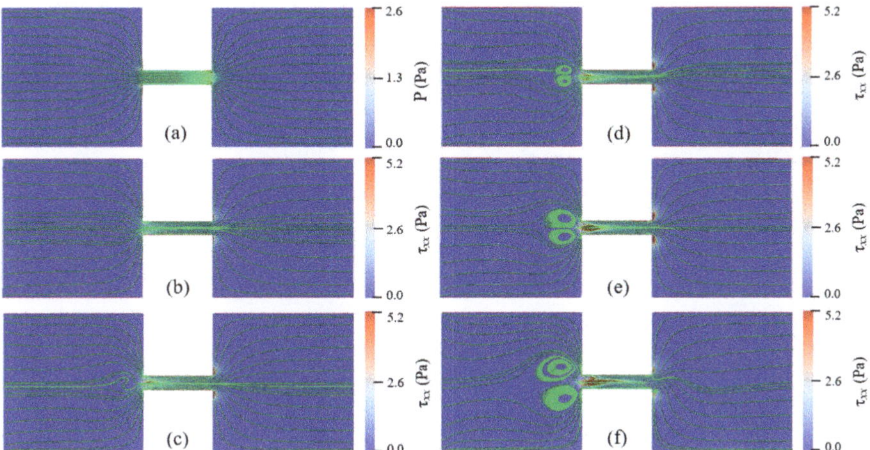

Figure 11. Streamlines of Newtonian fluid and PAA solutions with different concentrations under $E_{app} = 600$ V/cm at 1.70 s: (**a**) Newtonian fluid, (**b**) $c_p = 100$ ppm, (**c**) $c_p = 150$ ppm, (**d**) $c_p = 200$ ppm, (**e**) $c_p = 250$ ppm, and (**f**) $c_p = 500$ ppm. The color bar represents the elastic normal stress τ_{xx}.

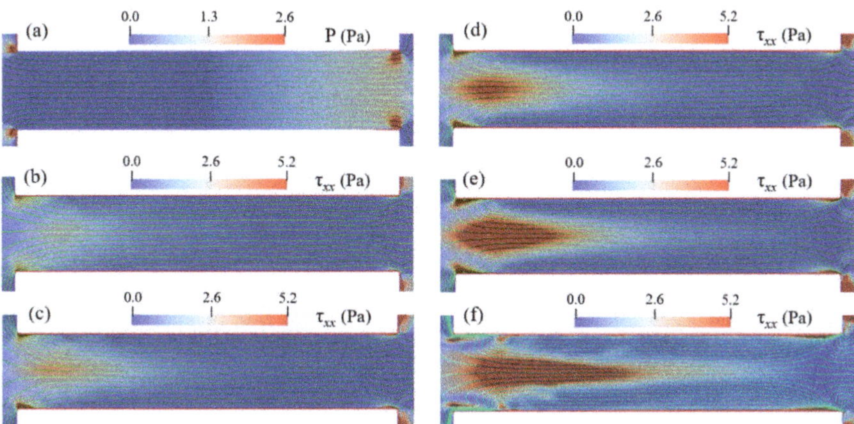

Figure 12. Streamlines in microchannel of Newtonian fluid and PAA solutions with different concentrations under $E_{app} = 600$ V/cm at 1.7 s: (**a**) Newtonian fluid, (**b**) $c_p = 100$ ppm, (**c**) $c_p = 150$ ppm, (**d**) $c_p = 200$ ppm, (**e**) $c_p = 250$ ppm, and (**f**) $c_p = 500$ ppm. The color bar represents the elastic normal stress τ_{xx}.

Figures 13 and 14 show the streamlines for $c_p = 150$ ppm when E_{app} is varied from 100 V/cm to 600 V/cm with the color bar representing the magnitude of τ_{xx}. At a relatively low electric field such as $E_{app} = 100$ V/cm, EOF of PAA solution is similar to the Newtonian fluid, and the flow is in a steady state and symmetric about channel centerline. In addition, the induced polymeric stress τ_{xx} in the constriction microchannel is relatively small. When E_{app} increases up to 400 V/cm, centerline streamlines start to show notable fluctuation and become asymmetric about the x-axis. At the entrance of the microchannel, a slight increase of τ_{xx} is observed, however, significant increase of τ_{xx} is observed near the microchannel walls and the downstream lips, as shown in Figure 14d. When E_{app} increases to 500 V/cm, a pair of upstream vortices are induced at upstream of the microchannel, and the size of vortices is about the height of constriction microchannel. Additionally, as shown in Figure 14e, significant τ_{xx} is induced near the entrance of the constriction microchannel. However, when E_{app} further increases to 600 V/cm, the size of the upstream vortices do not show notable increase in comparison with that of $E_{app} = 500$ V/cm. The results clearly show that increase of c_p and/or E_{app} can magnify the elastic instabilities of the viscoelastic EOF. However, the increase of c_p has a more significant enhancing effect on the elastic instabilities of the viscoelastic EOF than the increase of E_{app}.

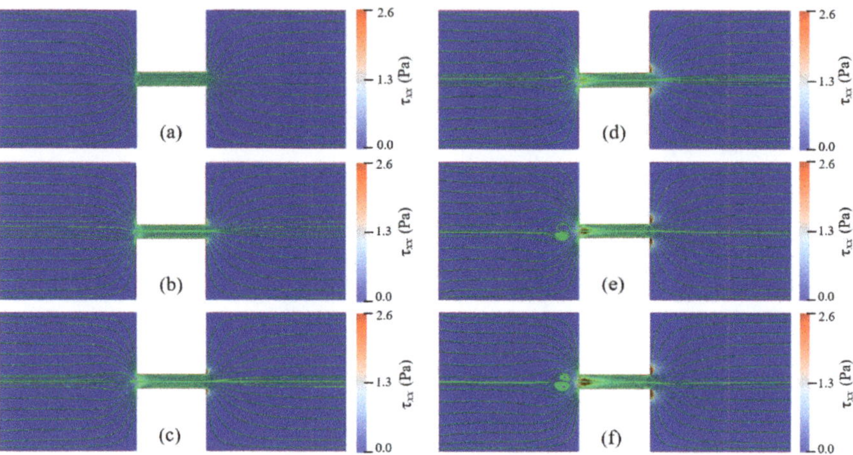

Figure 13. Streamlines for 150 ppm PAA solution under different E_{app} at 1.78 s: (**a**) 100 V/cm, (**b**) 200 V/cm. (**c**) 300 V/cm, (**d**) 400 V/cm, (**e**) 500 V/cm, (**f**) 600 V/cm. The color bar represents the elastic normal stress τ_{xx}.

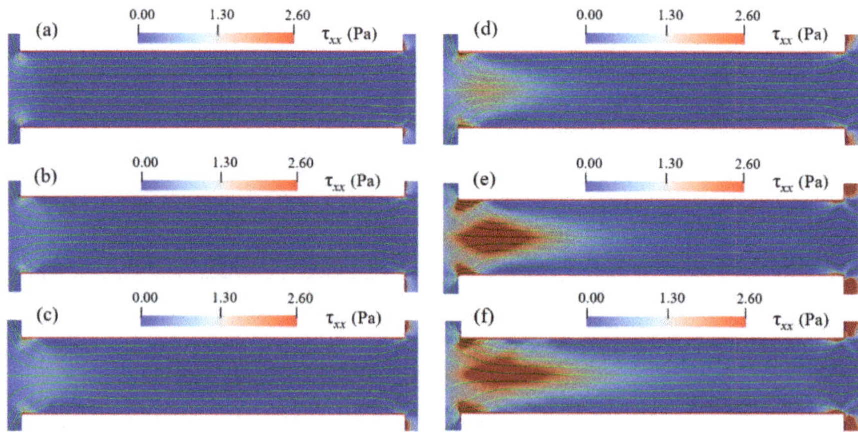

Figure 14. Streamlines in microchannel of 150 ppm PAA solution under different E_{app} at 1.78 s: (**a**) 100 V/cm, (**b**) 200 V/cm. (**c**) 300 V/cm, (**d**) 400 V/cm, (**e**) 500 V/cm, and (**f**) 600 V/cm. The color bar represents the elastic normal stress τ_{xx}.

Figure 15 depicts a flow map for the onset of vortices in unstable EOF as functions of c_p and E_{app}. At a fixed c_p (E_{app}), vortices and unstable EOF occur when E_{app} (c_p) exceeds a certain threshold value. For example, for $c_p = 200$ ppm, the flow becomes unstable with the occurrence of vortices when E_{app} is above 300 V/cm. At relatively low PAA concentration (i.e., $c_p = 100$ ppm), it requires a very high electric field (up to 850 V/cm) to yield unstable EOF with upstream vortices. In contrast, at relatively high c_p (i.e., $c_p = 500$ ppm), the onset of vortices occurs at E_{app} between 50 V/cm and 100 V/cm. An asymptotic curve fitting is implemented to illustrate the transition condition from no upstream vortices to the formation of upstream vortices, which is given as $E_{app} = 47.49 + 2892.25 \cdot 0.987^{c_p}$, where c_p represents the polymer concentration in ppm, and E_{app} is the apparent electric field in

V/cm. Above the curve in Figure 15, the EOF becomes time-dependent with upstream vortices, and no vortex forms under the conditions below the curve. Note that the flow map is only valid for the geometry considered in this study with the zeta potential of -110 mV. The instabilities of the viscoelastic EOF are dependent on the value of zeta potential. A comparison of the flow patterns of lower zeta potential (-70 mV) and higher zeta potential (-150 mV) for 150 ppm PAA solution under $E_{app} = 600$ V/cm in the Appendix A clearly shows that higher zeta potential triggers stronger instabilities under the same condition. In addition, the dimensionless numbers (Reynolds number and Weissenberg number) of the studied conditions corresponding to Figure 15 are given in the Appendix A.

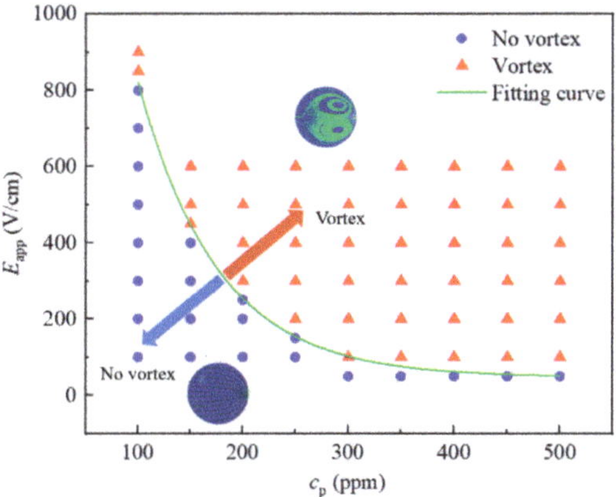

Figure 15. Flow map in c_p-E_{app} space for EOF of PAA solutions flowing through a 10:1:10 constriction/expansion microchannel. Up-right of the fitting curve are the conditions that trigger the vortex in the EOF.

Since the flow velocity is time-dependent, we first calculate the cross-sectional average velocity over a period of $\Delta t = t_2 - t_1$, and then take the time-average to obtain the averaged velocity as,

$$\overline{U} = \frac{\int_{t_1}^{t_2} \int_{-H_c/2}^{H_c/2} U(0,y) dy dt}{\Delta t \cdot H_c}, \tag{11}$$

We choose $\Delta t = 1$ s in the current study. Figure 16a shows the average velocity at the center of the constriction as a function of c_p at different values of E_{app}. In comparison, it also shows the result of Newtonian fluid whose viscosity is the same as the total viscosity of PAA solution under $E_{app} = 600$ V/cm. Under the same $E_{app} = 600$ V/cm, the average velocity in the constriction microchannel of the Newtonian flow is about 6–12% higher than that of the PAA solution. The decrease of the average velocity in PAA solution is attributed to the induced polymeric stress at the entrance of the constriction microchannel. For Newtonian fluid, the average velocity decreases as c_p increases, which is due to the increase of viscosity to make its viscosity be the same as that of PAA solution with the concentration of c_p. For PAA solutions, under the same E_{app} the average velocity exponentially decreases as the polymer concentration increases. One reason is attributed to the increase of total viscosity with the increase in c_p. In addition, the induced polymer stress within the constriction increases with the increase of polymer concentration, as shown in Figure 12, and the induced polymer stress slows down the flow.

Under the considered condition of $c_0 = 0.01$ mM, the EDL thickness is only 95 nm, which is much smaller than the height of the constriction. In Newtonian fluid, the EOF velocity can be approximated by the well-known Helmholtz–Smoluchowski velocity formula as described in Equation (10). We wonder if the average velocity of PAA solutions can be still approximated with the Helmholtz–Smoluchowski velocity formula. Since the actual local electric field within the constriction is much higher than the apparent electric field E_{app}, time-averaged electric field in the x-direction at the center of the constriction is used in the calculation of the Helmholtz–Smoluchowski velocity. Figure 16b shows the average velocity as a function of the apparent electric field under various PAA concentrations. The lines in Figure 16b represent the corresponding EOF velocity predicted by the Helmholtz–Smoluchowski velocity formula. At a fixed c_p, as expected, the EOF velocity increases with an increase in the applied electric field. In general, the Helmholtz–Smoluchowski formula over predicts the velocity, and at a fixed c_p the relative error increases with the increasing E_{app}. For example, for $c_p = 100$ ppm, the relative errors under $E_{app} = 100$ V/cm and 600 V/cm are, respectively, 1.2% and 8.6%. At a fixed E_{app}, the absolute error, which is the difference between the Helmholtz–Smoluchowski approximated velocity and the average velocity obtained from the full numerical simulation, increases with the increasing PAA concentration. However, the relative error shows no notable change with the increasing c_p. For example, at $E_{app} = 600$ V/cm, the relative errors for $c_p = 100$ ppm, 300 ppm, and 500 ppm are, respectively, 8.7%, 9.6%, and 9.0%. To evaluate the applicability of the Helmholtz–Smoluchowski formula to approximate the velocity of viscoelastic fluids for c_p ranging from 100 ppm to 500 ppm, the minimum, average, and maximum relative errors at different E_{app} are calculated as shown in Table 1. When $E_{app} \leq 300$ V/cm, the relative error is less than 5%. However, when $E_{app} \geq 400$ V/cm, the relative error is larger than 5%, and the Helmholtz–Smoluchowski formula failed to predict the velocity of the viscoelastic fluids accurately. In this study, the largest relative error is 9.6% when $c_p = 300$ ppm and $E_{app} = 600$ V/cm.

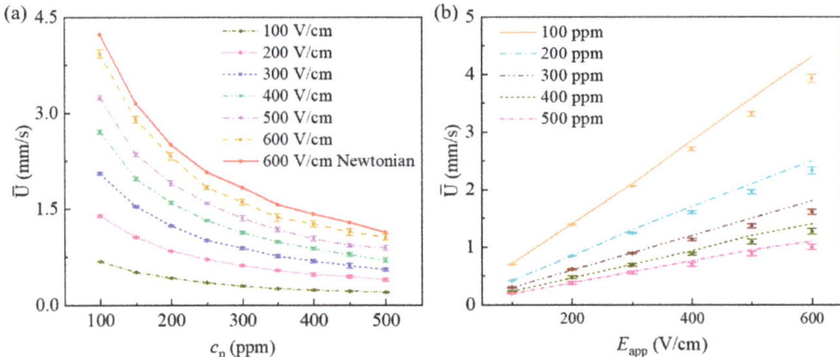

Figure 16. Time averaged cross-sectional average velocity at the center of the constriction microchannel ($x = 0$): (**a**) Average velocity, (**b**) comparison of average velocity (with deviation) and Helmholtz–Smoluchowski velocity (lines).

Table 1. Relative error between the Helmholtz–Smoluchowski velocity and the average velocity from the full mathematical model.

E_{app} (V/cm)	100	200	300	400	500	600
Minimum relative error	1.0%	1.3%	1.8%	5.5%	7.0%	8.5%
Average relative error	1.3%	1.6%	2.3%	5.9%	7.7%	8.9%
Maximum relative error	1.5%	1.9%	2.8%	6.5%	8.2%	9.6%

5. Conclusions

Electroosmotic flow (EOF) of viscoelastic fluid through a 10:1:10 constriction microchannel is numerically investigated as functions of the applied electric field and the polymer concentration. In the current study, we neglect the effect of the polymer concentration on the zeta potential of the channel walls. Comparing to the EOF of Newtonian fluid, the following distinct results for viscoelastic EOF through a 10:1:10 constriction microchannel are obtained:

(1) When polyacrylamide (PAA) concentration (applied electric field) exceeds a critical value, the EOF of viscoelastic fluid becomes time-dependent with upstream vortices occurring in the inlet reservoir near the entrance of the constriction microchannel. In contrast, EOF of Newtonian fluid is always in a steady state without vortices.

(2) For the viscoelastic EOF, significant polymer stress is induced near the entrance within the constriction and near the downstream lips of the constriction, causing the elastic instabilities of the viscoelastic EOF. The induced polymer stress is dramatically magnified with the increase of polymer concentration and applied electric field. However, the increase of polymer concentration shows a more significant enhancing effect on the polymer stress than the increase of applied electric field.

(3) The EOF velocity of viscoelastic fluid within the constriction becomes temporally and spatially dependent. Near the exit of the constriction, due to the extrudate swell effect of the polymers, the velocity at the centerline first decreases at the exit followed by an increase in the outlet reservoir.

(4) The velocity at the exit of the constriction is higher than that at the entrance of the constriction because of the formation of upstream vortices, which is in qualitative agreement with experimental observation obtained from the literature.

(5) Under the same total viscosity and applied electric field, the velocity of Newtonian fluid is higher than that of viscoelastic fluid, which is attributed to the induced polymeric stress within the constriction. When the applied electric field is less than 300 V/cm, the Helmholtz–Smoluchowski velocity formula can predict the cross-sectional average velocity of viscoelastic fluid with PAA concentration up to 500 ppm, and the relative error is less than 5%. At a fixed PAA concentration, in general the relative error of the Helmholtz–Smoluchowski approximation increases with an increase in the applied electric field.

Supplementary Materials: The following are available online at https://www.mdpi.com/article/10.3390/mi12040417/s1. Video S1: The development of vortices for 500 ppm PAA solution under 100 V/cm apparent electric field. $t_0 = 1.7\,\mathrm{s}$ and $\Delta t = 4.8\,\mathrm{ms}$ for each frame. Video S2: The development of vortices for 150 ppm PAA solution under 600 V/cm apparent electric field. $t_0 = 1.7\,\mathrm{s}$ and $\Delta t = 2.0$ ms for each frame.

Author Contributions: J.J. conducted simulations and drafted the manuscript; S.Q. supervised and revised the manuscript; and Z.L. analyzed and interpreted results. All authors have read and agreed to the published version of the manuscript.

Funding: This research was funded, in part, by the Foundation of State Key Laboratory of Coal Combustion (FSKLCCA1802).

Institutional Review Board Statement: Not applicable.

Informed Consent Statement: Not applicable.

Data Availability Statement: Not applicable.

Conflicts of Interest: The authors declare no conflict of interest.

Appendix A

Appendix A.1. Mesh Independence Study

Three different meshes are used to conduct the mesh independent study with $E_{app} = 600$ V/cm and $c_p = 100$ ppm. As shown in Figure A1, there are 135,192, 95,252, and 77,192 cells in mesh 1, mesh 2, and mesh 3, respectively. For mesh 1, the meshes near the charged wall are 7 nm, so there are 14 meshes within the EDL thickness. In mesh 2 and mesh 3, the meshes near the charged wall are 10 nm, and there are 10 meshes within the EDL thickness. However, there are less meshes within the two reservoirs in mesh 3 than in mesh 2. Figure A2 shows spatial distribution of the normal polymer stress and streamlines of three different meshes at $t = 1.78$ s. The normal polymer stress and streamlines of three meshes show no notable difference.

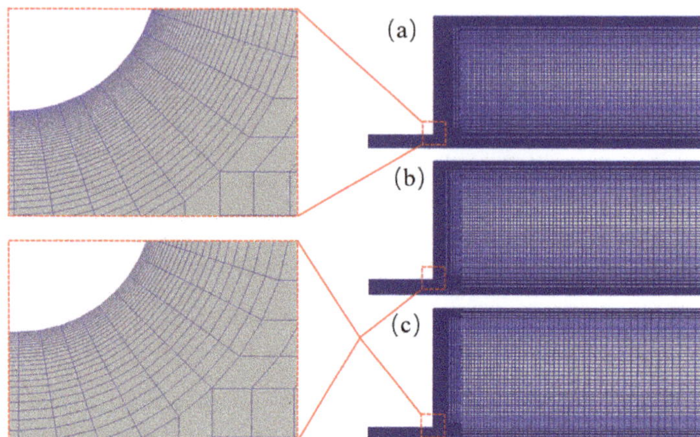

Figure A1. Three different meshes used for the mesh independence study. The meshes are symmetric with respect to the *x*-axis and *y*-axis, and only 1/4 of the total meshes are showed. (**a**) Mesh 1: 135.192 cells. (**b**) Mesh 2: 95.252 cells. (**c**) Mesh 3: 77.192 cells.

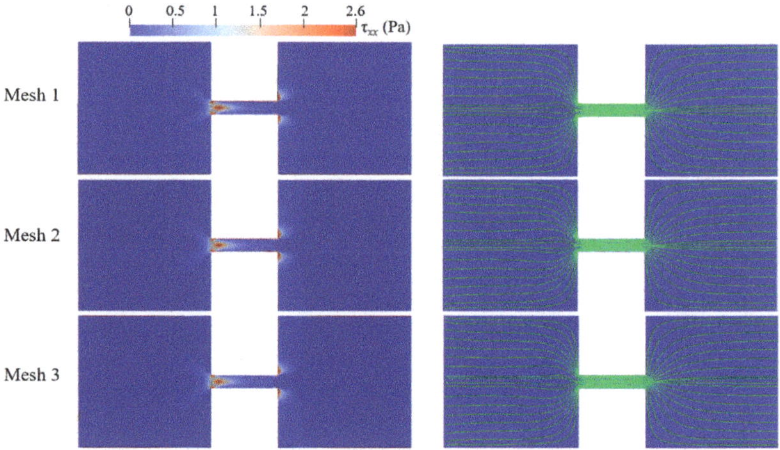

Figure A2. Spatial distribution of normal polymeric stress (**left**) and streamlines (**right**) for mesh 1 (**the top row**), mesh 2 (**the middle row**), and mesh 3 (**the bottom row**) at $t = 1.78$ s.

Figure A3 shows the spatial distribution of velocity magnitudes along $x = 0$ and $y = 0$ at $t = 1.78$ s. For velocity magnitude distribution along $x = 0$ (Figure A3a), the maximum relative error occurs at $y = 0$. The maximum relative error for mesh 2 is $|U_{mesh2} - U_{mesh1}|/U_{mesh1} = 0.9\%$, and the maximum relative error for mesh 3 is $|U_{mesh3} - U_{mesh1}|/U_{mesh1} = 1.1\%$. The average relative errors for all sampling points are 0.45% for mesh 2 and 0.64% for mesh 3. For velocity magnitude distribution along $y = 0$ (Figure A3b), comparing to mesh 1, the average relative errors for all sampling points are, respectively, 0.29% for mesh 2 and 0.72% for mesh 3.

Since the results from the above three meshes are in good agreement, we use mesh 3 to perform our other simulations.

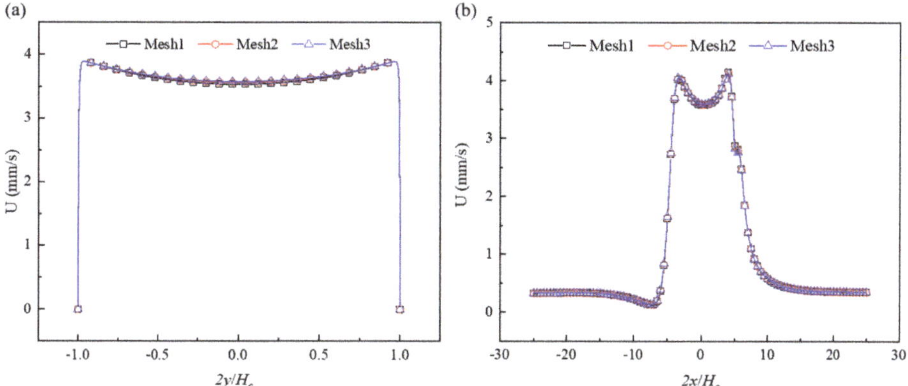

Figure A3. Spatial distribution of velocity magnitudes at t = 1.78 s. (**a**) Velocity magnitudes profile at *x* = 0. (**b**) Velocity magnitudes profile at *y* = 0.

Appendix A.2. Boundary Conditions

Boundaries:
1. Red dash line: Anode.
2. Black dash line: Cathode.
3. Green lines: Charged walls of reservoirs and microchannel.

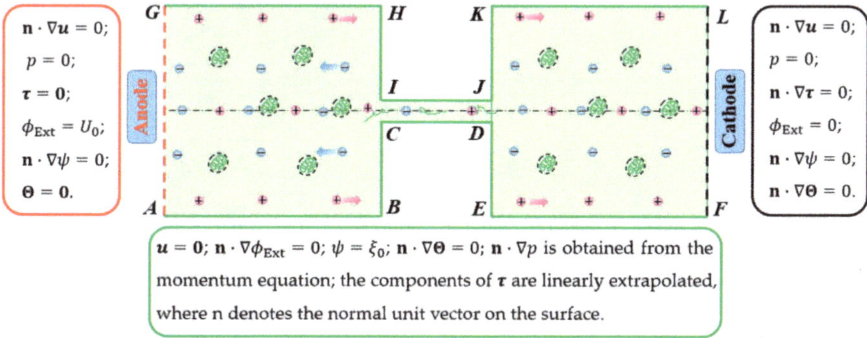

Figure A4. Boundary conditions with **n** denoting the normal unit vector on the surface.

Appendix A.3. Analytical Solution of Afonso et al

Based on PTT model, Afonso et al. [28] derived the analytical solution of mixed electro-osmotic/pressure driven flows of viscoelastic fluids between parallel plates based on the Debey–Hückel approximation. The analytical solution of the velocity profile across the height of the channel is given as:

$$u^E = \left(\frac{\epsilon \psi_0 E_x}{\eta} - 2\overline{C}\kappa^2 \varepsilon \lambda^2 \left[\frac{\epsilon \psi_0 E_x}{\eta} \right]^3 \right) (\overline{A} - 1) + \frac{2}{3}\kappa^2 \varepsilon \lambda^2 \left[\frac{\epsilon \psi_0 E_x}{\eta} \right]^3 \times (\overline{A}^3 - 1) \quad (A1)$$

where ϵ is the dielectric constant of the solution, $\overline{A} = ((\cosh(\kappa y))/(\cosh(\kappa H)))$, $\kappa^2 = ((2n_0 e^2 z^2)/(\epsilon \kappa_B T))$, κ_B is the Boltzmann constant, and n_0 is the ionic density.

For OB model, $\varepsilon = 0$. The Equation (A1) becomes $u^E = \left(\frac{\epsilon \psi_0 E_x}{\eta} \right) (\overline{A} - 1)$.

Appendix A.4. Summary of Current Studies

Table A1. Summary of current studies on elastic instabilities of viscoelastic EOF.

Geometry	Fluids	Electrical Parameters	Instabilities
Constriction channel [32] Contraction ratio: 2:1 Width: 200 μm and 100 μm Depth: 20 μm	PAA (18×10^6 and 5×10^6 Da) with 20:80 vol.% methanol: water mixture. 1–480 ppm.	Electro-osmotic mobility: $(5.6 \pm 0.5) \times 10^{-4}$, $(5.7 \pm 0.7) \times 10^{-4}$, and $(3.1 \pm 0.2) \times 10^{-4}$ cm^2/Vs for polymer free, high molecular weight, and low molecular weight 120 ppm PAA solutions. Electric field: 0–900 V/cm.	1. High molecular weight PAA solution leads to fluctuation above a critical flow rate. 2. The fluctuation is dependent on the polymer concentration.
Cross slot channel [34,35] Width: 100 μm Height: 50 μm	PAA, 5×106 Da, 300 and 1000 ppm.	Electro-osmotic mobility: $(6.4 \pm 0.2) \times 10^{-4}$, $7.2 \pm 0.1) \times 10^{-4}$cm^2/Vs for 1000 ppm and 300 ppm PAA solutions. $\Delta V = 20 - 140$ V.	1. For both 300 ppm and 1000 ppm PAA solution, transition from steady state to time-dependent state was observed at $\Delta V = 20$–40 V. 2. The fluctuation of the EOF does not enhance mixing effect.
T-shaped channel [37] Main-branch: Width: 200 μm Depth: 30 μm Side-branch 1: Width: 100 μm Depth: 50 μm Side-branch 2: Width: 100 μm Depth: 67 μm	Phosphate buffer-based aqueous polymer solutions. (500 ppm XG, 5% PVP, 2000 ppm PEO, 200 ppm PAA, 1000 ppm hyaluronic acid(HA)).	Electro-osmotic mobility: 3.5×10^{-4}, 0.1×10^{-4}, 4.6×10^{-4}, 0.1×10^{-4}, and 4.2×10^{-4} cm^2/Vs for XG, PVP, HA, PEO, PAA solutions. $\Delta V = 100$–500 V.	1. Fluid shear-thinning might be the primary cause of electro-elastic instabilities (fluctuation of velocity). 2. Fluid elasticity alone does not cause instability. 3. Threshold voltage for the onset of instability decreases with the increase of polymer concentration. 4. Increasing the buffer concentration causes a rise of threshold voltage.
Constriction channel [38] Contraction ratio: 10:1 Width: 400 μm and 40 μm Depth: 40 μm depth Length of constriction: 200 μm	1 mM phosphate buffer-based aqueous polymer solutions. (2000 ppm XG, 5% PVP, 3000 ppm PEO, and 200 ppm PAA.)	Electro-osmotic mobility: 0.54×10^{-4}, 0.75×10^{-4}, 0.06×10^{-4}, 0.14×10^{-4}, 1.82×10^{-4} cm^2/Vs for buffer, XG, PVP, PEO, PAA solutions. Electric field: 100–400 V/cm	1. Fluid elasticity alone does not cause instability. 2. Fluid shear-thinning effect alone caused counter-rotating circulations in weakly elastic XG solution. 3. No electric field-dependent phenomenon was observed in strong viscoelastic and shear-thinning PAA solutions.

Table A1. *Cont.*

Geometry	Fluids	Electrical Parameters	Instabilities
Hyperbolic/abrupt contraction with abrupt/hyperbolic expansion [39] Channel 1: (7.2:1) Width: 401 µm and 56 µm Depth: 100 µm Length of constriction: 382 µm Channel 2: (22.4:1) Width: 403 µm and 18 µm Depth: 100 µm Length of constriction: 128 µm	1 mM borate buffer with 0.05% wt. sodiumdodecylsulfate. 100, 300, 1000, and 10,000 ppm PAA solutions.	Electro conductivity: 20.2, 55.5, 178.3 and 161.8 µS/cm for 100, 300, 1000, and 10,000 ppm PAA solutions. $\Delta V = 2.5–90$ V.	1. For hyperbolic contraction, upstream vortices were observed on high contraction ratio geometry. The vortices grow with increasing electric field. 2. For abrupt contraction, small vortices were observed at the entrance of low contraction ratio geometry. 3. For abrupt contraction, small vortices and large flow circulation were observed near the entrance. Additionally, large downstream vortices were observed for 1000 ppm PAA solution.

Appendix A.5. Results for Zeta potentials of -70 mV and -150 mV

Two other different zeta potentials (-70 mV and -150 mV) are studied for 150 ppm PAA solution under $E_{app} = 600$ V/cm. Figure A5 shows the results of -70 mV zeta potential. Significant curvatures of the centerline streamlines are observed. However, at other places, no significant disturbance is observed. Figure A6 shows the results of -150 mV zeta potential. Similar to the results of -110 mV zeta potential, strong disturbance is induced in the viscoelastic EOF. Upstream vortices form and disappear. An increase of elastic normal stress is observed within the constriction microchannel. Figure A7 shows the velocity magnitude at the center of constriction channel (i.e., (0,0)). When the zeta potential is low (-70 mV), the velocity magnitude is almost steady state. However, when the zeta potential is high (-150 mV), the velocity magnitude shows strong fluctuation. The results for zeta potentials of -70 mV, -110 mV, and -150 mV show that the elastic instabilities of the EOF of PAA solutions are dependent on the value of zeta potential. Higher zeta potential induces larger electroosmotic velocity, and therefore stronger stretching of the polymers at the entrance of the constriction microchannel. Higher velocity and polymer normal stress lead to stronger instabilities of the viscoelastic flow.

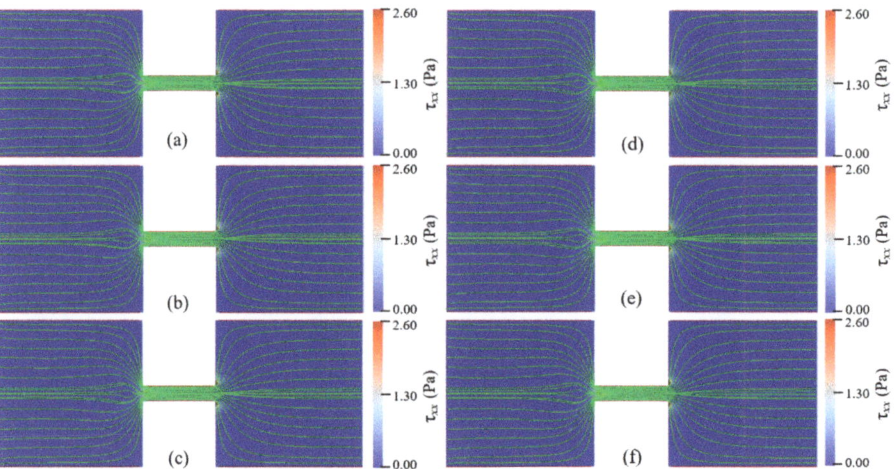

Figure A5. Streamlines of 150 ppm PAA solution under $E_{app} = 600$ V/cm at different times: (**a**) 1.70 s, (**b**) 1.72 s, (**c**) 1.74 s, (**d**) 1.76 s, (**e**) 1.78 s, (**f**) 1.80 s. Zeta potential is -70 mV. The color bar represents the elastic normal stress τ_{xx}.

Appendix A.6. Dimensionless Numbers

Dimensional analysis is a useful tool to fully characterize the flow and identify the dominant forces in complex flows of polymeric materials. Reynolds number (Re) is commonly used in rheological studies to determine whether the inertial force or the viscous force is dominating the flow, which is given by: $Re = \rho u l / \eta$, where ρ is the fluid density, u is the average velocity in the microchannel, l is the characteristic length scale (H_c), and η is the fluid viscosity. In addition, the Weissenberg number (Wi) is used to assess the flow elasticity of the PAA solutions, which is defined as: $Wi = u\lambda / l$, where λ is the relaxation time. Based on our results, the Re is nearly zero, indicating that the inertial force of the EOF is negligible. The Wi is on the order of 1, which indicates that the elastic effect in the flow is significant.

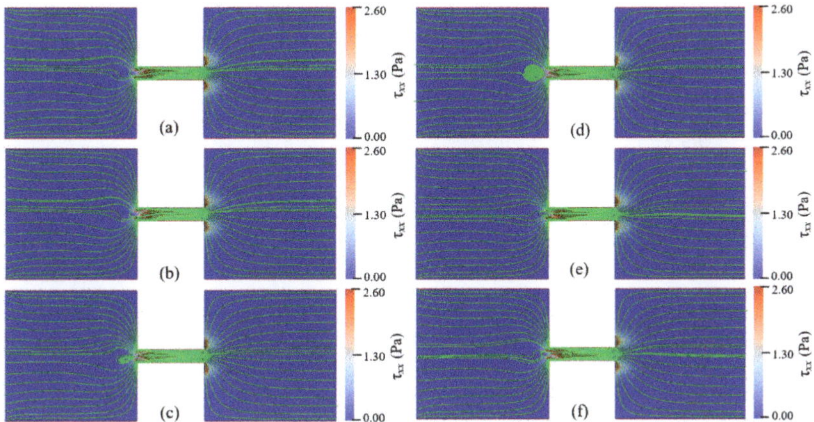

Figure A6. Streamlines of 150 ppm PAA solution under $E_{app} = 600$ V/cm at different times: (**a**) 1.70 s, (**b**) 1.72 s, (**c**) 1.74 s, (**d**) 1.76 s, (**e**) 1.78 s, (**f**) 1.80 s. Zeta potential is -150 mV. The color bar represents the elastic normal stress τ_{xx}.

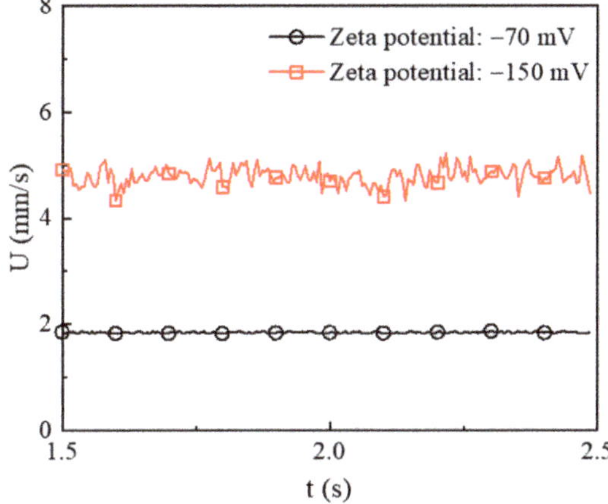

Figure A7. Velocity magnitudes at $(0,0)$ for 150 ppm PAA solution with zeta potentials of -70 mV and -150 mV under $E_{app} = 600$ V/cm.

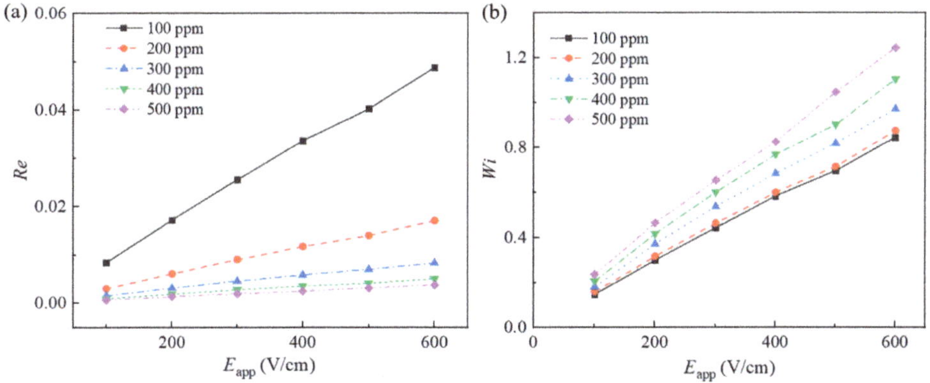

Figure A8. Reynolds number (**a**) and Weissenberg number (**b**) of the viscoelastic EOF under the conditions of Figure 15.

References

1. Gao, M.; Gui, L. A handy liquid metal based electroosmotic flow pump. *Lab Chip* **2014**, *14*, 1866–1872. [CrossRef] [PubMed]
2. Peng, R.; Li, D. Effects of ionic concentration gradient on electroosmotic flow mixing in a microchannel. *J. Colloid Interface. Sci.* **2015**, *440*, 126–132. [CrossRef]
3. Palyulin, V.V.; Ala-Nissila, T.; Metzler, R. Polymer translocation: The first two decades and the recent diversification. *Soft Matter.* **2014**, *10*, 9016–9037. [CrossRef]
4. Takamura, Y.; Onoda, H.; Inokuchi, H.; Adachi, S.; Oki, A.; Horiike, Y. Low-voltage electroosmosis pump for stand-alone microfluidics devices. *Electrophoresis* **2003**, *24*, 185–192. [CrossRef] [PubMed]
5. Li, L.; Wang, X.; Pu, Q.; Liu, S. Advancement of electroosmotic pump in microflow analysis: A review. *Anal. Chim. Acta* **2019**, *1060*, 1–16. [CrossRef]
6. Jiang, H.; Weng, X.; Chon, C.H.; Wu, X.; Li, D. A microfluidic chip for blood plasma separation using electro-osmotic flow control. *J. Micromech. Microeng.* **2011**, *21*, 085019. [CrossRef]
7. Ermann, N.; Hanikel, N.; Wang, V.; Chen, K.; Weckman, N.E.; Keyser, U.F. Promoting single-file DNA translocations through nanopores using electro-osmotic flow. *J. Chem. Phys.* **2018**, *149*, 163311. [CrossRef] [PubMed]
8. Huang, G.; Willems, K.; Soskine, M.; Wloka, C.; Maglia, G. Electro-osmotic capture and ionic discrimination of peptide and protein biomarkers with FraC nanopores. *Nat. Commun.* **2017**, *8*, 1–11. [CrossRef]
9. Buyukdagli, S. Facilitated polymer capture by charge inverted electroosmotic flow in voltage-driven polymer translocation. *Soft Matter.* **2018**, *14*, 3541–3549. [CrossRef] [PubMed]
10. Bello, M.S.; Besi, P.D.; Rezzonico, R.; Righetti, P.G.; Casiraghi, E. Electroosmosis of polymer solutions in fused silica capillaries. *Electrophoresis* **1994**, *15*, 623–626. [CrossRef]
11. Chang, F.M.; Tsao, H.K. Drag reduction in electro-osmosis of polymer solutions. *Appl. Phys. Lett.* **2007**, *90*, 194105. [CrossRef]
12. Kamişli, F. Flow analysis of a power-law fluid confined in an extrusion die. *Int. J. Eng. Sci.* **2003**, *41*, 1059–1083. [CrossRef]
13. Zimmerman, W.; Rees, J.; Craven, T. Rheometry of non-Newtonian electrokinetic flow in a microchannel T-junction. *Microfluid. Nanofluid.* **2006**, *2*, 481–492. [CrossRef]
14. Hakim, A. Mathematical analysis of viscoelastic fluids of White-Metzner type. *J. Math. Anal. Appl.* **1994**, *185*, 675–705. [CrossRef]
15. Das, M.; Jain, V.; Ghoshdastidar, P. Fluid flow analysis of magnetorheological abrasive flow finishing (MRAFF) process. *Int. J. Mach. Tool. Manu.* **2008**, *48*, 415–426. [CrossRef]
16. Oldroyd, J.G. On the formulation of rheological equations of state. *Proc. Math. Phys. Eng. Sci.* **1950**, *200*, 523–541.
17. Thien, N.P.; Tanner, R.I. A new constitutive equation derived from network theory. *J. Non-Newton. Fluid Mech.* **1977**, *2*, 353–365. [CrossRef]
18. Koszkul, J.; Nabialek, J. Viscosity models in simulation of the filling stage of the injection molding process. *J. Mater. Process. Technol.* **2004**, *157*, 183–187. [CrossRef]
19. Öztekin, A.; Brown, R.A.; McKinley, G.H. Quantitative prediction of the viscoelastic instability in cone-and-plate flow of a Boger fluid using a multi-mode Giesekus model. *J. Non-Newton. Fluid Mech.* **1994**, *54*, 351–377. [CrossRef]
20. Das, S.; Chakraborty, S. Analytical solutions for velocity, temperature and concentration distribution in electroosmotic microchannel flows of a non-Newtonian bio-fluid. *Anal. Chim. Acta* **2006**, *559*, 15–24. [CrossRef]
21. Zhao, C.; Zholkovskij, E.; Masliyah, J.H.; Yang, C. Analysis of electroosmotic flow of power-law fluids in a slit microchannel. *J. Colloid Interface Sci.* **2008**, *326*, 503–510. [CrossRef]
22. Zhao, C.; Yang, C. An exact solution for electroosmosis of non-Newtonian fluids in microchannels. *J. Non-Newton. Fluid Mech.* **2011**, *166*, 1076–1079. [CrossRef]
23. Zhao, C.; Yang, C. Joule heating induced heat transfer for electroosmotic flow of power-law fluids in a microcapillary. *Int. J. Heat Fluid Flow* **2012**, *55*, 2044–2051. [CrossRef]
24. Zhao, C.; Yang, C. Electroosmotic flows of non-Newtonian power-law fluids in a cylindrical microchannel. *Electrophoresis* **2013**, *34*, 662–667. [CrossRef]
25. Olivares, M.L.; Vera-Candioti, L.; Berli, C.L. The EOF of polymer solutions. *Electrophoresis* **2009**, *30*, 921–928. [CrossRef]
26. Tang, G.; Li, X.; He, Y.; Tao, W. Electroosmotic flow of non-Newtonian fluid in microchannels. *J. Non-Newton. Fluid Mech.* **2009**, *157*, 133–137. [CrossRef]
27. Park, H.; Lee, W. Effect of viscoelasticity on the flow pattern and the volumetric flow rate in electroosmotic flows through a microchannel. *Lab Chip* **2008**, *8*, 1163–1170. [CrossRef]
28. Afonso, A.; Alves, M.; Pinho, F. Analytical solution of mixed electro-osmotic/pressure driven flows of viscoelastic fluids in microchannels. *J. Non-Newton. Fluid Mech.* **2009**, *159*, 50–63. [CrossRef]
29. Afonso, A.; Alves, M.; Pinho, F. Electro-osmotic flow of viscoelastic fluids in microchannels under asymmetric zeta potentials. *J. Eng. Math.* **2011**, *71*, 15–30. [CrossRef]
30. Liu, Q.; Jian, Y.; Yang, L. Alternating current electroosmotic flow of the Jeffreys fluids through a slit microchannel. *Phys. Fluids* **2011**, *23*, 102001. [CrossRef]
31. Sousa, J.; Afonso, A.; Pinho, F.; Alves, M. Effect of the skimming layer on electro-osmotic-Poiseuille flows of viscoelastic fluids. *Microfluid. Nanofluidics* **2011**, *10*, 107–122. [CrossRef]
32. Bryce, R.; Freeman, M. Extensional instability in electro-osmotic microflows of polymer solutions. *Phys. Rev. E* **2010**, *81*, 036328. [CrossRef] [PubMed]

33. Bryce, R.; Freeman, M. Abatement of mixing in shear-free elongationally unstable viscoelastic microflows. *Lab Chip* **2010**, *10*, 1436–1441. [CrossRef] [PubMed]
34. Pimenta, F.; Alves, M. Stabilization of an open-source finite-volume solver for viscoelastic fluid flows. *J. Non-Newton. Fluid Mech.* **2017**, *239*, 85–104. [CrossRef]
35. Pimenta, F.; Alves, M. Electro-elastic instabilities in cross-shaped microchannels. *J. Non-Newton. Fluid Mech.* **2018**, *259*, 61–77. [CrossRef]
36. Song, L.; Jagdale, P.; Yu, L.; Liu, Z.; Li, D.; Zhang, C.; Xuan, X. Electrokinetic instability in microchannel viscoelastic fluid flows with conductivity gradients. *Phys. Fluids* **2019**, *31*, 082001.
37. Song, L.; Yu, L.; Li, D.; Jagdale, P.; Xuan, X. Elastic instabilities in the electroosmotic flow of non-Newtonian fluids through T-shaped microchannels. *Electrophoresis* **2020**, *41*, 588–597. [CrossRef]
38. Ko, C.H.; Li, D.; Malekanfard, A.; Wang, Y.N.; Fu, L.M.; Xuan, X. Electroosmotic flow of non-Newtonian fluids in a constriction microchannel. *Electrophoresis* **2019**, *40*, 1387–1394. [CrossRef]
39. Sadek, S.H.; Pinho, F.T.; Alves, M.A. Electro-elastic flow instabilities of viscoelastic fluids in contraction/expansion micro-geometries. *J. Non-Newton. Fluid Mech.* **2020**, *283*, 104293. [CrossRef]
40. Afonso, A.; Pinho, F.; Alves, M. Electro-osmosis of viscoelastic fluids and prediction of electro-elastic flow instabilities in a cross slot using a finite-volume method. *J. Non-Newton. Fluid Mech.* **2012**, *179*, 55–68. [CrossRef]
41. Huang, Y.; Chen, J.T.; Wong, T.; Liow, J.L. Experimental and theoretical investigations of non-Newtonian electro-osmotic driven flow in rectangular microchannels. *Soft Matter.* **2016**, *12*, 6206–6213. [CrossRef]
42. Ronshin, F.; Chinnow, E. Experimental characterization of two-phase flow patterns in a slit microchannel. *Exp. Therm. Fluid Sci.* **2019**, *103*, 262–273. [CrossRef]
43. Nito, F.; Shiozaki, T.; Nagura, R.; Tsuji, T.; Doi, K.; Hosokawa, C.; Kawano, S. Quantitative evaluation of optical forces by single particle tracking in slit-like microfluidic channels. *J. Phys. Chem. C* **2018**, *122*, 17963–17975. [CrossRef]
44. Sánchez, S.; Arcos, J.; Bautista, O.; Méndez, F. Joule heating effect on a purely electroosmotic flow of non-Newtonian fluids in a slit microchannel. *J. Non-Newton. Fluid Mech.* **2013**, *192*, 1–9. [CrossRef]
45. Varagnolo, S.; Filippi, D.; Mistura, G.; Pierno, M.; Sbragaglia, M. Stretching of viscoelastic drops in steady sliding. *Soft Matter.* **2017**, *13*, 3116–3124. [CrossRef]
46. Fattal, R.; Kupferman, R. Constitutive laws for the matrix-logarithm of the conformation tensor. *J. Non-Newton. Fluid Mech.* **2004**, *123*, 281–285. [CrossRef]
47. Fattal, R.; Kupferman, R. Time-dependent simulation of viscoelastic flows at high Weissenberg number using the log-conformation representation. *J. Non-Newton. Fluid Mech.* **2005**, *126*, 23–37. [CrossRef]
48. Walters, K.; Webster, M.F. The distinctive CFD challenges of computational rheology. *Int. J. Numer. Meth. Fluids.* **2003**, *43*, 577–596. [CrossRef]
49. Fogolari, F.; Brigo, A.; Molinari, H. The Poisson–Boltzmann equation for biomolecular electrostatics: A tool for structural biology. *J. Mol. Recognit.* **2002**, *15*, 377–392. [CrossRef]
50. Pimenta, F.; Alves, M.A. Numerical simulation of electrically-driven flows using OpenFOAM. *arXiv* **2018**, arXiv:1802.02843.
51. Alves, M.; Oliveira, P.; Pinho, F. A convergent and universally bounded interpolation scheme for the treatment of advection. *Int. J. Numer. Methods Fluids* **2003**, *41*, 47–75. [CrossRef]
52. Duarte, A.; Miranda, A.I.; Oliveira, P.J. Numerical and analytical modeling of unsteady viscoelastic flows: The start-up and pulsating test case problems. *J. Non-Newton. Fluid Mech.* **2008**, *154*, 153–169. [CrossRef]
53. Patankar, S.V.; Corp, H.P. *Numerical Heat Transfer and Fluid Flow*, 1st ed.; McGraw-Hill: New York, NY, USA, 1980.
54. Van Doormaal, J.P.; Raithby, G.D. Enhancements of the SIMPLE method for predicting incompressible fluid flows. *Numer. Heat Transf.* **1987**, *7*, 147–163.
55. López-Herrera, J.; Popinet, S.; Castrejón-Pita, A. An adaptive solver for viscoelastic incompressible two-phase problems applied to the study of the splashing of weakly viscoelastic droplets. *J. Non-Newton. Fluid Mech.* **2019**, *264*, 144–158. [CrossRef]
56. Sousa, P.C.; Vega, E.J.; Sousa, R.G.; Montanero, J.M.; Alves, M.A. Measurement of relaxation times in extensional flow of weakly viscoelastic polymer solutions. *Rheol. Acta* **2017**, *56*, 11–20. [CrossRef] [PubMed]
57. Martins, F.; Oishi, C.M.; Afonso, A.M.; Alves, M.A. A numerical study of the kernel-conformation transformation for transient viscoelastic fluid flow. *J. Comput. Phys.* **2015**, *302*, 653–673. [CrossRef]
58. Sze, A.; Erickson, D.; Ren, L.; Li, D. Zeta-potential measurement using the Smoluchowski equation and the slope of the current–time relationship in electroosmotic flow. *J. Colloid Interface Sci.* **2003**, *261*, 402–410. [CrossRef]
59. Sirisinha, C. A review of extrudate swell in polymers. *J. Sci. Soc. Thailand.* **1997**, *23*, 259–280. [CrossRef]
60. James, D.F. N1 stresses in extensional flows. *J. Non-Newton. Fluid Mech.* **2016**, *232*, 33–42. [CrossRef]
61. Bird, R.B.; Armstrong, R.C.; Hassager, O. Dynamics of polymeric liquids. In *Fluid Mechanics*, 2nd ed.; Wiley-Interscience: Hoboken, NJ, USA, 1987; Volume 1.
62. Latinwo, F.; Schroeder, C.M. Determining elasticity from single polymer dynamics. *Soft Matter.* **2014**, *10*, 2178–2187. [CrossRef]
63. Groisman, A.; Steinberg, V. Efficient mixing at low Reynolds numbers using polymer additives. *Nature* **2001**, *410*, 905–908. [CrossRef] [PubMed]
64. Grilli, M.; Vázquez-Quesada, A.; Ellero, M. Transition to turbulence and mixing in a viscoelastic fluid flowing inside a channel with a periodic array of cylindrical obstacles. *Phys. Rev. Lett.* **2013**, *110*, 174501. [CrossRef] [PubMed]

65. Burghelea, T.; Segre, E.; Steinberg, V. Elastic turbulence in von Karman swirling flow between two disks. *Phys. Fluids* **2007**, *19*, 053104. [CrossRef]
66. Pakdel, P.; McKinley, G.H. Elastic instability and curved streamlines. *Phys. Rev. Lett.* **1996**, *77*, 2459. [CrossRef] [PubMed]
67. Jovanović, M.R.; Kumar, S. Nonmodal amplification of stochastic disturbances in strongly elastic channel flows. *J. Non-Newton. Fluid Mech.* **2011**, *166*, 755–778. [CrossRef]

micromachines

Article

The Effects of Viscoelasticity on Droplet Migration on Surfaces with Wettability Gradients

Ying Jun Ren and Sang Woo Joo *

School of Mechanical Engineering, Yeungnam University, Gyeongsan 38541, Korea; renyingjun2003@gmail.com
* Correspondence: swjoo@yu.ac.kr; Tel.: +82-538102568

Abstract: A finite-volume method based on the OpenFOAM is used to numerically study the factors affecting the migration of viscoelastic droplets on rigid surfaces with wettability gradients. Parameters investigated include droplet size, relaxation time, solvent viscosity, and polymer viscosity of the liquid comprising droplets. The wettability gradient is imposed numerically by assuming a linear change in the contact angle along the substrate. As reported previously for Newtonian droplets, the wettability gradient induces spontaneous migration from hydrophobic to hydrophilic region on the substrate. The migration of viscoelastic droplets reveals the increase in the migration speed and distance with the increase in the Weissenberg number. The increase in droplet size also shows the increase in both the migration speed and distance. The increase in polymer viscosity exhibits the increase in migration speed but the decrease in migration distance.

Keywords: droplet migration; viscoelasticity; wettability gradient

Citation: Ren, Y.J.; Joo, S.W. The Effects of Viscoelasticity on Droplet Migration on Surfaces with Wettability Gradients. *Micromachines* **2022**, *13*, 729. https://doi.org/10.3390/mi13050729

Academic Editors: Lanju Mei, Shizhi Qian and Sung Sik Lee

Received: 7 March 2022
Accepted: 29 April 2022
Published: 30 April 2022

Publisher's Note: MDPI stays neutral with regard to jurisdictional claims in published maps and institutional affiliations.

1. Introduction

The motion of liquid droplets on solid surfaces is ubiquitous in nature and is associated with extremely broad applications in many different fields [1,2]. The manipulation of droplets by controlling the wettability of substrates in particular has been intensively investigated due to daily observations and various potential industrial applications. The methods for wettability control for transporting droplets on surfaces include creating temperature gradients, electrowetting, magnetic fields, and chemical or physical texture gradients [3–15]. When droplets are placed on a solid surface with a wettability gradient, they tend to move from regions of low wettability to high wettability due to the net driving force in the direction of increasing surface wettability. Greenspan and Brochard [16,17] studied the wettability gradient of a surface to night drop operation in detail. The net driving force is due to the difference in curvature between the front and back half of the droplet. Yang et al. [18] further discussed on the concept of manipulating droplets in the absence of an external factor environment. Li and Zhiguang [19] verified the spontaneous motion of droplets on solid surfaces. Moumen et al. [20] performed detailed experiments on droplet transport on horizontal solid surfaces using wettability gradients. Chowdhury et al. [21] verified the droplet transport mechanism for wettability gradient trajectories with different constraints. Liu et al. [22] conducted experiments on the motion of droplets on a surface with a wettability gradient and found that the velocity of the droplets increased with the surface wettability gradient. Subramanian et al. [23] verified the forces involved in the migration of droplets on solid surfaces, and demonstrated the forces and resistance provided by droplets by approximating the shape of the droplets as wedges, which is known as the wedge approximation. They also used the lubrication approximation to study the dynamics and resistance of fluids. Chaudhry et al. [24] verified that the velocity gradually decreased after increasing to a certain value along the direction of droplet movement. Xu and Qian [25] analyzed the motion of nanoscale droplets on a heated solid surface with a wettability gradient, and accurately simulated the phenomenon of rapid changes in solid-to-fluid temperature. They investigated the motion of evaporative droplets in a single-component fluid on a solid substrate with a wettability gradient. There are two main difficulties with fluid flow and heat flow near droplet contact lines on solid substrates: hydrodynamic (stress) singularities and thermal singularities. A continuum hydrodynamic model is proposed for the study of the motion of single-component droplets of fluids on solid substrates. The model can handle thermal singularities, which inevitably arise as the substrate temperature differs from the coexistence of liquid and gas. Liu and Xu [26] performed theoretical analysis and molecular dynamics simulations of droplet transport on surfaces with wettability gradients. A unified mechanical model is proposed that integrates the static configuration of the droplet at equilibrium and the dynamic configuration of the droplet during motion. Molecular dynamics (MD) simulations show that the configuration of water droplets on a solid surface relaxes during motion, and a dimensionless parameter is proposed to describe their dynamic contact area. In addition, the analysis showed that the friction coefficient of water droplets on the solid surface was significantly different from that of the water film, and a geometric factor related to the wettability of the solid surface was formulated to calibrate the kinetic friction of water droplets. Full-velocity trajectories of droplet motion are extracted, and the predictions are in good agreement with extensive MD simulations across the entire surface wettability gradient, from superhydrophobic to superhydrophilic. Raman et al. [27] used a phase-field-based Boltzmann method (LMB) to simulate the dynamics of droplet aggregation on a wettability-gradient surface and observed that when the droplet impinges on the wettability gradient surface, the droplet shape is not necessarily spherical, resulting in different droplet morphologies near the droplet junction area. Huang et al. [28] conducted a 2D numerical simulation of droplet transport on a surface with a stepwise wettability gradient by considering the contact-angle hysteresis (CAH) on the droplet surface. They used the Lattice Boltzmann Method (LBM) and found that the velocity of the droplet has a strong dependency on viscosity ratio, wetting gradient magnitude, and CAH. Ahmadlouydarab and Feng [29] used wettability gradient and external flow to numerically study the movement and coalescence of droplets, and analyzed the transport of droplets on a surface with wettability gradient, making comparisons with the results of Moumen et al. [20] In addition to the work of Nasr et al. [30], Chaudhry et al. [21,31] studied the migration of droplets on surfaces with linear wettability gradients, and concluded that the droplet shape was found to evolve over time to maintain a minimum energy state. Even with different wettability gradients, the surface energy of a droplet can be the same at a specific dimensionless time. Droplets, located at different locations and times, can be identical in shape. Paul Ch. Zielke et al. [32] reported that velocity increases with the droplet size.

Although the research on Newtonian fluids has made great progress, the research on non-Newtonian fluids (viscoelastic fluids) is still very scarce. In the study of non-Newtonian fluids [33–38] most of the studies are on flows in microchannels. Although the spontaneous migration of Newtonian droplets due to wettability gradients has been widely studied, that for viscoelastic droplets is very limited. Bai et al. [39] used the OpenFOAM to numerically analyze viscoelastic droplet migration on surfaces with linear wettability gradients. They showed that the migration speed increases monotonically with the increase in fluid elasticity until it saturates for high enough Weissenberg number. The migration distance, however, was not obtained because the cases reported did not contain migrations coming to an end, which is observed in experiments. Li et al. [40] proposed a dynamically controlled particle separation by employing viscoelastic fluids in deterministic lateral displacement (DLD) arrays. The process of deceleration and termination of viscoelastic droplet migration due to the growing viscous force with droplet deformation, however, is omitted in the report. Zhang et al. [41] studied the transient flow response of viscoelastic fluids to different external forces. Damped harmonic oscillation and periodic oscillation are induced and modulated depending on the fluid intrinsic properties such as viscosity and elasticity. External body forces, such as constant force, step force, and square wave force, are applied at the inlet of the channel. It is revealed that the oscillation damping originates from the fluid viscosity, while the oscillation frequency is dependent on the fluid elasticity. An innovative way is also developed to characterize the time relaxation of the viscoelastic fluid by modulating the frequency of the square wave force. Zhang et al. [42] investigated temporal-pulse flow mixing of Newtonian and viscoelastic fluids at different pulse frequencies and showed that viscoelastic fluids are more mixed than Newtonian fluids. Despite these findings on effects of the viscoelasticity, the finite migration distance of droplets affected by the viscoelasticity is yet to be reported. In this work the effect of viscoelasticity on the droplet dynamics is studied parametrically, and the changes in migration speed and distance due to viscoelasticity are revealed for the first time.

2. Numerical Simulation

The volume-of-fluid (VOF) method is a simulation technique used to track and locate free-form surfaces or fluid interfaces in computational fluid dynamics. It uses static and migrating mesh to accommodate the evolution of the interface shape. It is based on the Eulerian formulation. In this paper, the VOF solver tracking interface included in OpenFOAM is used to calculate the volume fraction in the gas/liquid two-phase flow: α is the volume fraction of a liquid, and the value of α in the grid varies between 0 and 1. When a grid is completely filled with liquid, the value of α is 1. When there is no liquid in the grid, the value of α is 0. The continuity equation is then written as

$$\frac{\partial \alpha}{\partial t} + (U \cdot \nabla)\alpha = 0 \tag{1}$$

where U is the fluid velocity vector. The transport properties of this fluid are obtained by a volume average of the equations:

$$\rho = \alpha\rho_1 + (1 - \alpha)\rho_2 \tag{2}$$

$$\mu = \alpha\mu_1 + (1 - \alpha)\mu_2 \tag{3}$$

where ρ and μ represent the density and dynamic viscosity of the two liquids, respectively. The interactive reaction between the two phases of the fluid can be calculated on the surface tension by the following equation:

$$\Delta p = \sigma k \widehat{n} \tag{4}$$

where Δp is the pressure difference across the interface, σ represents the surface tension coefficient, k is the curvature of the surface, and $\widehat{n}$ is the unit outward normal on the surface. The surface tension is included in the Navier–Stokes equation as a source term. Based on the case of incompressible fluids, the governing equations of viscoelastic fluids and the conservation of mass and momentum can be expressed as

$$\Delta \cdot U = 0 \tag{5}$$

$$\frac{\partial \rho U}{\partial t} + U \cdot (\rho U) = -\nabla p + \nabla(\tau_s + \tau_p) + \sigma\kappa\nabla\alpha + \rho g \tag{6}$$

where p is the pressure, ρ is the fluid density, and g is gravitational acceleration, the stress tensor can be expressed as

$$\tau = \tau_s + \tau_p \tag{7}$$

where stress τ is divided into that contributed by the Newtonian solvent τ_s, and the viscoelastic polymer τ_p. To focus on the effect of viscoelasticity without the complications of shear-thinning the Oldroyd-B viscoelastic constitutive model is adopted, which can be expressed as

$$\tau_p + \lambda\tau_p^{\nabla} = 2\eta_p\left[\nabla U + (\nabla U)^T\right] \tag{8}$$

where λ is the relaxation time, η_p is the polymer viscosity, τ_p^{∇} is the derivative on the elastic stress tensor:

$$\tau_p^{\nabla} = \frac{\partial \tau_p}{\partial t} + \nabla \cdot (U\tau_p) - (\nabla U)^T \cdot \tau_p - \tau_p \cdot (\nabla U) \tag{9}$$

The relation of the Oldroyd-B constitutive model can be expressed as

$$\tau_p = \eta_p / (\lambda(C - I)) \tag{10}$$

where C is the conformational tensor of the polymer molecule, and a symmetry tensor I is a unit tensor. Equation (6) can then be simplified as

$$\frac{\partial U}{\partial t} + U \cdot \nabla U = -\frac{1}{\rho_c}\nabla p + \frac{\beta \eta_c}{\rho_c}\nabla^2 U + \frac{\eta_c}{\rho_c \lambda_c}(1 - \beta)\nabla \cdot C + \frac{\sigma k \nabla \alpha}{\rho_c} + g \tag{11}$$

where $\beta = \eta_s / \eta_p = \eta_s / (\eta_s + \eta_p)$ and $\eta_c = \eta_s + \eta_p$, The viscosity and density fields depend on the order parameter:

$$\eta_c = \alpha \eta_L + (1 - \alpha)\eta_G, \ \rho_c = \alpha \rho_L + (1 - \alpha)\rho_G \tag{12}$$

where η_L and ρ_L denote the viscosity and density of the liquid. The viscosity and density of the gas is denoted by η_G and ρ_G. The transport equation of the deformation rate tensor is expressed as

$$\frac{\partial C}{\partial t} + \nabla \cdot (UC) - (\nabla U)^T \cdot C - C \cdot (\nabla U) = \frac{1}{\lambda}(1 - C) \tag{13}$$

If we set the droplet radius as a and the total substrate length as L, and the dimensionless droplet radius in units of L is

$$R = \frac{a}{L} \tag{14}$$

The spatial and temporal variables then are nondimensionalized as

$$x^* = \frac{x}{L}, \ y^* = \frac{y}{L}, \ T = \frac{tv}{a^2} \tag{15}$$

where the kinematic viscosity $v = \eta_c / \rho_L$.

For viscoelastic droplets, the Weissenberg number Wi is an important parameter, the measure of fluid elasticity:

$$Wi = \frac{v\lambda}{a^2} \tag{16}$$

If we take the center point x_m of the droplet as a reference for droplet location, dimensionless droplet location M and the dimensionless migration distance are expressed as

$$M = \frac{x_m}{L} \text{ and } M_f = \frac{x_f}{L} \tag{17}$$

where x_f is the final value of x_m when the droplet ceases to move.

In this paper, the OpenFOAM software is used for computations. Initially a viscoelastic droplet is placed in a rectangular area with a length L of 10 mm and a height H of 1.5 mm. The contact angle along the substrate decreases from the superhydrophobic region in the left side to the hydrophilic region in the right side, and the droplet migration is observed as in Figure 1. The boundary conditions are set to no-slip on the substrate and atmospheric conditions on other boundaries, as available in the OpenFOAM. The spatial resolution of the calculation is determined by grid-independent studies to ensure an absolute error bound of 10^{-6} on the calculation of the droplet migration distance. The change in the contact angle decreases from $160°$ at the initial droplet location to $0°$ at the right end of the substrate. As shown in Figure 2, the deformation that occurs with droplet migration is consistent with [24].

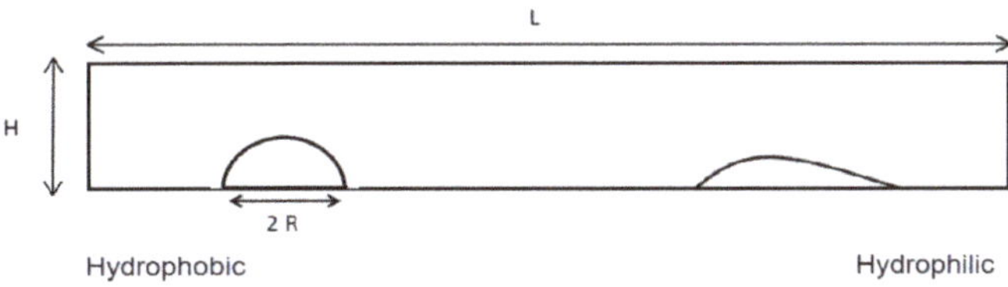

Figure 1. The computational domain of the simulation. Semicircle: The droplets move from the superhydrophobic side to the hydrophilic side.

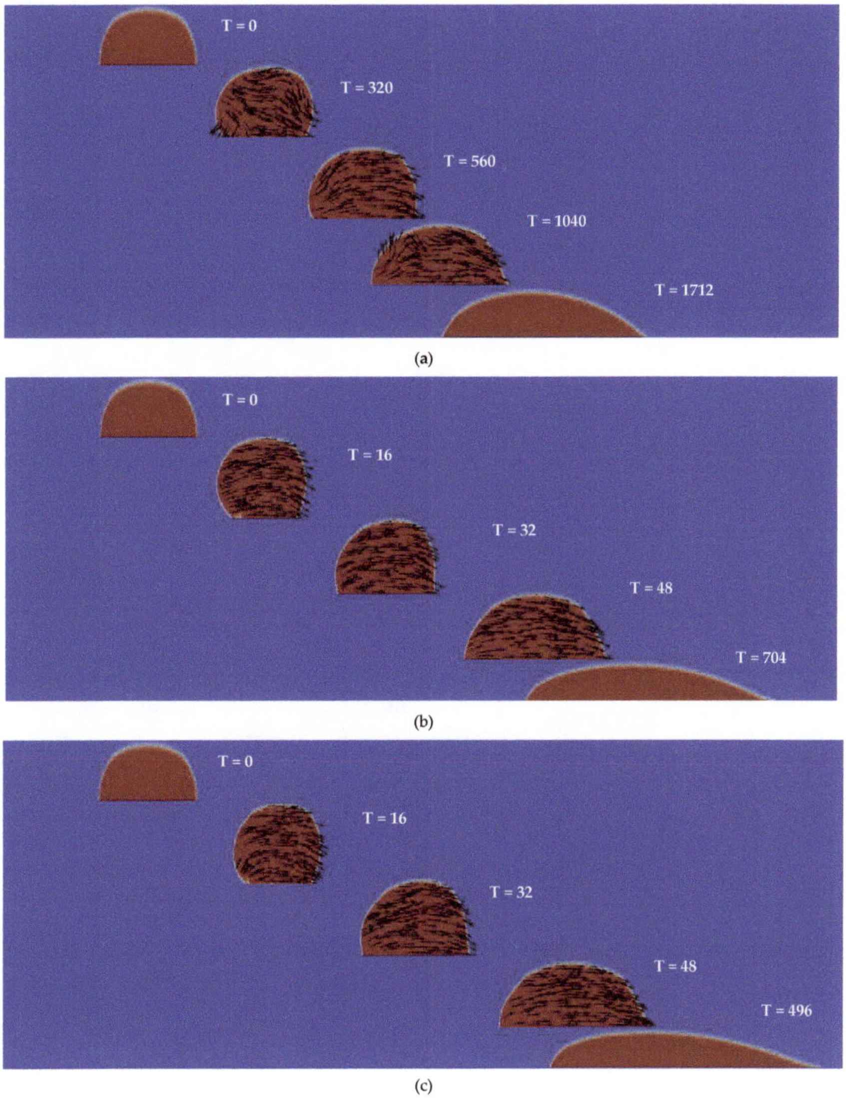

Figure 2. (**a**) Migration of a Newtonian droplet (**b**) Migration a of viscoelastic droplet with $Wi = 16$ (**c**) with $Wi = 40$.

The contact angle model we use is the dynamic contact angle model, and the equation of the dynamic contact angle model is:

$$\theta_d = \begin{cases} \theta_a, & U_w \geq 0 \\ \theta_r, & U_w \geq 0 \end{cases} \tag{18}$$

where U_w is the velocity near the wall. The dynamic contact angle model based on Open-FOAM is

$$\theta = \theta_e + (\theta_a - \theta_r)\tanh\left(\frac{U_w}{U_\theta}\right) \tag{19}$$

where U_θ is the characteristic velocity scale. θ_a, θ_r, θ_e, respectively, are the advancing, the receding, and the balance angle.

Density of the Oldroyd-B and the Newtonian liquid is set identically to $\rho = 1000 \text{ kg/m}^3$, while polymer and solvent viscosities are set to $\eta_p = 0.36 \text{ Pa·s}$ and $\eta_s = 0.04 \text{ Pa·s}$, respectively. The initial relaxation time is set as $\lambda = 0.01 \text{ s}$ [43]. For the gas phase, we set $\rho = 1 \text{ kg/m}^3$ and $\eta_s = 1 \times 10^{-5} \text{ Pa·s}$, with the surface tension $\sigma = 0.073 \text{ N/m}$, as listed in Table 1. The migration of droplets on the substrate with a contact angle distribution ranging from 160° to 0° is investigated.

Table 1. Liquid properties used.

Fluid	ρ (kg/m^3)	η_p (Pa·s)	η_s (Pa·s)	λ (s)	σ (N/m)
Oldroyd-B	1000	0.36	0.04	0.01	0.073
Newtonian liquid	1000		0.04		0.073
Newtonian gas	1		1×10^{-5}		0.073

Figure 2 shows representative cases of Newtonian and viscoelastic droplet migration obtained by the OpenFOAM simulations described above. The red and blue regions, respectively, represent liquid and air phase, with the arrows indicating the local velocity vector. As in Figure 1, the initial droplet shape is set to a semicircle, with which the droplet radius and volume can be clearly specified.

3. Results and Discussion

Figure 3a shows the time-dependent location of droplet center M for droplets with identical viscosity, as a droplet migrates from the superhydrophobic region to the hydrophilic region of the substrate. The slope of each line indicates instantaneous migration speed, which eventually becomes zero for all droplets shown. It is thus seen that droplets start to move due to the wettability gradient, decelerate, and cease to move due to the viscous dissipation. The migration speed in the early stages of the motion and the migration distance in the final stage both increase with the Wi. The time spent to reach the final stationary state is seen to decrease with Wi. It can thus be deduced that more elastic droplets migrate faster and farther and stop sooner. Figure 3b shows the location of droplets with different volumes, with an enlarged scale provided in the inset. With the fixed wettability gradient along the substrate, bigger droplets would experience bigger differences in the contact angle between the advancing and receding side of the droplet. It is thus seen that droplets with bigger initial radius migrate faster and farther. Since the difference in the migration distance is more pronounced than the migration speed, the time required to reach the stationary state increases as well with the initial droplet radius.

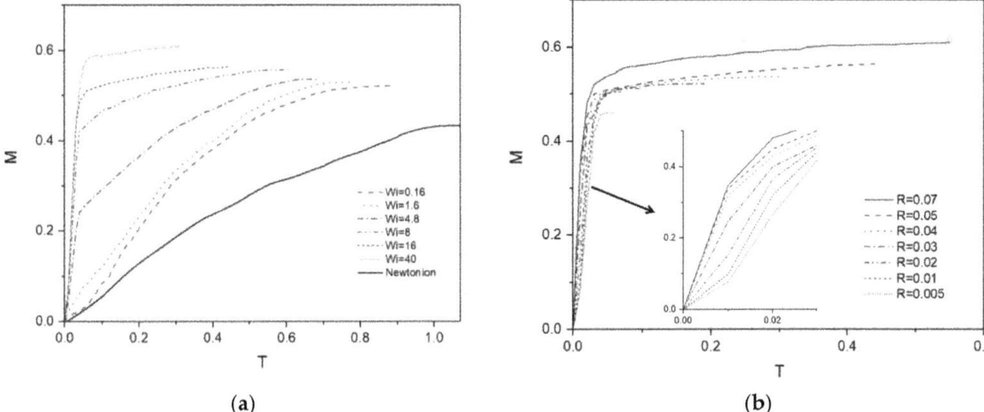

Figure 3. (**a**) Migration of Newtonian droplet and viscoelastic droplets with different *Wi* numbers against dimensionless time. (**b**) Migration of a viscoelastic droplet (*Wi* = 16) of different initial radii.

Figure 4a shows the location of viscoelastic droplets with time, depending on the solvent viscosity η_s for an identical polymer viscosity η_p. For high solvent viscosity, $\eta_s = 25$ Pa·s, the viscoelastic droplet migrates slowly and for a relatively short distance. Low solvent viscosity, $\eta_s = 0.04$ Pa·s, gives faster and longer migration, as can be easily understood. In Figure 4b three different polymer viscosities, $\eta_p = 0.36$, 1.9, 4.2 Pa·s, are tested with the solvent viscosity kept identically at $\eta_s = 0.04$ Pa·s. In early stages, the difference in migration speed is not conspicuous, but eventually the differences in migration distance and migration time are obvious. Droplets with higher polymer viscosity show shorter migration distance and smaller migration time.

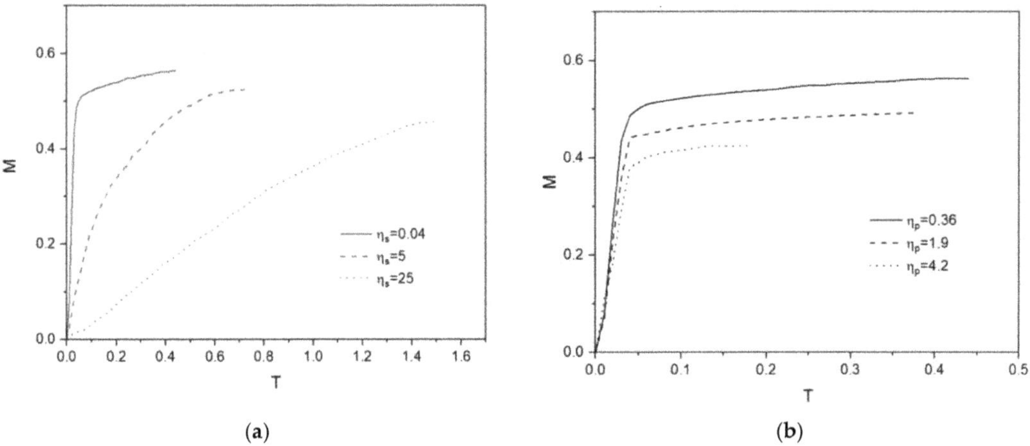

Figure 4. (**a**) Droplet migration for different η_s. (**b**) Droplet migration for different η_p.

Figure 5a shows the dimensionless migration distance M_f depending on different viscoelasticity. The final migration distance of the droplets increases with the *Wi* number. The Newtonian droplet, *Wi* = 0, shows shortest migration distance. The increase in the migration distance with the increase in *Wi* is monotonic. Near *Wi* = 80, the increase seems minimized, but the terminal equilibrated value for high *Wi*, if any, cannot be verified due to difficulties in computations for high enough *Wi*. As seen in Figure 3, a longer migration distance is accompanied by a shorter migration time. With the increase in *Wi* the total migration time decreases monotonically. The viscoelasticity seems to promote the droplet migration due to the wettability gradient both in terms of migration velocity and migration distance. The exact mechanism for this promotion needs to be analyzed. The increase in the migration distance with respect to the droplet radius is shown in Figure 5b. For all values of *Wi*, the migration distance increases monotonically with the droplet radius. In contrast to the migration-distance increase due to the elasticity, however, the total migration time also increases with the increase in the migration distance. It is to be noted that with in present length scale increase in the droplet radius is analogous to that in the wettability gradient.

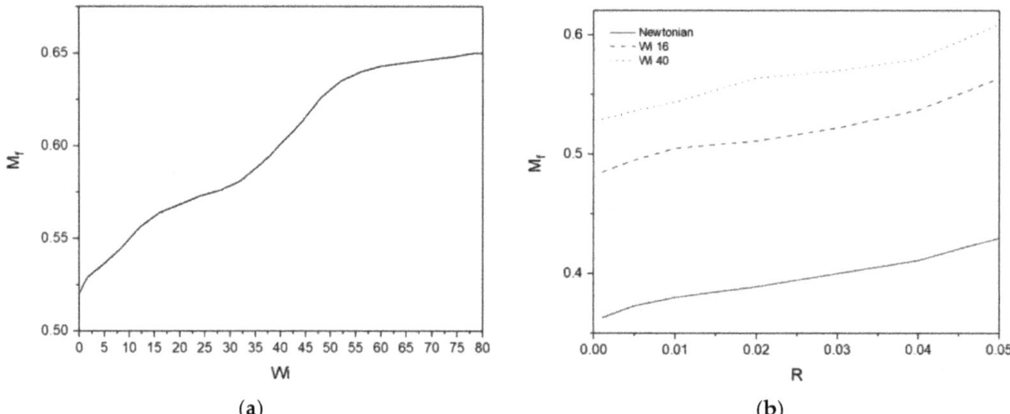

(a)

(b)

Figure 5. (a) Migration distance of droplets for different *Wi*; (b) Migration distance of droplets with different initial radius.

4. Concluding Remarks

Based on the VOF method, the effect of viscoelasticity on the migration distance, speed, and time for spontaneous droplet motion due to wettability gradients is analyzed for the first time. It is found that as the fluid elasticity increases, the farther and the faster the viscoelastic droplet migrates, but the sooner it stops. Increase in the droplet size also makes it migrate farther and faster, but total migration time becomes longer. Increase in the viscosity of the solvent causes the droplet to move more slowly and over a shorter distance, as does the change in the viscosity of the polymer due to elastic effects. The viscoelasticity seems to promote the droplet migration due to the wettability gradient both in terms of migration velocity and migration distance. The exact mechanism for this promotion needs to be analyzed. It is to be noted that in the present length scale used, increase in droplet radius is analogous to that in the wettability gradient. This work focuses on the effect of elasticity on the droplet migration without other complications of viscoelastic fluids, and so the Oldroyd-B model is chosen. With the features embedded in the OpenFOAM, it is straightforward to extend the work to other constitutive models with other desired effects. Here, the migration-promoting effect of elasticity is reported quantitatively. Its difference with the effect of wettability-gradient increase is revealed.

Author Contributions: Conceptualization, Y.J.R.; methodology, Y.J.R.; investigation, Y.J.R.; writing—original draft preparation, Y.J.R.; writing—review and editing, S.W.J.; supervision, S.W.J.; project administration, S.W.J.; funding acquisition, S.W.J. All authors have read and agreed to the published version of the manuscript.

Funding: This work is funded by the Grant NRF-2022R1A2C2002799 of the National Research Foundation of Korea.

Conflicts of Interest: The authors have declared no conflict of interest.

References

1. Yao, X.; Song, Y.; Jiang, L. Applications of bio-inspired special wettablesurfaces. *Adv. Mater.* **2011**, *23*, 719–734. [CrossRef] [PubMed]
2. Darhuber, A.A.; Valentino, J.P.; Troian, S.M.; Wagner, S. Thermocapillary actuation of droplets on chemically patterned surfaces by programmablemicroheater arrays. *J. Microelectromech. Syst.* **2003**, *12*, 873–879. [CrossRef]
3. Mettu, S.; Chaudhury, M.K. Motion of drops on a surface induced bythermal gradient and vibration. *Langmuir* **2008**, *24*, 10833–10837. [CrossRef] [PubMed]
4. Foroutan, M.; Fatemi, S.M.; Esmaeilian, F.; Fadaei Naeini, V.; Baniassadi, M. Contact angle hysteresis and motion behaviors of a water nanodroplet on suspended graphene under temperature gradient. *Phys. Fluids* **2018**, *30*, 052101. [CrossRef]
5. Jaiswal, V.; Harikrishnan, A.; Khurana, G.; Dhar, P. Ionic solubility and solutal advection governed augmented evaporation kinetics of salt solution pendant droplets. *Phys. Fluids* **2018**, *30*, 012113. [CrossRef]
6. Cho, S.K.; Moon, H.; Kim, C.-J. Creating, transporting, cutting, and merging liquid droplets by electrowetting-based actuation for digital microfluidic circuits. *J. Microelectromech. Syst.* **2003**, *12*, 70–80. [CrossRef]
7. Latorre, L.; Kim, J.; Lee, J.; De Guzman, P.-P.; Lee, H.; Nouet, P.; Kim, C.-J. Electrostatic actuation of microscale liquid-metal droplets. *J. Microelectromech. Syst.* **2002**, *11*, 302–308. [CrossRef]
8. Kunti, G.; Mondal, P.K.; Bhattacharya, A.; Chakraborty, S. Electrothermally modulated contact line dynamics of a binary fluid in a patterned fluidicenvironment. *Phys. Fluids* **2018**, *30*, 092005. [CrossRef]
9. Kunti, G.; Bhattacharya, A.; Chakraborty, S. Electrothermally actuatedmoving contact line dynamics over chemically patterned surfaces with resistive heaters. *Phys. Fluids* **2018**, *30*, 062004. [CrossRef]
10. Long, Z.; Shetty, A.M.; Solomon, M.J.; Larson, R.G. Fundamentals ofmagnet-actuated droplet manipulation on an open hydrophobic surface. *Lab Chip* **2009**, *9*, 1567–1575. [CrossRef]
11. Ichimura, K.; Oh, S.-K.; Nakagawa, M. Light-driven motion of liquids on aphotoresponsive surface. *Science* **2000**, *288*, 1624–1626. [CrossRef] [PubMed]
12. Zheng, Y.; Cheng, J.; Zhou, C.; Xing, H.; Wen, X.; Pi, P.; Xu, S. Droplet Motion on a Shape Gradient Surface. *Langmuir* **2017**, *33*, 4172–4177. [CrossRef] [PubMed]
13. Morrissette, J.M.; Mahapatra, P.S.; Ghosh, A.; Ganguly, R.; Megaridis, C.M. Rapid, Self-driven Liquid Mixing on Open-Surface Microfluidic Platforms. *Sci. Rep.* **2017**, *7*, 1800. [CrossRef] [PubMed]
14. Dhiman, S.; Jayaprakash, K.S.; Iqbal, R.; Sen, A.K. Self-Transport and Manipulation of Aqueous Droplets on Oil-Submerged Diverging Groove. *Langmuir* **2018**, *34*, 12359–12368. [CrossRef] [PubMed]
15. Petrie, R.J.; Bailey, T.; Gorman, C.B.; Genzer, J. Fast directed motion of fakir droplets. *Langmuir* **2004**, *20*, 9893–9896. [CrossRef]
16. Greenspan, H.P. On the motion of a small viscous droplet that wets a surface. *J. Fluid Mech.* **1978**, *84*, 125–143. [CrossRef]
17. Brochard, F. Motions of droplets on solid surfaces induced by chemical or thermal gradients. *Langmuir* **1989**, *5*, 432–438. [CrossRef]
18. Yang, J.-T.; Chen, J.C.; Huang, K.-J.; Yeh, J.A. Droplet manipulation on a hydrophobic textured surface with roughened patterns. *J. Microelectromech. Syst.* **2006**, *15*, 697–707. [CrossRef]
19. Li, J.; Guo, Z. Spontaneous Directional Transportations of Water Droplets on Surfaces Driven by Gradient Structures. *Nanoscale* **2018**, *10*, 13814–13831. [CrossRef]
20. Moumen, N.; Subramanian, R.S.; McLaughlin, J.B. Experiments on the Motion of Drops on a Horizontal Solid Surface Due to a Wettability Gradient. *Langmuir* **2006**, *22*, 2682–2690. [CrossRef]
21. Chowdhury, I.U.; Mahapatra, P.S.; Sen, A.K. Self-driven droplet transport: Effect of wettability gradient and confinement. *Phys. Fluids* **2019**, *31*, 042111. [CrossRef]
22. Liu, C.; Sun, J.; Li, J.; Xiang, C.; Chenghao, X.; Wang, Z.; Zhou, X. Long-range spontaneous droplet self-propulsion on wettability gradient surfaces. *Sci. Rep.* **2017**, *7*, 7552. [CrossRef] [PubMed]
23. Subramanian, R.S.; Moumen, A.N.; McLaughlin, J.B. Motion of a Drop on a Solid Surface Due to a Wettability Gradient. *Langmuir* **2005**, *21*, 11844–11849. [CrossRef] [PubMed]
24. Chaudhury, M.K.; Chakrabarti, A.; Daniel, S. Generation of Motion of Drops with Interfacial Contact. *Langmuir* **2015**, *31*, 9266–9281. [CrossRef]
25. Xu, X.; Qian, T. Droplet motion in one-component fluids on solid substrates with wettability gradients. *Phys. Rev. E* **2012**, *85*, 051601. [CrossRef]
26. Liu, Q.; Xu, B. A unified mechanics model of wettability gradient-driven motion of water droplet on solid surfaces. *Extreme Mech. Lett.* **2016**, *9*, 304–309. [CrossRef]
27. Raman, K.A. Dynamics of simultaneously impinging drops on a dry surface: Role of inhomogeneous wettability and impact shape. *J. Colloid Interface Sci.* **2018**, *516*, 232–247. [CrossRef]
28. Huang, J.; Shu, C.; Chew, Y. Numerical investigation of transporting droplets by spatiotemporally controlling substrate wettability. *J. Colloid Interface Sci.* **2008**, *328*, 124–133. [CrossRef]
29. Ahmadlouydarab, M.; Feng, J.J. Motion and coalescence of sessile drops driven by substrate wetting gradient and external flow. *J. Fluid Mech.* **2014**, *746*, 214–235. [CrossRef]

30. Nasr, H.; Ahmadi, G.; McLaughlin, J.; Jia, X. Drop motion simulation on a surface due to a wettability gradient. In Proceedings of the ASME 2006 2nd Joint US-European Fluids Engineering Summer Meeting Collocated with the 14th International Conference on Nuclear Engineering, Miami, FL, USA, 17–20 July 2006; American Society of Mechanical Engineers: New York, NY, USA; pp. 547–550.
31. Chowdhury, I.U.; Mahapatra, P.S.; Sen, A.K. Shape evolution of drops on surfaces of different wettability gradients. *Chem. Eng. Sci.* **2021**, *229*, 116136. [CrossRef]
32. Zielke, P.C.; Subramanian, R.S.; Szymczyk, J.A.; McLaughlin, J.B. Movement of drops on a solid surface due to a contact angle gradient. *Proc. Appl. Math. Mech.* **2003**, *2*, 390–391. [CrossRef]
33. Ji, J.; Qian, S.; Liu, Z. Electroosmotic Flow of Viscoelastic Fluid through a Constriction Microchannel. *Micromachines* **2021**, *12*, 417. [CrossRef] [PubMed]
34. Casas, L.; Ortega, J.A.; Gómez, A.; Escandón, J.; Vargas, R.O. Analytical Solution of Mixed Electroosmotic/Pressure Driven Flow of Viscoelastic Fluids between a Parallel Flat Plates Micro-Channel: The Maxwell Model Using the Oldroyd and Jaumann Time Derivatives. *Micromachines* **2020**, *11*, 986. [CrossRef] [PubMed]
35. Escandón, J.; Torres, D.; Hernández, C.; Vargas, R. Start-Up Electroosmotic Flow of Multi-Layer Immiscible Maxwell Fluids in a Slit Microchannel. *Micromachines* **2020**, *11*, 757. [CrossRef]
36. Mei, L.; Zhang, H.; Meng, H.; Qian, S. Electroosmotic Flow of Viscoelastic Fluid in a Nanoslit. *Micromachines* **2018**, *9*, 155. [CrossRef]
37. Mei, L.; Qian, S. Electroosmotic Flow of Viscoelastic Fluid in a Nanochannel Connecting Two Reservoirs. *Micromachines* **2019**, *10*, 747. [CrossRef]
38. Omori, T.; Ishikawa, T. Swimming of Spermatozoa in a Maxwell Fluid. *Micromachines* **2019**, *10*, 78. [CrossRef]
39. Bai, F.; Li, Y.; Zhang, H.; Joo, S.W. A numerical study on viscoelastic droplet migration on a solid substrate due to wettability gradient. *Electrophoresis* **2019**, *40*, 851–858. [CrossRef]
40. Li, Y.; Zhang, H.; Li, Y.; Li, X.; Wu, J.; Qian, S.; Li, F. Dynamic control of particle separation in deterministic lateral displacement separator with viscoelastic fluids. *Sci. Rep.* **2018**, *8*, 3618. [CrossRef]
41. Zhang, M.; Zhang, W.; Wu, Z.; Shen, Y.; Wu, H.; Cheng, J.; Zhang, H.; Li, F.; Cai, W. Modulation of viscoelastic fluid response to external body force. *Sci. Rep.* **2019**, *9*, 9402. [CrossRef]
42. Zhang, M.; Wu, Z.; Shen, Y.; Chen, Y.; Lan, C.; Li, F.; Cai, W. Comparison of Micro-Mixing in Time Pulsed Newtonian Fluid and Viscoelastic Fluid. *Micromachines* **2019**, *10*, 262. [CrossRef] [PubMed]
43. Figueiredo, R.A.; Oishi, C.M.; Cuminato, J.A.; Azevedo, J.C.; Afonso, A.M.; Alves, M.A. Numerical investigation of three dimensional viscoelastic free surface flows: Impacting drop problem. In Proceedings of the 6th European Conference on Computational Fluid Dynamics (ECFD VI), Barcelona, Spain, 20–25 July 2014; Volume 5.

micromachines

Article

Generation and Dynamics of Janus Droplets in Shear-Thinning Fluid Flow in a Double Y-Type Microchannel

Fan Bai [1,2], Hongna Zhang [1,*], Xiaobin Li [1], Fengchen Li [1] and Sang Woo Joo [2,*]

[1] School of Mechanical Engineering, Tianjin University, Tianjin 300072, China; baifan@ynu.ac.kr (F.B.); lixiaobin@tju.edu.cn (X.L.); lifc@tju.edu.cn (F.L.)
[2] School of Mechanical Engineering, Yeungnam University, Gyeongsan 38541, Korea
* Correspondence: hongna@tju.edu.cn (H.Z.); swjoo@yu.ac.kr (S.W.J.)

Abstract: Droplets composed of two different materials, or Janus droplets, have diverse applications, including microfluidic digital laboratory systems, DNA chips, and self-assembly systems. A three-dimensional computational study of Janus droplet formation in a double Y-type microfluidic device filled with a shear-thinning fluid is performed by using the multiphaseInterDyMFoam solver of the OpenFOAM, based on a finite-volume method. The bi-phase volume-of-fluid method is adopted to track the interface with an adaptive dynamic mesh refinement for moving interfaces. The formation of Janus droplets in the shear-thinning fluid is characterized in five different states of tubbing, jetting, intermediate, dripping and unstable dripping in a multiphase microsystem under various flow conditions. The formation mechanism of Janus droplets is understood by analyzing the influencing factors, including the flow rates of the continuous phase and of the dispersed phase, surface tension, and non-Newtonian rheological parameters. Studies have found that the formation of the Janus droplets and their sizes are related to the flow rate at the inlet under low capillary numbers. The rheological parameters of shear-thinning fluid have a significant impact on the size of Janus droplets and their formation mechanism. As the apparent viscosity increases, the frequency of Janus droplet formation increases, while the droplet volume decreases. Compared with Newtonian fluid, the Janus droplet is more readily generated in shear-thinning fluid due to the interlay of diminishing viscous force, surface tension, and pressure drop.

Keywords: microfluidics; Janus droplet; OpenFOAM; volume of fluid method; adaptive dynamic mesh refinement; shear-thinning fluid

Citation: Bai, F.; Zhang, H.; Li, X.; Li, F.; Joo, S.W. Generation and Dynamics of Janus Droplets in Shear-Thinning Fluid Flow in a Double Y-Type Microchannel. *Micromachines* **2021**, *12*, 149. https://doi.org/10.3390/mi12020149

Received: 7 January 2021
Accepted: 30 January 2021
Published: 3 February 2021

Publisher's Note: MDPI stays neutral with regard to jurisdictional claims in published maps and institutional affiliations.

1. Introduction

Droplet-based microfluidic technology has many advantages in biomedicine, chemical analysis, material science and microreactions due to its capability to produce high surface-volume ratios in large quantities with low reagent use, rapid reaction, and independent control of each droplet [1–7]. Many studies have been carried out to develop efficient and stable methods for conventional single-phase and multiphase droplet formation and movements [8]. Janus droplet, composed of two adhering immiscible drops of different fluids in a third phase, can offer a wide range of applications that cannot be realized with single-attribute structures, due to their centrally asymmetrical structure. The flow mechanism of Janus droplets in a microfluidic environment is diverse, including wettability gradient, magnetic force, and electrical force, among others [9,10]. The formation of bi-phase structures of droplets is motivated by the minimization of free energy at the interface. Controlling the interface energy between the three liquid phases is crucial for achieving flexible droplet shape changes. It is foreseen that a wider range of imaginable applications of liquid combinations based on two distinct chemical properties can be developed.

Due to its significant advantages in manipulating the hydrodynamics of micro-droplets, droplet-based microfluidics technology has become a promising method for the preparation of Janus droplets at the (sub)-micron level. The microfluidic device can realize single-step production of high-yield monodisperse Janus droplets, and has great flexibility to control the droplet size and anisotropic shape [11]. In the microfluidic device, each droplet provides a microreactor in which species migration or reaction can occur [12–15]. Various microfluidic devices, such as cross-flow microchannels, focused fluidics, embedded channels, and co-flow devices, can produce droplets in the microchannel by shearing the dispersed phase in a continuous liquid stream [16–20]. Two mutually adhering immiscible dispersed phase streams enter the microfluidic focusing device, and are sheared into Janus droplets by another stream at the orifice [21,22]. Nisisako et al. used a direct method to provide two-sided star particles for various systems, and explained the advantage of producing highly dispersed two-sided star hemispherical droplets [22,23]. The idea of preparing Janus droplets in a flow-focusing device has been extended to various systems by many researchers [21,24,25]. The size and shape of droplets are controlled by changing the flow rate of the continuous or the dispersed phase. Gupta et al. studied the formation mechanism of droplets in a T-junction channel, and proposed a power-law correlation to predict their size [26]. Wu Ping et al. analyzed the formation of microfluidic bubbles in a cross-junction and the transition mechanism from extrusion to dripping [27]. Wang et al. used a 3-dimensional lattice-Boltzmann method to simulate the formation of Janus droplets in Newtonian fluid in the Y-junction channel and its transition from extrusion to dripping [7]. Fu et al. also proposed the flow ratio and the capillary number (Ca) scaling law to estimate the bubble size of different systems [28]. Chen et al. studied the influence of liquid viscosity and surface tension on the formation of droplets, and showed that the slug formation period increases with the increase in surface tension [29]. Li et al. used a fluid-volume method to study the formation of single-phase droplets in a cross-junction microchannel, which was used to guide the design of microchips for droplet formation [30]. Raj et al. used the volume of fluid method (VOF) to study the formation of droplets in T-junction and Y-junction microchannel, and analyzed the influence of flow ratio, liquid viscosity, surface tension, channel size and wall adhesion characteristics on the length of Newtonian fluid slugs [31]. Fu et al. studied the formation of oil droplets in flow-focusing microchannels, and proposed a scaling law for predicting the length of oil droplets [32].

There are many existing reports on Janus droplet/particle preparation methods, but more fundamental studies on underlying mechanism and new phenomena associated can be of great consequence and use, especially those involving non-Newtonian fluids [33,34]. It is well known that the nonlinear rheology in non-Newtonian fluids has a profound influence on the flow dynamics [35]. In biological applications, many fluids exhibit non-Newtonian behaviours, which are amplified in combination with the small length scale in microfluidics. Abate et al. studied the formation of monodisperse particles in a flow-focusing device in a non-Newtonian polymer solution [33]. Arratia et al. reported the thinning of polymer filaments and the rupture of Newtonian and viscoelastic liquids in flow-focusing microchannels [36]. The results show that the decomposition mechanism of Newtonian liquid and polymer liquid with the same viscosity is drastically different. This phenomenon is due to the rheological difference between the two liquids. Qiu et al. numerically studied the droplet formation of non-Newtonian liquids in cross-flow microchannels [37]. From their findings, it is obvious that the rheological parameters of non-Newtonian fluids significantly affect the formation mechanism and size of droplets. Aytouna et al. examined the pinch-off dynamics, yield stress, and shear-thinning fluid of droplets in Newton through experiments [38]. Sontti et al. studied the flow pattern of Newtonian fluid and non-Newtonian fluid using T-junction microchannels [39]. Although the dynamics of droplet breaking in Newtonian fluids has been systematically studied [40,41], there is still lack of in-depth understanding of droplet formation and flow in non-Newtonian fluids at the microscale. In this work, we focus on the shear-thinning of non-Newtonian fluids among other features, and investigate for the first time the microfluidic formation of shear-thinning Janus droplets and their subsequent evolution.

The process of Janus droplet generation in a microfluidic device involves a complex mechanism, which stems from the force competition between surface tension, viscous shear, pressure drop and possible disturbances outside the system. These forces depend on fluid properties, flow geometry and flow conditions. When Ca is fixed, the droplet length increases with the increase in Q, or the velocity of the dispersed phase liquid increases or the aspect ratio increases [7,42,43]. Garstecki et al. proposed that the droplet size decreases with the increase in Ca, and is almost independent of the fluid properties [42]. In addition to the influence of Ca, the viscosity contrast after conversion is more important than before. Tice et al. and Sang et al. introduced the viscosity effect to the Ca, and announced that more viscous fluid produced greater resistance in the main channel and made the force system unbalanced [44,45]. Sontti et al. obtained a similar power-law of droplet volume from extrusion to dripping [39]. Independent experiments by Xu et al. have also verified these power-law that vary with Ca or the flow ratio [46]. Numerical simulations have been successfully used to study the performance of droplet formation in multiphase microfluidics [47,48]. Shardt et al. showed the phenomenon of Janus particle transport in shear flow [49]. Daghighi et al. used a 3D multiphysics model to study the transient motion of Janus droplets in a microchannel [50]. However, the formation of Janus droplets in microfluidic devices has not attracted much attention. The volume-of-fluid (VOF) method is very useful in simulating two-phase flow requiring interface tracking. An adaptive dynamic mesh based on the VOF method further enhances accuracy in capturing the evolution of moving interfaces, and has been used to simulate multiphase flow in microchannels.

Here numerical simulations for the Janus droplet formation and ensuing dynamics in a double Y-junction microfluidic device filled with a shear-thinning fluid are performed by a multiphase model of the OpenFOAM, based on a finite-volume method. The bi-phase VOF is adopted to track the interface with the aforementioned adaptive dynamic mesh. The power-Law model is adopted as a simple constitutive model of the shear-thinning fluid. We describe the effects of rheological properties, surface tension, and velocity ratio on droplet characteristics, including formation mechanism, droplet size, and velocity. These understandings can greatly contribute to controlling the preparation of Janus droplets with non-Newtonian liquids.

2. Numerical Simulation and Validation

2.1. Governing Equations and Computational Scheme

The open-source CFD software OpenFOAM is used with the multiphase flow solver implemented using the VOF method [51]. Conservation equations are solved for a fluid mixture with distributed concentrations rather than a single fluid for each phase. By solving the transport equation of volume fraction, the interface between the phases is realized. The volume fraction, α_i specifies the volume of a phase in each calculation unit. In a two-phase system, a cell that completely fills one phase is expressed as $\alpha_i = 1$, and that for the other phase as $\alpha_i = 0$. Obviously, the interface between the two phases is expressed as $0 < \alpha_i < 1$. For a two-phase system, only the volume fraction of one phase needs to be determined because the other is obtained by $\alpha_2 = 1 - \alpha_1$.

The fluid viscosity and density for each unit are defined as a function of volume fraction as:

$$\eta = \alpha_{CP}\eta_{CP} + (1 - \alpha_{CP})\eta_{DP} \tag{1}$$

$$\rho = \alpha_{CP}\rho_{CP} + (1 - \alpha_{CP})\rho_{DP} \tag{2}$$

where η_{CP} and ρ_{CP} denote the viscosity and the density of the continuous phase, while η_{DP} and ρ_{DP} denote those of the disperse phase, respectively.

The volume fraction of the entire region is determined by solving the transport equation, which is expressed for α_1 of multiphase flow in OpenFOAM as [52]:

$$\frac{\partial \alpha_1}{\partial t} + \nabla \cdot (\mathbf{U}\alpha_1) + \nabla \cdot (\mathbf{U}\alpha_1(1 - \alpha_1)) = 0 \tag{3}$$

where **U** represents the fluid velocity [53]. For laminar incompressible flow, the conservation of mass and momentum are written as:

$$\frac{\partial \rho}{\partial t} + \nabla \cdot (\rho \mathbf{U}) = 0 \tag{4}$$

$$\frac{\partial \rho \mathbf{U}}{\partial t} + \nabla \cdot (\rho \mathbf{U} \mathbf{U}) = -\nabla p + \nabla \left(\eta \left(\nabla \mathbf{U} + \nabla \mathbf{U}^T \right) \right) + \frac{\rho \sigma k \nabla \alpha}{\frac{1}{2} (\rho_{DP} + \rho_{CP})} + \rho \mathbf{g} \tag{5}$$

where ρ, σ, k, p, and $\mathbf{g}$ represent the fluid density, fluid surface tension, interface curvature, pressure, and gravitational acceleration, respectively.

2.2. Flow Geometry

A double Y-shaped microfluidic channel is designed to prepare Janus droplets, as shown in Figure 1. The three-dimensional model has a total of four inlets and one outlet. The inlets of the dispersed phase forms a 45° Y-shaped channel. After they merge into one channel, it forms a 45° Y-shaped channel with the other two continuous phase inlets, beyond which Janus droplets are generated and flow through a sufficiently long downstream channel. As shown in Figure 1b, the inlet width of the first Y-shaped channel is 50 µm, and that after the intersection is 100 µm. The inlet width of the second Y-shaped channel is 100 µm, and the width after the intersection is 200 µm, while the depth of this channel is uniformly 100 µm. As shown in Table 1, the two dispersed phase A and B phase are chosen with matching densities, 1,6-hexanediol diacrylate (HDDA) and silicon oil. The two organic phases are incompatible and have an obvious interface. In order to compare differences between Newtonian and shear-thinning fluids, water and 0.25% carboxymethylcellulose (CMC) are selected, respectively. γ_{ab} represents the interfacial tension of two dispersed phases, γ_{ac} represents the interfacial tension of dispersed phase A and continuous phase C, and γ_{bc} represents the interfacial tension of dispersed phase B and continuous phase C. Uniform velocity and pressure are set at each of the four inlets, and the no-slip condition and the contact angle of 165° are imposed on channel walls. The channel upstream to the second Y-shaped channel orifice is filled with A and B phase initially.

2.3. Validation

2.3.1. Mesh Independence

The interface width of a real physical system is much smaller and sharper than the typical diffusive interface thickness adopted in most VOF simulations. Considering the dependence of the VOF method on the mesh size, solution convergence must be ensured with extreme care. The computational reconstruction of the droplet interface is directly related to the reliability of the simulation. The interface capturing method of VOF has a great relationship with the construction of mesh. A suitable mesh can form a clear and thin interface, making the simulation results more reliable. Generally speaking, the denser the mesh, the clearer the interface will be, but excessively dense mesh can also cause divergence. Here an adaptive dynamic mesh is used to refine the two-phase interface of the droplet by changing the maximum number of subdivision layers (maxRefinementInterface) that a cell in OpenFOAM can experience, as shown in Figure 2. Max Refinement (MR) is the maximum number of subdivision levels that a cell can go through. When the refinement level specifies a power of 2, for example, the maximum refinement is $2^{-3} = 0.125$ times the original cell size. The subdivision is relative to the size of the current cell, and the typical value is in the range of 2–4. The flow rates of the dispersed and continuous phases are 18 µL/h and 1800 µL/h, respectively, with $Ca_{CP} = 0.0103$ and $We = 0.051$. When the index is equal to 1, although Janus droplets are generated, the interface is very blurred. For MR = 2, the satellite droplets are not obvious; for MR = 3, a clear interface and satellite droplets are formed. In the case of MR = 4 and MR = 5, problematic divergence is encountered as shown in Table 2. Thus, in most computations performed for the parameter values chosen in this study, MR = 3 is employed.

(a)

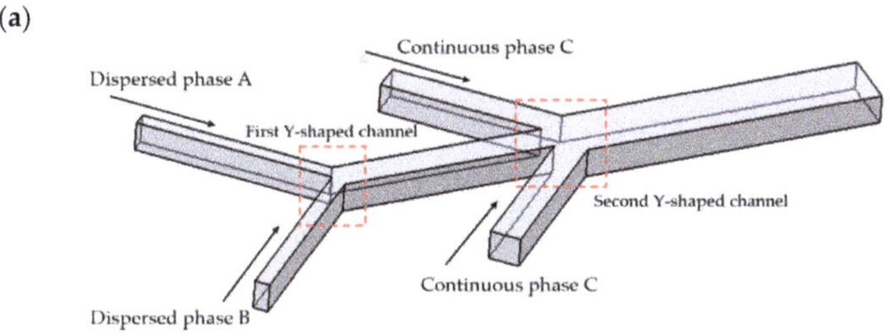

(b)

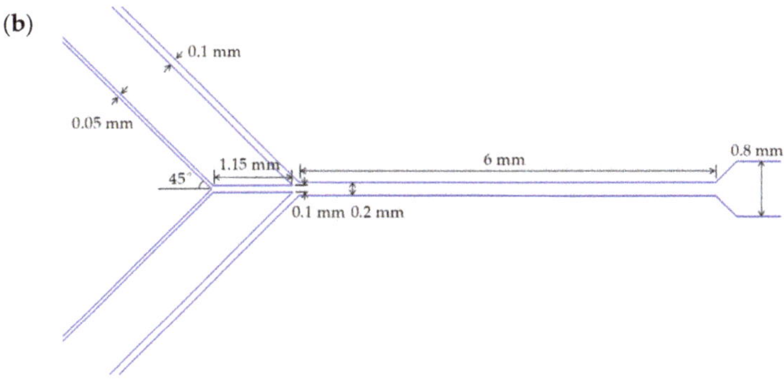

Figure 1. (**a**) Schematic illustration of the double Y-shaped channel used in the manufacture of Janus droplets; (**b**) Specific dimensions of each channel.

Table 1. Physical properties of various liquid-liquid systems used.

Phase	Density (kg/m^3) ρ	Dynamic Viscosity (Pa·s) ν	Surface Tension (mN/m) γ
Aqueous (CP)	1000	0.00101	$\gamma_{ab} = 2.2$
0.25% CMC (CP)	999.5	0.0489	$\gamma_{ab} = 2.8$
Silicon oil (DP)	960	0.00934	$\gamma_{bc} = 5.3$
HDDA (DP)	1000	0.00635	$\gamma_{ac} = 4.5$

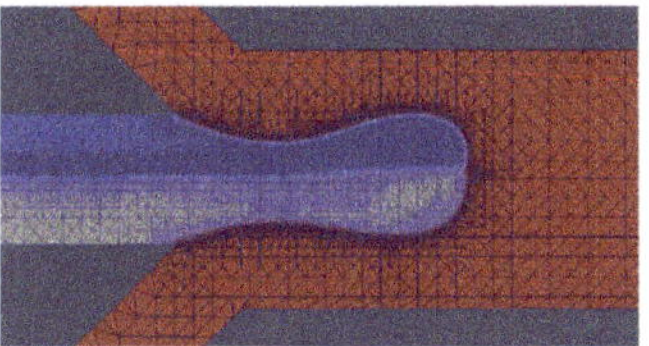

Figure 2. Schematic illustration of adaptive dynamic mesh of Janus droplet at the orifice.

Table 2. Results with different max refinement.

Name	Max Refinement	Results
Mesh 1	1	Obscure boundary
Mesh 2	2	The satellite droplet is not obvious
Mesh 3	3	Janus droplet and satellite droplet are both obvious
Mesh 4	4	Interface divergence
Mesh 5	5	Interface divergence

2.3.2. Model Validation

To validate the numerical model, the numerical results of droplet migration are compared with the experiment result by Nisisako et al. [22]. The standard deviation error bar for three different fixed Q_{DP} is studied. At different flow rates, by comparing the characteristic length of the droplets, it is found that the simulation results are consistent with the experimental results. Table 3 shows that the error rate relative to the experiment increases with the increase of the capillary number, $Ca = \rho U / \gamma$, where U and γ are inlet velocity of continuous phase and surface tension, respectively. In order to show the comparison results of experiment and simulation more comprehensively, error bar is also compared as shown in Figure 3. It can be seen that when $Ca < 0.04$, the droplet diameter is slightly smaller than the reference, but there is no significant difference, and it conforms to the law of change. However, as the number of Ca increases, the variance of the Janus droplets generated gradually increases, and it can be seen that the formation of droplets is not stable. Especially, when the $Ca > 0.1$, the change is more obvious. It can be seen that the simulation in this paper is reliable when the Ca is less than 0.1.

Table 3. Comparisons of droplet diameter obtained by the current simulation and previous experiment result.

Parameter	Droplet Diameter (m) $Q_{DP} = 1.4$ mL h^{-1}, $Ca = 0.02$	Droplet Diameter (m) $Q_{DP} = 1$ mL h^{-1}, $Ca = 0.04$	Droplet Diameter (m) $Q_{DP} = 0.4$ mL h^{-1}, $Ca = 0.1$
Empirical value by Nisisako et al.	1.82×10^{-4}	1.5×10^{-4}	1×10^{-4}
Present simulation	1.7×10^{-4}	1.3×10^{4}	8×10^{-5}
Rate of deviation	1.07	1.15	1.25

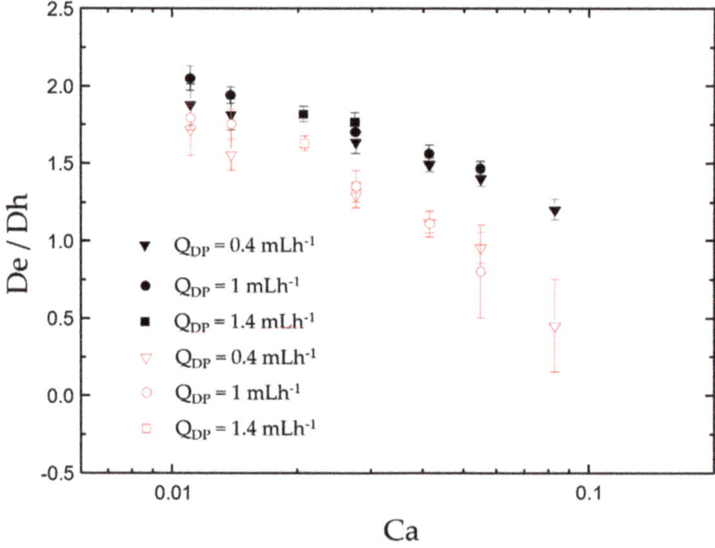

Figure 3. Schematic illustration of error bar of the comparison results of experiment and simulation more comprehensively. The solid icons represent the experimental results, and the open ones are the simulation results.

3. Results and Discussion

3.1. Phase Diagram

The phase diagram of flow states under different Ca of the continuous phase and the dispersed phase was obtained, as shown in Figure 4. Diverse working conditions can be obtained by changing the inlet flow ratio. The formation of Janus droplets in the shear-thinning fluid is characterized in five different states: Jetting and tubbing, the fluid neck is not completely contracted and flows downstream in the form of a thread without forming droplets. Jetting has U_{CP} much larger than U_{DP}, and thread width is smaller than the aperture. While tubbing means that U_{CP} is much smaller than U_{DP}, and the thread width is larger than the aperture. Dripping, through continuous phase cutting forms obvious and continuous Janus droplets, which can subdivide into three situations, the Janus droplets are large enough to touch the squeezing of the inner wall of the channel, single dripping droplets and dripping droplets accompanied by satellite droplets. Unstable dripping, the strong force of the continuous phase causes an unstable drip state, which cannot maintain a stable spherical droplet flowing downstream. Intermediate, Janus droplets pinch off an extended thread which maintains the connection with the fluid in the port through the fluid neck at the same time. When the flow rate of the continuous phase further increases, the intermediate state will be a transition to dripping state. Under the situation of a high shear rate, the viscosity of the shear-thinning continuous phase decreases as the number of shear stress increases. Therefore, the viscous stress of continuous phase imposed on the dispersed phase is lower, and the liquid thread will not be pulled downstream further. When Ca_{DP} is greater than 0.04, jetting will occur because the large inertial force. When Ca_{DP} is less than 0.01, there will be a transition from jetting to dripping with satellite droplets. Inertial force is necessary to induce satellite droplets, because when the viscous force is greater than the inertial force, the formation of droplets will be inhibited; while jetting will appear by the very large inertial force, and the filament will break into Janus droplets further downstream. For the formation of Janus droplets and satellite droplets, the elongated center droplet fluctuates and squeezes at numerous locations then producing a series of minor satellite droplets at a small viscosity ratio. In contrast, when the viscosity is quite prohibitive, the internal flow that causes the rupture is weakened, resulting in a decrease in the formation of satellite droplets.

Viscosity is one of the important factors affecting the formation of Janus droplets. The study of the formation of Janus droplets in non-Newtonian fluids, compared with the Newtonian multiphase system, has important academic and industrial significance. Under the interaction of viscous stress and the surface tension between the two fluids, the formation of Janus droplets can be controlled. The viscous stress of shear-thinning fluid related to the shear rate and the shear rate varies greatly with the change of position. The shear rate has an effect on the viscosity in all directions in the three-dimensional flow. Due to the influence of the shear rate, the viscosity of the solvent becomes smaller, the dispersed phase is more likely to accumulate at the orifice, and the shear force acting on the interface becomes smaller. Under the same conditions, compared with Newtonian fluid, the shear-thinning solution resulting in Janus droplets are larger and have longer intervals.

The control variable method was used to compare the migration of Janus droplets in shear-thinning fluid and Newtonian fluid, where the parameters of each fluid were the same, except for the dynamic viscosity. As shown in Table 4, when $Ca_{DP} = 0.0007$ and 0.0014, the Janus droplet will generate successfully in shear-thinning fluid. However, a series of small Janus droplets were generated in a Newtonian fluid in an intermediate state. When Ca_{DP} is greater than 0.07, the frequency of Janus droplets generated in the shear-thinning fluid will be faster due to the effect of smaller viscous force, but the flow state will change from intermediate to jetting in the Newtonian fluid. With the increase of the shear force of the shear-thinning fluid, the viscous force becomes smaller, so that more dispersed phase liquid will inject in the main channel and eventually form larger Janus droplets.

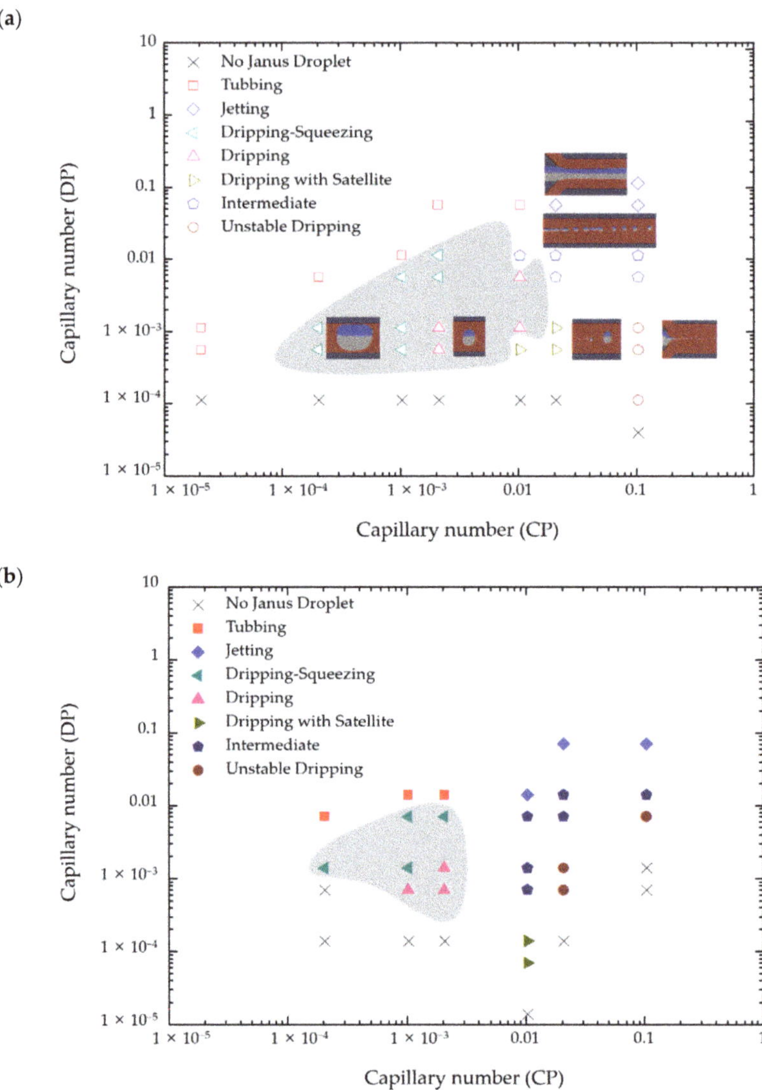

Figure 4. Schematic illustration of the phase diagram. (**a**) The phase diagram in shear-thinning fluid; (**b**) The phase diagram in Newtonian fluid. The gray area represents the dripping state.

Using an effective model, the formation mechanism and behavior of droplets in non-Newtonian fluids are systematically studied. The following chapters will elaborate on the influence of various rheological parameters, namely the power-law index (n), consistency index (K), surface tension, and the flow rate of continuous phase (Q_{CP}) and the flow rate of dispersed phase (Q_{DP}).

Table 4. The difference between Newtonian fluid and shear-thinning fluid under the same conditions.

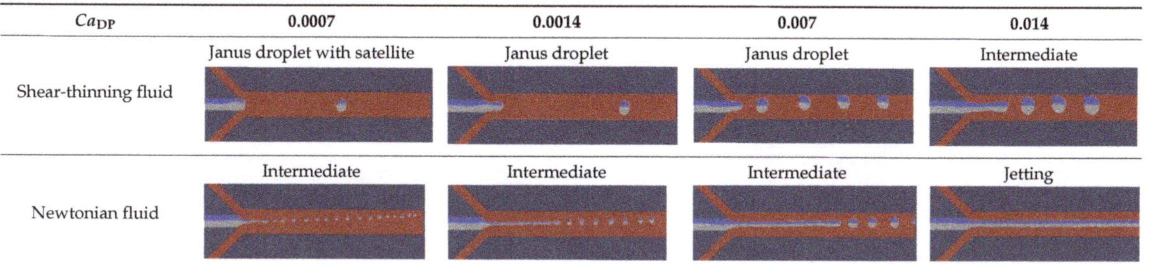

Ca_{DP}	0.0007	0.0014	0.007	0.014
	Janus droplet with satellite	Janus droplet	Janus droplet	Intermediate
Shear-thinning fluid				
	Intermediate	Intermediate	Intermediate	Jetting
Newtonian fluid				

3.2. Effect of Power-Law Index and Consistency Index

In this section, the power-law model, $\eta = Kx^{n-1}$, was chosen to describe shear-thinning fluid. The influence of the consistency index and the power-law index on the formation mechanism, size, and velocity of the droplet is discussed systematically [39].

In general, the forces acting on the dispersed phases at the cross junction are viscous force, pressure drop, and surface tension. The viscous force is caused by the viscous stress acting on the liquid-liquid interface, which is proportional to the area of the dispersed phase with the velocity gradient. When the K value is 0.0489 Pa·s^n, $n > 1$ represents shear-thickening fluid that shows the intermediate state that the dispersed phase is pulled downstream of the main channel before the droplet ruptures due to the main surface tension as shown in Figure 5a. For $n < 1$, which is shear-thinning fluid, the dripping state that resulting in larger and more stable Janus droplets will be achieved due to the low viscous force. As n increases, the effective viscosity of the continuous phase increases, resulting in higher viscous resistance, which helps the rapid separation of droplets. In the case of shear-thinning fluid, surface tension plays a dominant role and delay the separation of dispersed phases at the cross junction. For high-n shear-thinning fluid, the droplet length hardly changes. With the increase of n, in Newtonian fluids and shear-thickened liquids, due to the increase in viscosity and shear force, droplets are formed quickly before they rise to the top wall of the channel. It can be seen that as the flow state changes from dripping to jetting, the shape of the droplet changes from plug shape to spherical shape.

The generation of Janus droplets in power-law fluids can also be realized by changing the consistency index value. When $n = 0.83$, the various droplets generation state can be obtained in the different continuous phase by changing K from 0.01 Pa·s^n to 0.1 Pa·s^n as shown in Figure 5b. The flow state changes from dripping to intermediate due to the increase in effective viscosity. In addition, the size of Janus droplet decreases with the increase of the consistency index because the interaction between the viscous force and interface force.

In order to realize the influence of the power-law index on the body viscosity and overall viscosity, the effective viscosity of various liquids was estimated from the simulation and compared with the result of Equation (6) [54]:

$$\eta_{\text{eff}} = K\left(\frac{3n+1}{4n}\right)^n \left(\frac{8\mathbf{U}_L}{W_c}\right)^{n-1} \tag{6}$$

As shown in Figure 6a that the effective viscosity increases with the increase of n, and the CFD calculation results agree well with the theoretical values. In the middle of the microchannel, the effective viscosity of the shear-thinning fluid increases, which is the hallmark of the typical non-Newtonian fluid flow characteristics.

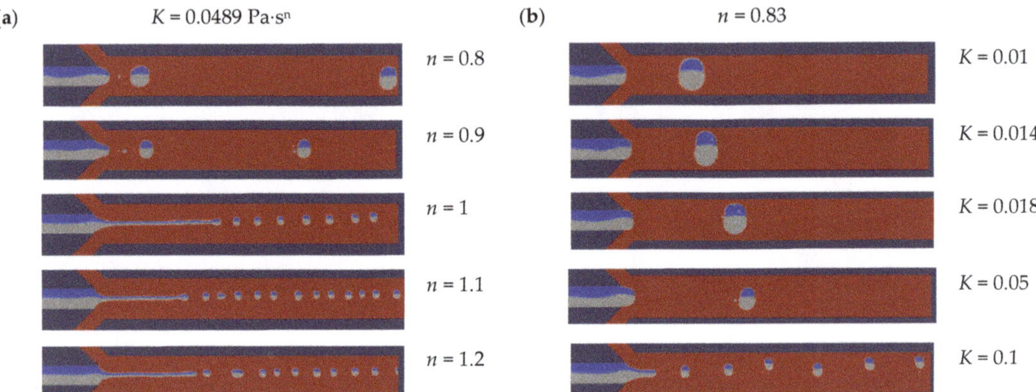

Figure 5. Schematic illustration of the effect of consistency index and liquidity index on droplet length at Ca_{CP} = 0.0103, Ca_{DP} = 0.0014. (**a**) When K = 0.0489 Pa·s^n, the flow states under different power-law index; (**b**) When n = 0.83, the flow states under different consistency index.

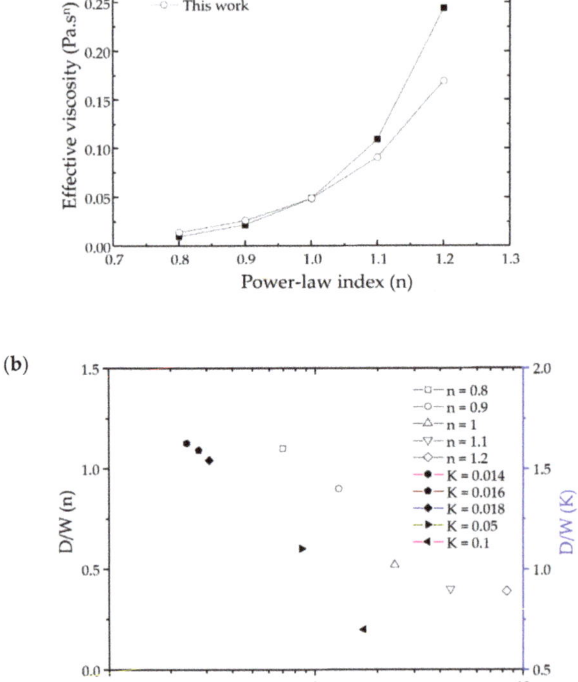

Figure 6. (**a**) Comparison of the calculated result with our simulation result. The solid icons represent the experimental results, and the open ones are the simulation results; (**b**) The scaling relation for the non-dimensional droplet size with the modified *Ca'* for various power-law liquids at different conditions. The solid icons represent the *K* results, and the open ones are the n results.

The dimensionless droplet size, D/W where D is diameter of Janus droplet and W is the width of channel, is scaled using the Ca' as a power-law relationship, as shown in Figure 6b, where $Ca' = KU_L{}^n W_c{}^{1-n}/\sigma$. When $K = 0.0489$ Pa·s^n, n change from 0.8 to 1.2, the size of Janus droplet decrease with the increase of modified Ca'. While $n = 0.83$, K change from 0.014 to 0.1, it is obvious that the size of the Janus droplet decreases with the increase of modified Ca'.

3.3. Effect of Surface Tension

In order to understand the effect of surface tension on the formation of Janus droplets, a series of simulations were carried out with additional properties unchanged, where $\gamma_{ac} + \gamma_{bc} = 10$ mN/m ($S = \gamma_{bc} - (\gamma_{ac} + \gamma_{ab}) < 0$). It can be seen from Figure 7 that when γ_{ac} (blue) is changed, as the surface tension of the shear-thinning fluid increases, the formation of Janus droplets changes from unstable state to dripping, and finally due to the difference between being increasing and the Janus droplet cannot be formed. As the surface tension of the one side dispersed phase increases to 9 mN/m, the dispersed phase (blue) blocks the front Y-type channel, and its velocity decreases. It can be observed that the size of the Janus droplet increases with the increase of the surface tension. When the surface tension between 3 mN/m and 7 mN/m, the flow state is dripping, and Janus droplets are the most stable when surface tension is 5 mN/m. When the surface tension is 1 mN/m, the phenomena that are more complex can be observed. As the surface tension increases, the satellite droplets will decrease or even disappear. Because the Weber number that determines the inertial force of the droplet is small enough, it can be ignored in this work.

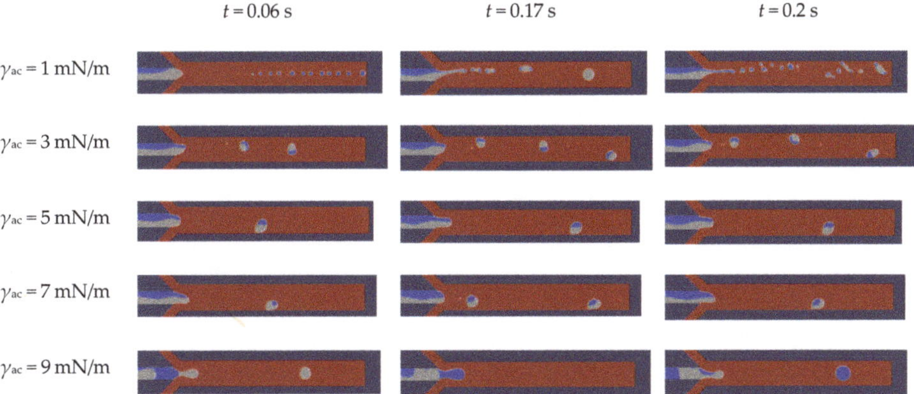

Figure 7. Effect of surface tension on droplet generation at $Ca_{CP} = 0.0103$, $Ca_{DP} = 0.0014$, $\gamma_{bc} = 5$ mN/m, and $\gamma_{ab} = 2.2$ mN/m.

Owing to the change of the curvature of the dispersed phase caused by a continuous phase, the pressure on the droplet increases with the increase of the surface tension. However, the pressure drop over the entire channel length is constant. With the increase of pressure drop on the dispersed phase, the dispersed phase will be cut off faster to form Janus droplets.

3.4. Effect of Flow Rate of Continuous Phase and Dispersed Phase

Under the fixed conditions of $K = 0.048$ Pa·s^n, $n = 0.83$, the influence of Q_{CP} on the droplet generation mechanism and size was studied. As the flow rate of dispersed phase $Q_{DP} = 45$ μL/h, the flow rate of continuous phase Q_{CP} changed from 720 μL/h to 36,000 μL/h, as shown in Figure 8a. It was observed that the viscosity decreased as the Q_{CP} increased, and the flow state changes from dripping to intermediate, and finally reaches unstable dripping. When Q_{CP} is small, the resistance of the continuous phase is weak, such that the dispersed phase easily pushes into the main channel and completely blocks at the cross-section. With the shear stress increases, the fluid neck of the dispersed phase gradually decreases, and finally separates when the neck reaches critical thickness. Moreover, as the flow rate of the continuous phase increases, the increase in viscosity and inertial force will cause the droplets to separate quickly in the neck, showing an unstable flow state.

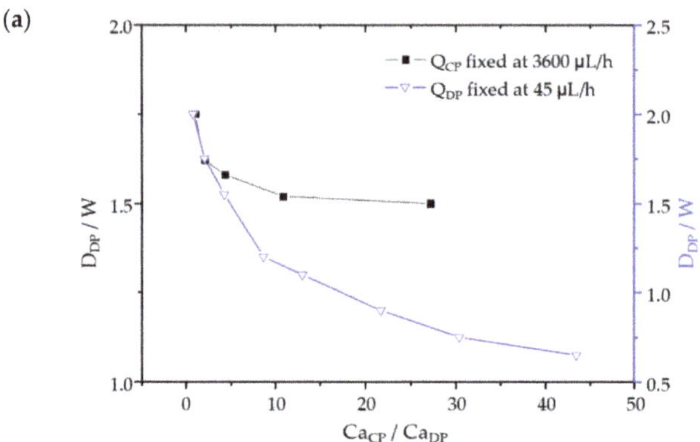

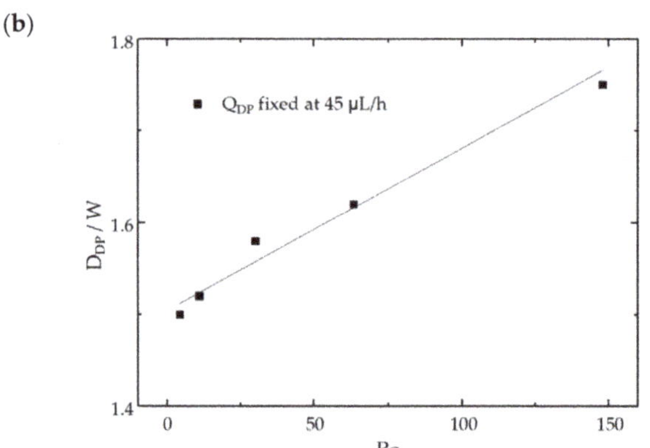

Figure 8. (**a**) Diagram of the characteristic length and *Ca* ratio under different flow rates of continuous phase and dispersed phase; (**b**) Diagram of characteristic length and Re of dispersed phase.

At the condition of Q_{CP} = 3600 μL/h, the Q_{DP} varies between 7.2–180 μL/h to understand its influence on the droplet generation characteristics. The result can be explained by the Reynolds number of the dispersed phase. It can be observed in Figure 8a that in the case of shear-thinning fluid, with the increase of Q_{DP}, there is a squeezing state ($D_{DP}/W > 1.5$). Compared with the Newtonian fluids, due to the increase of the dispersed phases entering the main channel continuously, the flow state changes from a dripping to squeezing with the increase of Q_{DP}. In the shear-thinning fluid, as the Re of the dispersed phase increases, the droplet size increases significantly, as shown in Figure 8b because the inertial force is sufficient to resist the opposing continuous phase shear stress. However, the shear stress and viscous stress of the continuous phase suppress this inertial effect in Newtonian fluids, making the droplet length change very small.

3.5. Mechanism for Janus Droplet Generation

For shear-thinning fluid, after the dispersed phase invades the main channel, it grows slowly under the balance of surface tension, shear force, and pressure drop, and is finally pinched to Janus droplets. Before detachment, the schematic diagram of the force on the emerging droplet is shown in Figure 9, which is surface tension, shear stress on the interface, and hydrostatic pressure difference on both sides of the droplet [48]. With the continuous injection of the dispersed phase at the orifice, the pressure gradient on both sides of the droplet in the continuous phase increases under the resistance of surface tension. The pressure caused by the pressure drop is:

$$F_p = (p_r - p_t) \cdot A_c = Q_c \frac{k\mu_c L_g}{h(w - L_g)^3} L_g h = \frac{k\mu_c Q_c L_g^2}{(w - L_g)^3} \tag{7}$$

where $Q = \Delta p / R_{hydro}$. In shear-thinning fluid, additional Laplace pressure is generated because the arc interface discharges the continuous fluid around itself. It is always affected by the continuous phase on the interface and express as the total pressure along the channel axis. Under the dripping state, the flow velocity between the interface and the wall is large, and the viscous force acting on the interface becomes diluted, the dispersed phase separates quickly by the dual action of surface tension and pressure drop. Viscous shear force can be represented by [55]:

$$F_\tau = 2\tau \cdot A = 2\cos 45 \mu_c \frac{Q_c}{h(w - L_g)(w - L_g)} L_g h = \frac{\sqrt{2}\mu_c Q_c L_g}{(w - L_g)^2} \tag{8}$$

where $\tau = \mu \, du/dy = \mu \, dQ/A\,dy$. The surface tension between the two phases is expressed as:

$$\begin{aligned}
F_\sigma &= \Delta p \cdot A_c \\
&= \left[\sigma_1\left(\tfrac{1}{h/2} + \tfrac{1}{L_g}\right) - \sigma_1\left(\tfrac{1}{h/2} + \tfrac{1}{L_g/2}\right)\right]\tfrac{L_g}{2}h + \left[\sigma_2\left(\tfrac{1}{h/2} + \tfrac{1}{L_g}\right) - \sigma_2\left(\tfrac{1}{h/2} + \tfrac{1}{L_g/2}\right)\right]\tfrac{L_g}{2}h \\
&= -\tfrac{1}{2}h(\sigma_1 + \sigma_2)
\end{aligned} \tag{9}$$

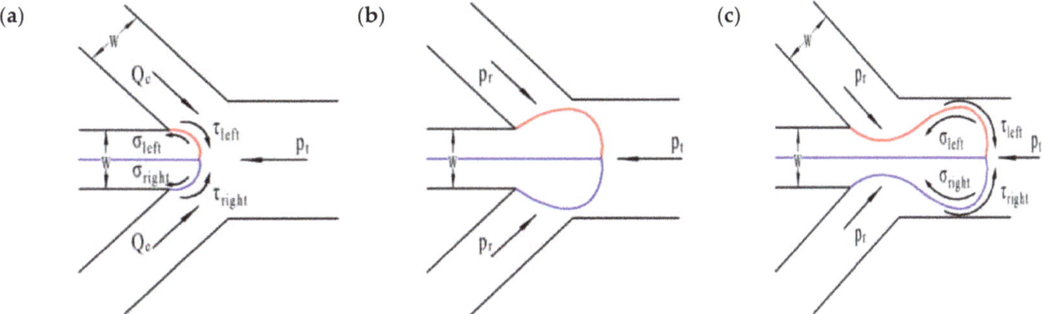

Figure 9. Sketch of emerging droplet prior to detachment and the related force competition. (**a**) Droplet just extruded from the orifice; (**b**) The extruded dispersed phase continuously accumulates at the orifice; (**c**) The dispersed phase was concave toward the centerline under the force of the continuous phase.

In summary, the total force is expressed as:

$$F_{total} = \frac{\sqrt{2}\mu_c Q_c L_g}{(w - L_g)^2} - \sigma h + \frac{k\mu_c Q_c L_g^2}{(w - L_g)^3} \tag{10}$$

Therefore, under the pressure accumulation and shear of the continuous phase fluid, the formation of droplets is a dynamic process, and the continuous generation is periodic, as are the changes in local pressure and velocity.

3.6. Mechanism for Janus Droplet Migration

Janus droplets are formed only within a limited range of $Q_{DP} = Q_A + Q_B$ at a particular value of Q_{CP} under low Ca. The three surface tensions can be balanced at equilibrium, as shown in Figure 10a, only if $S = \gamma_{ac} - (\gamma_{ab} + \gamma_{bc}) < 0$, where S is the spreading parameter. If $S > 0$, the A phase will spread entirely across the B phase to form a core–shell geometry. According to Young's equation, the surface tension at the interface must satisfy the following relationship to form Janus droplets [56]:

$$\gamma_{ab}\cos_\alpha + \gamma_{bc}\cos_\beta \geq \gamma_{ac} \tag{11}$$

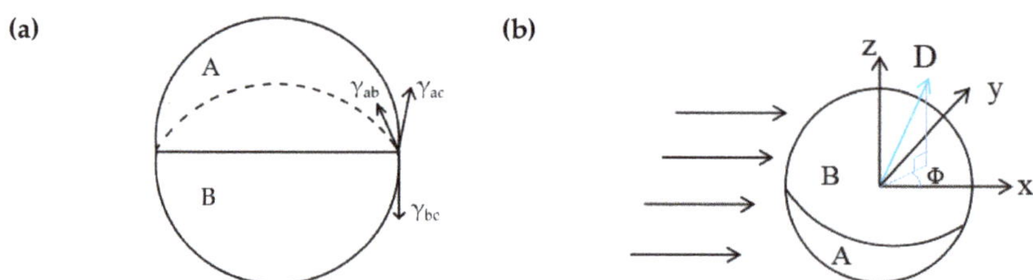

Figure 10. (**a**) 2D illustration of the equilibrium of surface tension in continuous phase; (**b**) 3D illustration of the direction of a Janus droplet.

In a simple shear flow, the direction of Janus droplets can only be characterized by the polar angle φ. The analysis here completely ignores the rotational dynamics of the Janus droplet before it reaches its stable direction, because in the case of spherical particles in the expanding flow, regardless of the initial position, they can be aligned with one of the principal axes [55]. The problem can be simply analyzed by looking for the torque-free direction D, as shown in Figure 10b. The process of determining the torque and force on the Janus droplet follows the classic direct method, that is integrating the fluid stress on the outer surface of the droplet. The integration of the surface gravitational force on the surface of the outer droplet ($r = 1$) produces the hydrodynamic force on the Janus droplet:

$$F = \int_{S_d} n \cdot \sigma dS_d \tag{12}$$

where S_d is the surface area of the outer droplet. The torque is given by the first moment of surface traction:

$$T = \int_{S_d} x \times (n \cdot \sigma) dS_d \tag{13}$$

4. Conclusions

The VOF method is used to study the formation and migration of Janus droplets in a double Y-type microchannel using Newtonian fluid/shear-thinning fluid two-phase microsystem. A multiphase fluid flow solver based on the open-source software OpenFOAM, MultiphaseInterDyMFoam, is used to solve the problem. The traditional interface-tracking method VOF depends on the mesh of simulation domain, while more accurate results can be obtained by selecting a dynamic adaptive mesh. A new understanding of the formation process of Janus droplets in non-Newtonian liquids is obtained. Under the shear-thinning fluid, the dynamic characteristics are used to explain the formation and flow state of Janus droplets under different flow conditions. The viscosity on the droplet interface is significantly reduced by the shear-thinning fluids in continuous phase. Compared with the continuous phase of Newtonian fluid under the same environment, the viscosity effect is weaker, resulting in faster separation and a larger droplet size.

In the observed extrusion, jetting, tubing, dripping, unstable dripping and parallel flow under various flow conditions, the power-law index, consistency index, flow ratio, surface tension, and rheology have significant effects on the formation and size of Janus droplets. As the power-law index and consistency index increase, the size of droplet decreases, as the effective viscosity increases. Emulsions produced with non-Newtonian fluids are commonly used for drug delivery and other biochemical applications. Accurate dosage must be ensured for a reliable operation, which requires fine control of Janus droplet size. We have also determined the correlation between the droplet size and the number of flow rate, which may provide a higher degree of control over the Janus droplet size produced by shear-thinning fluids. As the flow rate of the continuous phase increases, the Janus droplet also decreases. On the contrary, the size of the Janus droplet increases with the flow rate of the dispersed phase increases. As with Newtonian media, droplet size increases with the surface tension increases in all cases in non-Newtonian liquids. However, the size of Janus droplets changes from a small bead connected by a liquid thread to a Janus droplet in the case of shear-thinning fluid.

The development of microfluidic methods for generating and manipulating mono-dispersed droplets has brought more potentially interesting applications. Although computational research cannot completely eliminate the necessity of being exhaustive and expensive, a fully verified CFD model can certainly supplement all aspects of physical phenomena that can be obtained by experiments. The present results presented are called upon to provide a better understanding and experimental guidance for the control parameters of forming ideal Janus droplets of different shapes and sizes in non-Newtonian fluid.

Author Contributions: Conceptualization, S.W.J. and H.Z.; methodology, H.Z.; software, F.B.; validation, F.B.; formal analysis, F.B., X.L., and F.L.; investigation, F.B.; resources, S.W.J. and H.Z.; data curation, F.B.; writing—original draft preparation, F.B.; writing—review and editing, S.W.J.; visualization, F.B.; supervision, S.W.J., H.Z., X.L., and F.L.; project administration, S.W.J.; funding acquisition, S.W.J. and H.Z. All authors have read and agreed to the published version of the manuscript.

Funding: This research was funded by the National Research Foundation of Korea, grant number NRF-2018R1A2B3001246 and the National Natural Science Foundation of China, grant number NSFC-51976238.

Informed Consent Statement: Informed consent was obtained from all subjects involved in the study.

Conflicts of Interest: The authors declare no conflict of interest.

References

1. Basova, E.Y.; Foret, F. Droplet microfluidics in (bio) chemical analysis. *Analyst* **2015**, *140*, 22–38. [CrossRef]
2. Hattori, S.; Tang, C.; Tanaka, D.; Yoon, D.H.; Nozaki, Y.; Fujita, H.; Akitsu, T.; Sekiguchi, T.; Shoji, S. Development of microdroplet generation method for organic solvents used in chemical synthesis. *Molecules* **2020**, *25*, 5360. [CrossRef]
3. Kung, C.-T.; Gao, H.; Lee, C.-Y.; Wang, Y.-N.; Dong, W.; Ko, C.-H.; Wang, G.; Fu, L.-M. Microfluidic synthesis control technology and its application in drug delivery, bioimaging, biosensing, environmental analysis and cell analysis. *Chem. Eng. J.* **2020**, *125748*. [CrossRef]
4. Liu, L.; Xiang, N.; Ni, Z. Droplet-based microreactor for the production of micro/nano-materials. *Electrophoresis* **2020**, *41*, 833–851. [CrossRef]
5. Wang, Y.; Chen, Z.; Bian, F.; Shang, L.; Zhu, K.; Zhao, Y. Advances of droplet-based microfluidics in drug discovery. *Expert Opin. Drug Discov.* **2020**, *15*, 969–979. [CrossRef] [PubMed]
6. Boufadel, M.C.; Socolofsky, S.; Katz, J.; Yang, D.; Daskiran, C.; Dewar, W. A review on multiphase underwater jets and plumes: Droplets, hydrodynamics, and chemistry. *Rev. Geophys.* **2020**, *58*, e2020RG000703. [CrossRef]
7. Wang, H.; Fu, Y.; Wang, Y.; Yan, L.; Cheng, Y. Three-dimensional lattice boltzmann simulation of janus droplet formation in y-shaped co-flowing microchannel. *Chem. Eng. Sci.* **2020**, *225*, 115819. [CrossRef]
8. Zhu, P.; Wang, L. Passive and active droplet generation with microfluidics: A review. *Lab Chip* **2017**, *17*, 34–75. [CrossRef] [PubMed]
9. Li, J.; Hou, Y.; Liu, Y.; Hao, C.; Li, M.; Chaudhury, M.K.; Yao, S.; Wang, Z. Directional transport of high-temperature janus droplets mediated by structural topography. *Nat. Phys.* **2016**, *12*, 606–612. [CrossRef]
10. Sun, X.-T.; Yang, C.-G.; Xu, Z.-R. Controlled production of size-tunable janus droplets for submicron particle synthesis using an electrospray microfluidic chip. *RSC Adv.* **2016**, *6*, 12042–12047. [CrossRef]
11. Sun, X.-T.; Liu, M.; Xu, Z.-R. Microfluidic fabrication of multifunctional particles and their analytical applications. *Talanta* **2014**, *121*, 163–177. [CrossRef]

12. Niu, G.; Ruditskiy, A.; Vara, M.; Xia, Y. Toward continuous and scalable production of colloidal nanocrystals by switching from batch to droplet reactors. *Chem. Soc. Rev.* **2015**, *44*, 5806–5820. [CrossRef]

13. Plouffe, P.; Roberge, D.M.; Sieber, J.; Bittel, M.; Macchi, A. Liquid–liquid mass transfer in a serpentine micro-reactor using various solvents. *Chem. Eng. J.* **2016**, *285*, 605–615. [CrossRef]

14. Shen, Z.; Cao, M.; Zhang, Z.; Pu, J.; Zhong, C.; Li, J.; Ma, H.; Li, F.; Zhu, J.; Pan, F. Efficient ni2co4p3 nanowires catalysts enhance ultrahigh-loading lithium–sulfur conversion in a microreactor-like battery. *Adv. Funct. Mater.* **2020**, *30*, 1906661. [CrossRef]

15. Zhang, L.; Niu, G.; Lu, N.; Wang, J.; Tong, L.; Wang, L.; Kim, M.J.; Xia, Y. Continuous and scalable production of well-controlled noble-metal nanocrystals in milliliter-sized droplet reactors. *Nano Lett.* **2014**, *14*, 6626–6631. [CrossRef] [PubMed]

16. Huerre, A.; Miralles, V.; Jullien, M.-C. Bubbles and foams in microfluidics. *Soft Matter* **2014**, *10*, 6888–6902. [CrossRef]

17. Li, S.; Xu, J.; Wang, Y.; Luo, G. Controllable preparation of nanoparticles by drops and plugs flow in a microchannel device. *Langmuir* **2008**, *24*, 4194–4199. [CrossRef] [PubMed]

18. Ren, Y.; Liu, Z.; Shum, H.C. Breakup dynamics and dripping-to-jetting transition in a newtonian/shear-thinning multiphase microsystem. *Lab Chip* **2015**, *15*, 121–134. [CrossRef] [PubMed]

19. Rahimi, M.; Yazdanparast, S.; Rezai, P. Parametric study of droplet size in an axisymmetric flow-focusing capillary device. *Chin. J. Chem. Eng.* **2020**, *28*, 1016–1022. [CrossRef]

20. Yao, J.; Lin, F.; Kim, H.S.; Park, J. The effect of oil viscosity on droplet generation rate and droplet size in a t-junction microfluidic droplet generator. *Micromachines* **2019**, *10*, 808. [CrossRef]

21. Nie, Z.; Li, W.; Seo, M.; Xu, S.; Kumacheva, E. Janus and ternary particles generated by microfluidic synthesis: Design, synthesis, and self-assembly. *J. Am. Chem. Soc.* **2006**, *128*, 9408–9412. [CrossRef] [PubMed]

22. Nisisako, T.; Torii, T. Formation of biphasic janus droplets in a microfabricated channel for the synthesis of shape-controlled polymer microparticles. *Adv. Mater.* **2007**, *19*, 1489–1493. [CrossRef]

23. Nisisako, T.; Torii, T.; Takahashi, T.; Takizawa, Y. Synthesis of monodisperse bicolored janus particles with electrical anisotropy using a microfluidic co-flow system. *Adv. Mater.* **2006**, *18*, 1152–1156. [CrossRef]

24. Shepherd, R.F.; Conrad, J.C.; Rhodes, S.K.; Link, D.R.; Marquez, M.; Weitz, D.A.; Lewis, J.A. Microfluidic assembly of homogeneous and janus colloid-filled hydrogel granules. *Langmuir* **2006**, *22*, 8618–8622. [CrossRef]

25. Zhao, L.; Pan, L.; Zhang, K.; Guo, S.; Liu, W.; Wang, Y.; Chen, Y.; Zhao, X.; Chan, H. Generation of janus alginate hydrogel particles with magnetic anisotropy for cell encapsulation. *Lab Chip* **2009**, *9*, 2981–2986. [CrossRef] [PubMed]

26. Gupta, A.; Kumar, R. Effect of geometry on droplet formation in the squeezing regime in a microfluidic t-junction. *Microfluid. Nanofluid.* **2010**, *8*, 799–812. [CrossRef]

27. Wu, P.; Luo, Z.; Liu, Z.; Li, Z.; Chen, C.; Feng, L.; He, L. Drag-induced breakup mechanism for droplet generation in dripping within flow focusing microfluidics. *Chin. J. Chem. Eng.* **2015**, *23*, 7–14. [CrossRef]

28. Fu, T.; Ma, Y.; Funfschilling, D.; Zhu, C.; Li, H.Z. Squeezing-to-dripping transition for bubble formation in a microfluidic t-junction. *Chem. Eng. Sci.* **2010**, *65*, 3739–3748. [CrossRef]

29. Chen, S.; Liu, K.; Liu, C.; Wang, D.; Ba, D.; Xie, Y.; Du, G.; Ba, Y.; Lin, Q. Effects of surface tension and viscosity on the forming and transferring process of microscale droplets. *Appl. Surf. Sci.* **2016**, *388*, 196–202. [CrossRef]

30. Li, D.-Y.; Li, X.-B.; Li, F.-C. Experimental and numerical study on the droplet formation in a cross-flow microchannel. *J. Nanosci. Nanotechnol.* **2015**, *15*, 2964–2969. [CrossRef] [PubMed]

31. Raj, R.; Mathur, N.; Buwa, V.V. Numerical simulations of liquid−liquid flows in microchannels. *Ind. Eng. Chem. Res.* **2010**, *49*, 10606–10614. [CrossRef]

32. Fu, T.; Wu, Y.; Ma, Y.; Li, H.Z. Droplet formation and breakup dynamics in microfluidic flow-focusing devices: From dripping to jetting. *Chem. Eng. Sci.* **2012**, *84*, 207–217. [CrossRef]

33. Abate, A.R.; Kutsovsky, M.; Seiffert, S.; Windbergs, M.; Pinto, L.F.; Rotem, A.; Utada, A.S.; Weitz, D.A. Synthesis of monodisperse microparticles from non-newtonian polymer solutions with microfluidic devices. *Adv. Mater.* **2011**, *23*, 1757–1760. [CrossRef]

34. Yıldırım, Ö.E.; Basaran, O.A. Dynamics of formation and dripping of drops of deformation-rate-thinning and-thickening liquids from capillary tubes. *J. NonNewton. Fluid Mech.* **2006**, *136*, 17–37. [CrossRef]

35. Li, S.; Zhang, H.; Cheng, J.; Li, X.; Cai, W.; Li, Z.; Li, F. A state-of-the-art overview on the developing trend of heat transfer enhancement by single-phase flow at micro scale. *Int. J. Heat Mass Transf.* **2019**, *143*, 118476. [CrossRef]

36. Arratia, P.E.; Gollub, J.P.; Durian, D.J. Polymeric filament thinning and breakup in microchannels. *Phys. Rev. E* **2008**, *77*, 036309. [CrossRef] [PubMed]

37. Qiu, D.; Silva, L.; Tonkovich, A.L.; Arora, R. Micro-droplet formation in non-newtonian fluid in a microchannel. *Microfluid. Nanofluid.* **2010**, *8*, 531–548. [CrossRef]

38. Aytouna, M.; Paredes, J.; Shahidzadeh-Bonn, N.; Moulinet, S.; Wagner, C.; Amarouchene, Y.; Eggers, J.; Bonn, D. Drop formation in non-newtonian fluids. *Phys. Rev. Lett.* **2013**, *110*, 034501. [CrossRef]

39. Sontti, S.G.; Atta, A. Cfd analysis of microfluidic droplet formation in non−newtonian liquid. *Chem. Eng. J.* **2017**, *330*, 245–261. [CrossRef]

40. Andrew, S.; Utada, A.F.-N.; Stone, H.A.; David, A. Weitz. Dripping to jetting transitions in coflowing liquid streams. *Phys. Rev. Lett.* **2007**, *99*, 094502.

41. Eggers, J.; Villermaux, E. Physics of liquid jets. *Rep. Prog. Phys.* **2008**, *71*, 036601. [CrossRef]

42. Garstecki, P.; Fuerstman, M.J.; Stone, H.A.; Whitesides, G.M. Whitesides. Formation of droplets and bubbles in a microfluidic t-junction—scaling and mechanism of break-up. *Lab Chip* **2006**, *6*, 437–446. [CrossRef]

43. Nisisako, T.; Torii, T.; Higuchi, T. Novel microreactors for functional polymer beads. *Chem. Eng. J.* **2004**, *101*, 23–29. [CrossRef]

44. Tice, J.D.; Lyon, A.D.; Ismagilov, R.F. Effects of viscosity on droplet formation and mixing in microfluidic channels. *Anal. Chim. Acta* **2004**, *507*, 73–77. [CrossRef]

45. Sang, L.; Hong, Y.; Wang, F. Investigation of viscosity effect on droplet formation in t-shaped microchannels by numerical and analytical methods. *Microfluid. Nanofluid.* **2009**, *6*, 621–635. [CrossRef]

46. Xu, J.H.; Li, S.W.; Tan, J.; Luo, G.S. Correlations of droplet formation in t-junction microfluidic devices: From squeezing to dripping. *Microfluid. Nanofluid.* **2008**, *5*, 711–717. [CrossRef]

47. Guo, F.; Chen, B. Numerical study on taylor bubble formation in a micro-channel t-junction using vof method. *Microgravity Sci. Technol.* **2009**, *21*, 51–58. [CrossRef]

48. Wang, W.; Liu, Z.; Jin, Y.; Cheng, Y. LBM simulation of droplet formation in micro-channels. *Chem. Eng. J.* **2011**, *173*, 828–836. [CrossRef]

49. Shardt, O.; Derksen, J.J.; Mitra, S.K. Simulations of janus droplets at equilibrium and in shear. *Phys. Fluids* **2014**, *26*, 012104. [CrossRef]

50. Daghighi, Y.; Gao, Y.; Li, D. 3d numerical study of induced-charge electrokinetic motion of heterogeneous particle in a microchannel. *Electrochim. Acta* **2011**, *56*, 4254–4262. [CrossRef]

51. Hirt, C.W.; Nichols, B.D. Volume of fluid (vof) method for the dynamics of free boundaries. *J. Comput. Phys.* **1981**, *39*, 201–225. [CrossRef]

52. Henrik, R. Computational Fluid Dynamics of Dispersed Two-Phase Flows at High Phase Fractions. Ph.D. Thesis, University of London, Londone, UK, 2002.

53. Berberović, E.; van Hinsberg, N.P.; Jakirlić, S.; Roisman, I.V.; Tropea, C. Drop impact onto a liquid layer of finite thickness: Dynamics of the cavity evolution. *Phys. Rev. E* **2009**, *79*, 036306. [CrossRef] [PubMed]

54. Rhodes, M.J. *Introduction to Particle Technology*; John Wiley & Sons: Hoboken, NJ, USA, 2008.

55. Li, X.-B.; Li, F.-C.; Yang, J.-C.; Kinoshita, H.; Oishi, M.; Oshima, M. Study on the mechanism of droplet formation in t-junction microchannel. *Chem. Eng. Sci.* **2012**, *69*, 340–351. [CrossRef]

56. De Gennes, P.-G.; Brochard-Wyart, F.; Quéré, D. *Capillarity and Wetting Phenomena: Drops, Bubbles, Pearls, Waves*; Springer Science & Business Media: New York, NY, USA, 2013.

micromachines

Article

The Effect of Surface Wettability on Viscoelastic Droplet Dynamics under Electric Fields

Bo Sen Wei [ID] and Sang Woo Joo *

School of Mechanical Engineering, Yeungnam University, Gyeongsan 38541, Korea; wbs1996919@naver.com
* Correspondence: swjoo@yu.ac.kr; Tel.: +82-53-810-2568

Abstract: The effects of surface wettability and viscoelasticity on the dynamics of liquid droplets under an electric field are studied experimentally. A needle-plate electrode system is used as the power source to polarize a dielectric plate by the corona discharge emitted at the needle electrode, creating a new type of steerable electric field realized. The dynamics of droplets between the dielectric plate and a conductive substrate include three different phenomena: equilibrium to a stationary shape on substrates with higher wettability, deformation to form a bridge between the top acrylic plate and take-off on the substrates with lower wettability. Viscoelastic droplets differ from water in the liquid bridge and takeoff phenomena in that thin liquid filaments appear in viscoelastic droplets, not observed for Newtonian droplets. The equilibrated droplet exhibits more pronounced heights for Newtonian droplets compared to viscoelastic droplets, with a decrease in height with the increase in the concentration of the elastic constituent in the aqueous solution. In the take-off phenomenon, the time required for the droplet to contact the upper plate decreases with the concentration of the elastic constituent increases. It is also found that the critical voltage required for the take-off phenomenon to occur decreases as the elasticity increases.

Keywords: droplet deformation; viscoelasticity; wettable surface; dielectric field

Citation: Wei, B.S.; Joo, S.W. The Effect of Surface Wettability on Viscoelastic Droplet Dynamics under Electric Fields. *Micromachines* **2022**, *13*, 580. https://doi.org/10.3390/mi13040580

Academic Editors: Lanju Mei and Shizhi Qian

Received: 7 March 2022
Accepted: 5 April 2022
Published: 7 April 2022

Publisher's Note: MDPI stays neutral with regard to jurisdictional claims in published maps and institutional affiliations.

1. Introduction

In recent years, applications of new microfluidic technologies have grown tremendously, especially in biological, chemical, optoelectronic tweezers technology [1,2], biomedical, and other thermofluidic operations. Controlling microfluidic operations with better efficiency and accuracy has become a key part in recent years [3,4]. An important driving mechanism of fluids in microfluidics is by use of the electricity, as is common in many processes, such as electrohydrodynamic (EHD) atomization [5], electrospinning [6], inkjet printing [7], dielectrophoresis [8–11], electrowetting [12], polymer patterning preparation [13], and electrostatic spraying [14], among others. As a representative example, the deformation of thin films or droplets caused by an applied electric field is of scientific interest and a practical importance that has been studied experimentally and theoretically for many decades. The deformation of droplets before reaching the Rayleigh limit [15–18] and the droplet ejection after rupture [19,20] both have been enticing subjects of investigations. Depending on the physical properties of droplets, phenomena such as jetting and electrospraying can occur [21,22]. When a droplet is subject to an applied electric field, it undergoes deformation due to the electrostatic stresses exerted on the interface.

Swan [23] discovered the charge-induced stress in resins and viscous compounds of resins and oils. Cheng and Miksis [24] investigated the shape and the stability of droplets on conducting planes in electric fields. Basaran and Scriven [25] studied the relative importance of the electricity and the gravity compared to the surface tension. Wohlhuter and Basaran [26] found that, regardless of the ratio of the permittivity of the droplet to that of the surrounding fluid κ, the droplet shape exhibits a conical tip as its deformation develops. Three types of behaviors are found, depending on the value of κ. For $\kappa < 20.25$, the droplet deformation grows without bound as the field strength rises. On the other hand, for $21.75 \geq \kappa > 20.25$, families of equilibrium droplet shapes become unstable at turning points with respect to the field strength. For $\kappa > 21.75$, results predict that droplet deformations exhibit hysteresis. Reznik et al. [27] evolution of small droplets attached to a conducting surface and subjected to a relatively strong electric field has been investigated experimentally and numerically. Three different droplet-shape evolution scenarios are distinguished based on a numerical solution of the Stokes equation for perfectly conducting droplets. In a sufficiently weak (subcritical) electric field, the droplet is stretched by Maxwell electrical stress, and acquires a steady-state shape, where equilibrium is achieved with the action of the surface tension. In a stronger (supercritical) electric field, the Maxwell stress overcomes the surface tension, and for static (initial) contact angle of the droplet with the conducting electrode $\alpha_s < 0.8\pi$ an ejection starts from the tip of the droplet. In this case, the base of the jet acquires a quasi-stable, nearly conical shape with a vertical half angle $\beta \leq 30°$, which is significantly smaller than the Taylor cone ($\beta_T = 49.3°$). Finally, in a supercritical electric field acting on a droplet with a contact angle in the range of $0.8\pi < \alpha_s < \pi$, there is no ejection but the entire droplet jumps off, as reported by Mugele and Baret [28]. Corson et al. [29] investigated the electric field-induced deformation of a nearly hemispherical conducting droplet theoretically and temporally. Tsakonas et al. [30] studied the electric field-induced deformation of hemispherical sessile droplets of ionic liquid. Sessile droplets of an ionic liquid with contact angles close to 90° were subjected to an electric field $E = V/h$ inside a capacitor with plate separation h and potential difference V. For small field induced deformations of the droplet shapes the change in maximum droplet height $\Delta H = H(E) - H(0)$ was found to be virtually independent of the plate separation provided that $h > 3H(0)$.

In the study of droplet behavior on electric field-controlled superhydrophobic surfaces, A. Glière [31] studied the complex lift-off process caused by the competition between gravity, electricity and capillary forces. The results of B. Traipattanakul [32] show that with the increase in the plate electrode gap width, both the voltage threshold and the electric field threshold increase, while the droplet charge decreases. Christos Stamatopoulos [33] manipulated the droplet discharge by changing the wettability, and illustrated the difference between the strength of the applied electric field and the deformed shape of the droplet. Arshia Merdasi [34] analyzed electrowetting-induced droplet hopping from substrates with conical geometric heterogeneity, and compared the results with those of planar substrates with different wettability and hydrophobicity. The results show that droplet dynamics can be enhanced by applying topographical heterogeneity. However, increasing the height of the cone does not always provide better conditions for jumping, and there is an optimal value for the height of the cone. This enhancement is due to the fact that more liquid flow affects the pressure gradient within the droplet, resulting in higher jump velocities. For flat surfaces, most of the kinetic energy can be converted into oscillations of the droplet during retraction and does not promote droplet jumping.

Although the research on Newtonian fluids has made great progress, the research on non-Newtonian fluids (viscoelastic fluids) is still very little. In the study of non-Newtonian fluids [35–41] most of the studies are on motion in microchannels. There is no study on the dynamics of viscoelastic droplets between a dielectric plate and on substrates of different wettability under the corona discharge emitted by a needle-plate electrode system. It has a diverse applicability due to its flexibility and maneuverability, and we examine through experiments with droplets of PEO aqueous solutions the entire process of droplet deformations by varying the PEO concentration, electric field strength, and surface wettability of the conducting substrate. Notably, we make use of a novel steerable system that uses a pin-plate electrode to polarize the dielectric to create a steerable electric field. We can adjust the position of the needle to make the electric field appear diagonally above the droplet, so that the lateral migration of the droplet can be achieved by the force of the electric field.

2. Materials and Methods

The experimental apparatus is composed of an iron needle of diameter 1 mm, a copper plate (10 cm × 10 cm × 1 mm), an acrylic plate (5 cm × 5 cm × 2 mm), an ITO conductive glass (3 cm × 3 cm × 2 mm), and a high voltage DC power supply, as shown in Figure 1. The distance between the needle and the acrylic plate is 1 mm (The horizontal position of the acrylic plate does not affect the movement of the water droplets. As the acrylic plate is a homogeneous dielectric, moving the acrylic plate horizontally will not affect the magnitude of the electric field, so it will not affect the movement of the droplet. The vertical position of the acrylic plate affects the movement of the water droplets, because the spacing of the needle from the acrylic plate affects the polarization of the acrylic plate and thus the magnitude of the electric field). The distance between the acrylic plate and the ITO plate was 5 mm. The droplets used in this experiment were all extracted by a pipette (Nichipet EX-Plus II pipette, volume 2–20 μL), and the volume was 5 μL. The experimental results are taken by a high-speed digital camera (MotionXtra N4), the lens is a macro lens (DG Macro Lens 105 mm), and the light source used is a high-power LED (Cyclops 1). During shooting, the total number of frames is 1912 frames, with the frame rate of 300 FPS and the shooting time of 6.37 s. The acrylic plate is irradiated by the needle-tip corona discharge, generating a local electric field. The electric field can be controlled by adjusting the position of the needle to control the movement of the droplet [42]. The ITO conductive glass plate supporting liquid droplets is placed above the copper plate. The surface treatment of the glass plate is performed to obtain different contact angles. The contact angle for the original untreated plate is 70°. With a thin PDMS layer of thickness less than 1 mm placed on the glass plate, the contact angle is measured as 105°. A superhydrophobic surface can be prepared by spraying a thin layer of hydrophobic nanoparticles (Glaco Mirror Coat Zero, Soft 99, Osaka, Japan) on silicon wafers to form loose and porous structures, so it has an ultra-low surface energy. Here, superhydrophobic surface was prepared by spraying with Glaco nanoparticles 1 and 3 times to obtain contact angles of 135 and 150 degrees, respectively. We prepared adhesives by dispersing 1 M poly(ethylene oxide) (PEO with average molecular weight Mw = 1 × 10^6 g mol^{-1}, Sigma Aldrich, Burlington, MA, USA) in pure water (18.4 MΩ cm, Millipore Synergy, Darmstadt, Germany) elastic solution. To determine the effect of liquid viscoelasticity on droplet dynamics, we prepared a low and a high-concentration PEO aqueous solutions of 0.2% and 1%, respectively (at a molecular weight of 1 × 10^4 g mol^{-1}, the solution at 2% concentration was non-viscous elastic gel, so 1% concentration of Mw = 1 × 10^6 g mol^{-1} molecular weight can be regarded as high concentration of PEO), and compared with the 0% pure aqueous solution.

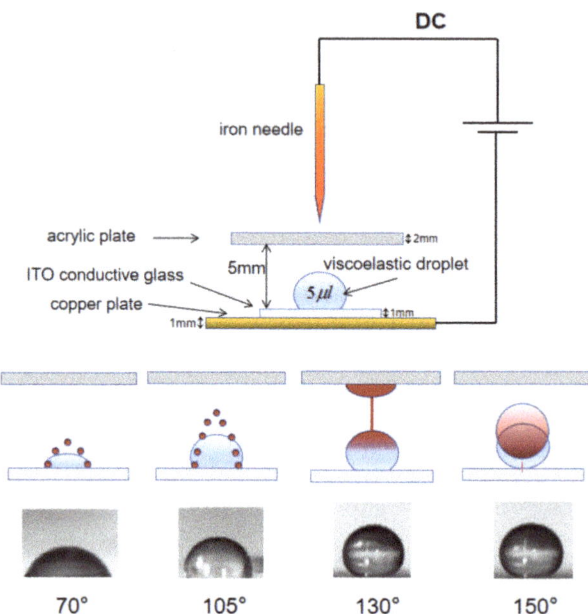

Figure 1. Deformation of viscoelastic droplets between a needle-plate electrode system. The electrodes are composed by an iron needle and a copper plate, and the acrylic plate is used as a platform to receive the corona discharge, and is polarized to generate an electric field, making the ITO glass conduct electricity and droplets on it deform. Behaviors of droplets for different contact angles of 70, 105, 130, and 150 degrees. The droplet deformation reaches an equilibrium for contact angles of 70 and 105 degrees, whereas for 130 degrees the droplet is stretched to the top acrylic plate, forming a liquid bridge a thin filament in the middle. For 150 degrees, the droplet can take off toward the top acrylic plate.

3. Results and Discussions

Figure 2a shows droplet shapes on the ITO glass substrate before and after applying the electrical field in the pin-plate electrode system. The initial droplet shape, shown on the left, varies depending on the wettability of the substrate. For a droplet of volume 5 µL, the initial height h_0 is measured as h_0 = 1.3 mm, 1.8 mm, 2.08 mm and 2.12 mm, respectively, for contact angles 70°, 105°, 130° and 150°. After applying the voltage (11 kV), the droplet starts to deform soon after a time lag for the polarization of the top acrylic plate. For all three scenarios shown in Figure 2, evolutions to an equilibrium, to a bridge, and to a take-off are not monotonic, but with transient oscillations due to competing effects of gravitational, capillary, and electrostatic forces. Additional effects of viscoelasticity seem to amplify (for bridge and take-off) or weaken (for equilibrium) these oscillations depending on the scenario, but always shorten the evolution time. On hydrophilic substates droplets have relatively bigger footprints with lower height, and equilibration occurs after the oscillatory transiency with the balance of forces involved. On hydrophobic substrates droplets can touch the top acrylic plate, eliminating the possibility of equilibration. It is observed that regardless of the wettability of the substrate the time lag is longer for Newtonian droplets than those with nonzero PEO concentrations. The droplet deformation is accompanied by the change in its height from the initial value h_0 to h_t with time. Figure 2b–d show the change in the droplet height against time in measures of its initial value:

$$H = \frac{h_t}{h_0}$$

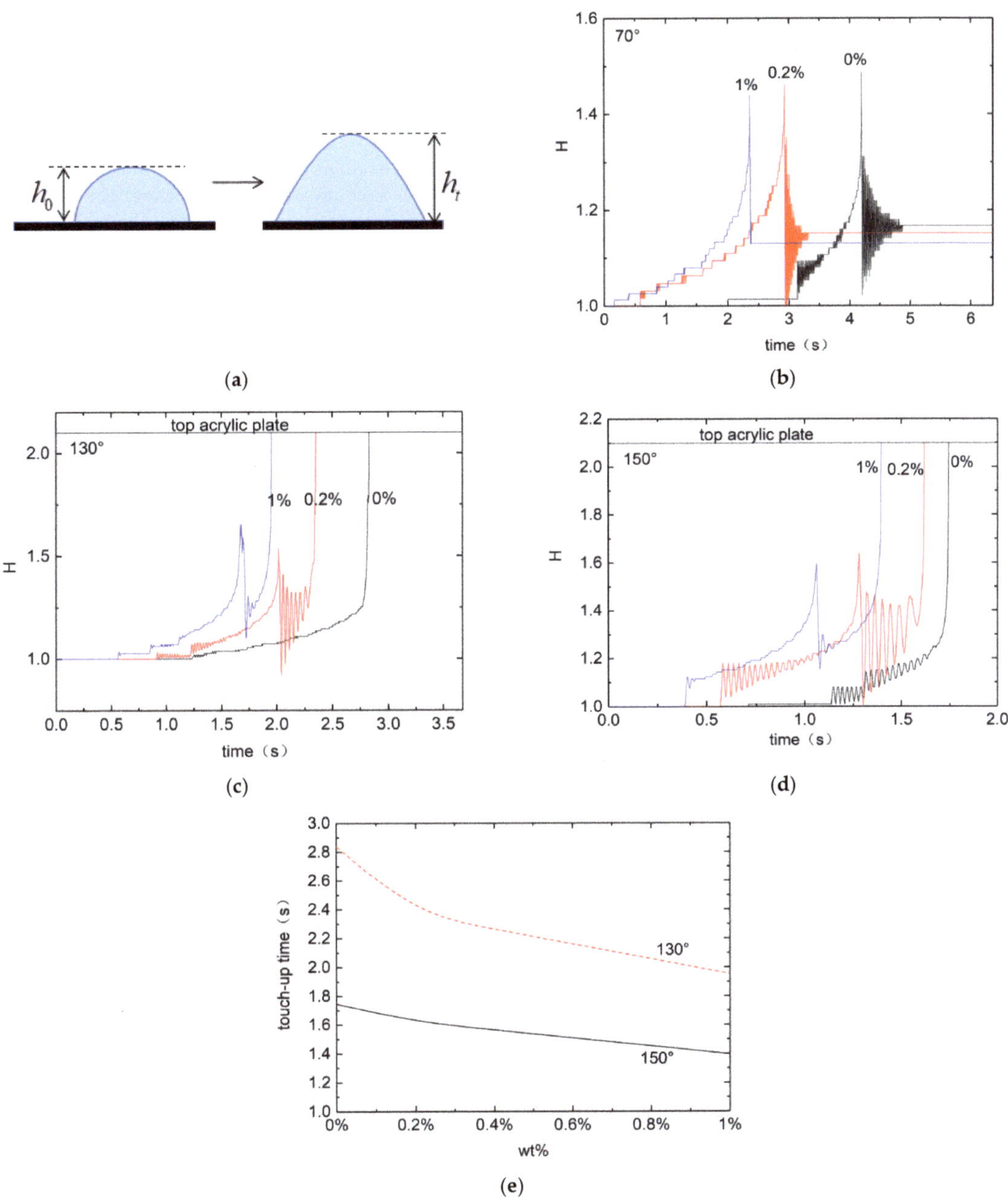

Figure 2. Deformation process of water, low-concentration PEO aqueous solution, and high-concentration PEO aqueous solution at different contact angles. (**a**) Initial and deformed droplet configuration. (**b–d**) Dimensionless droplet height evolution for contact angle of 70°, 130° and 150°. (**e**) Dimensionless time required for droplet tip to reach the top acrylic plate against PEO concentration. Voltage applied through the pin-plate electrode system is 11 kV.

On the hydrophilic surface (contact angle of 70°) droplets are stretched to higher heights due to the electric field. For all three PEO concentrations transient small-amplitude oscillations are obvious before stationary equilibrium droplet shapes are reached. It is seen that both the equilibrium height and the time required for the equilibration decrease with the PEO concentration, whereas the time required for the equilibration. It can be speculated that on more hydrophobic substrate an equilibrated droplet can reach the top acrylic plate. On the superhydrophobic surface (contact angle of 150°) even more pronounced transient oscillations observed, followed by droplet take-offs from the substrate. The take-off time decreases with the PEO concentration. Viscoelastic droplets thus take shorter time for take-off, but with more complicated transiency of droplet bouncing. Upon take-off viscoelastic droplets also create a thin filament tail attached to the substrate. Neither the bouncing before take-off nor the tail after take-off is observed with water droplets. Figure 2e shows the total time after the application of the electric field required for a launching droplet from the superhydrophobic substrate to touch the top acrylic plate against the PEO concentration. The higher the elasticity due to increase in the PEO concentration, the shorter becomes the droplet to reach the top acrylic plate.

It is observed that on a hydrophilic substrate droplets tend to equilibrate to a final stationary state. For low voltage applied the equilibration is monotonic, whereas for high enough voltage rather severe transient vibrations exist before eventual equilibration, as shown in Figure 2b. Figure 3 shows the final equilibrium droplet height beyond the transient vibrations depending on the PEO concentration for an applied voltage of 10 kV. It is found that the final equilibrium height H_f decreases with the increase in PEO concentration, or the increase in the elasticity of the droplet.

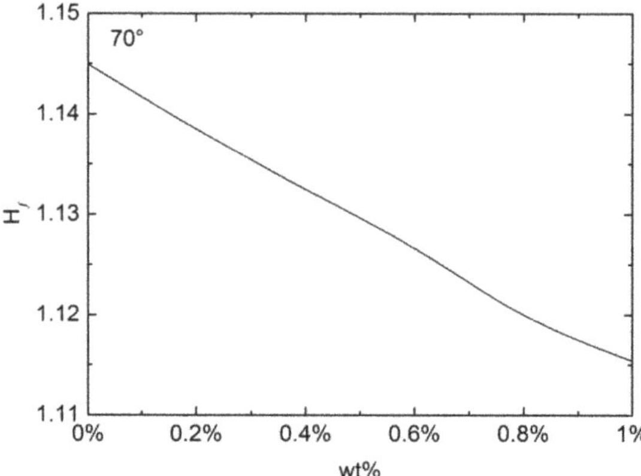

Figure 3. Equilibrium droplet height with respect to PEO concentration at a voltage of 10 kV and a contact angle of 70°.

Figure 4 shows droplet evolution on a hydrophobic substrate with a contact angle of 105°, which is not high enough to exhibit the bridge with the top acrylic plate or the take off. As with the hydrophilic substrate, for the high enough voltage (11 kV) applied the droplet goes through severe transient oscillations before reaching the final equilibrium state. Both the maximum transient peak H_{max} for the droplet height and the final equilibrium height are seen to decrease with the PEO concentration. The contact angle observed is insensitive to the PEO concentration tested in this study. The initial droplet height thus is independent of the PEO concentration. Regardless of PEO concentration droplets on a hydrophobic substrate would stand higher than those on a hydrophilic substrate. Figure 4b shows the transient maximum and final equilibrium droplet height with respect to the PEO concentration. Decrease in these heights with the PEO concentration is obvious, as seen also on a hydrophilic substrate. While the transient maximum is consistently higher on the hydrophobic surface, however, the equilibrium height observed is higher on the hydrophilic substrate. This reverse in the height is caused by the electrowetting present only with the dielectric PDMS substrate used for the contact angle 105° [43].

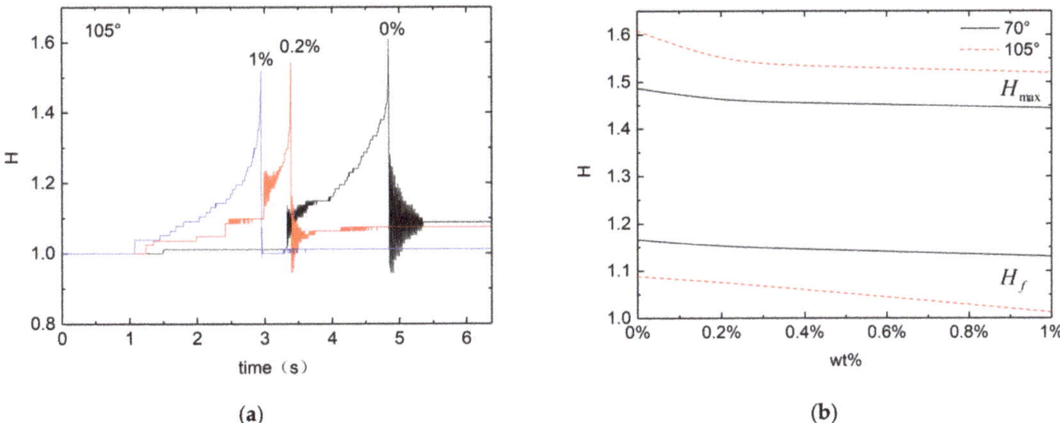

Figure 4. Droplet deformation on a dielectric PDMS substrate with contact angle 105°. (**a**) Droplet height against time; and (**b**) equilibrium height against PEO concentration.

The phenomena shown above, including the severe transient oscillations before equilibration, bridging with top acrylic plate, and the take-off from the substrate, require high enough electric field to be applied. It is thus of great importance to identify the critical voltage for each phenomenon, which should vary with the wettability and the elasticity of the droplet. In Figure 5, the critical voltage is shown for three different contact angles and PEO concentrations for a droplet volume of 5 μL on the ITO glass. The severe transiency with final equilibrium shown on a hydrophilic substrate (contact angle 70°) is to be observed for voltages exceeding 10.9 kV, 10.6 kV, and 10.4 kV, respectively, for PEO concentration of 0%, 0.2%, and 1%. The critical voltage is seen to be maximum for the Newtonian droplet, and to decrease with the increase in the elasticity. Below these critical voltages, droplets evolve monotonically to an equilibrium without complicated transiency. The bridging phenomenon of droplet reaching the top acrylic plate is observed for voltages higher than 7.9 kV, 7.68 kV, and 7.6 kV, respectively, for PEO concentration of 0%, 0.2%, and 1%. Again, more elastic droplets require lower voltage for the bridging. Below the critical voltage, droplets saturate to an equilibrium, as shown in Figure 5. The critical voltage for droplets to take off from the superhydrophobic substrate also decreases with the PEO concentration, as shown for the case of 150° contact angle. For all cases shown, droplets reach an equilibrium shape for a low applied voltage. The critical voltage to escape this monotonic equilibrium decreases with the contact angle and the PEO concentration.

Figure 6 shows that at a contact angle of 130° the droplet is stretched to the top acrylic plate to form a liquid bridge. Here, Newtonian and elastic droplets exhibit different phenomena. After the water droplet touches the top acrylic plate, the resulting liquid bridge quickly disappears, and so the liquid filament is not sustained. The PEO aqueous solution on the other hand shows sustained liquid filament, more stable with higher PEO concentration. The liquid filament gradually becomes thinner, and reacts to the electric field. When the vertical electric field is weakened by the PEO aqueous solution attached to the top acrylic plate, the PEO liquid filament will be attracted by the nearby strong electric field because the top acrylic plate is uniformly polarized under the electric field. The liquid filament thus moves with a circular trajectory.

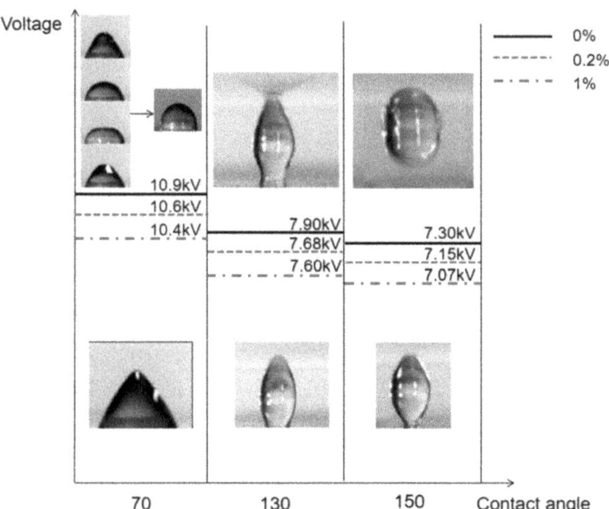

Figure 5. Critical voltage to overcome monotonic equilibration to a static droplet on hydrophilic, hydrophobic, and superhydrophobic substrate for three different PEO concentrations.

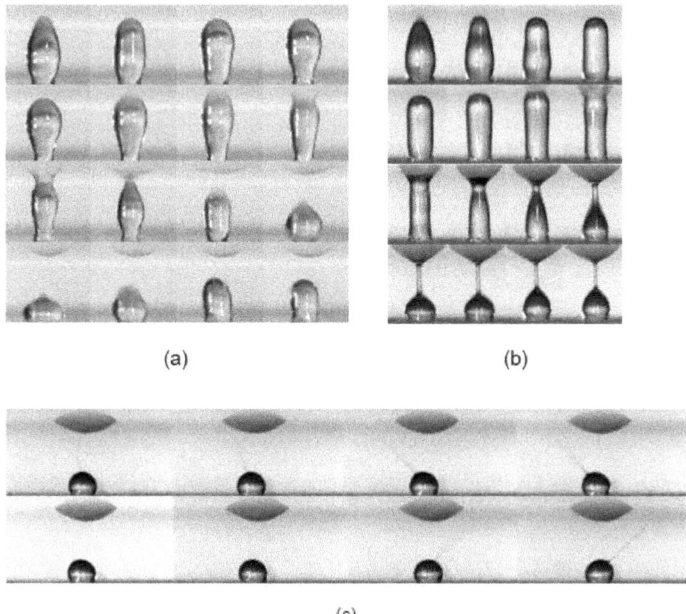

(a)

(b)

(c)

Figure 6. The liquid bridge phenomenon at a contact angle of 130°. (**a**) After the water contacts the top acrylic plate, a liquid bridge is formed for a short time, and then disappears; (**b**) PEO aqueous solution produced obvious liquid filaments after contacting the top acrylic plate; and (**c**) a phenomenon similar to electrospinning appears after the liquid filaments becomes thinner.

In the take-off phenomenon of droplets detaching from the superhydrophobic substrate, water and PEO aqueous solution also produced different behaviors. As shown in Figure 7, after the electric field (11 kV) is applied, the water droplet is stretched for a whole, and then the bottom is detached from the substrate, forming an oblate sphere, flying towards the top acrylic plate. The PEO aqueous solution takes a long time in the stretching process, and flies to the top acrylic plate in a spherical shape when separated from the wall, with a liquid filament appearing at the bottom during the flight. During the take-off process of the high-concentration PEO aqueous solution, the top is slightly stretched into a cone, and there is no obvious deformation in the process of flying to the top acrylic plate, but with a very thin liquid filament formed at the bottom.

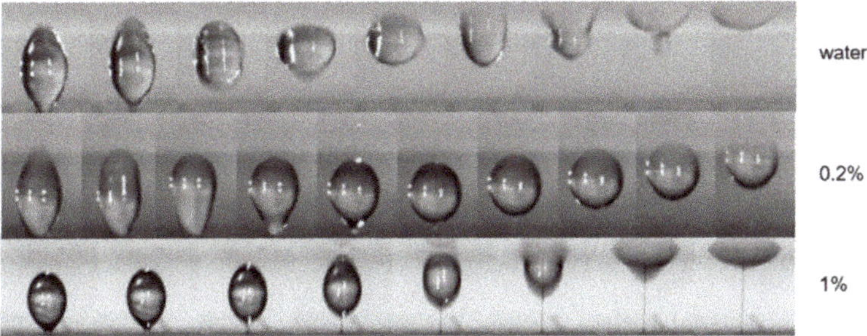

Figure 7. The sequence of water, 0.2% PEO aqueous solution and 1% PEO aqueous solution droplet taking off from superhydrophobic substrate. The water droplet shows no filament formation, while PEO droplets do.

For the process of droplet takeoff in Figure 8, comparisons are made for Newtonian and 0.2% PEO droplets. We compared the changes of position and velocity during the take-off of the droplet, respectively, using the previous time interval in which the severe deformation occurred as a reference target. We determined the position of the droplet by observing the value of H of the droplet, we found that the H value of the droplet before the violent deformation (initial position) was larger than that of the PEO droplet, and the bottom of the droplet collided with the top because the bottom of the droplet was detached from the bottom plate, this results in a faster drop rate for water droplets and a slower, more gradual drop for PEO droplets. In terms of speed, the water droplets fly much faster than PEO droplets, and touch the top acrylic plate in a shorter time. Negative velocities for a time interval are generated as the droplets collide and fuse in the air. Here, the recorded time interval is 33 ms.

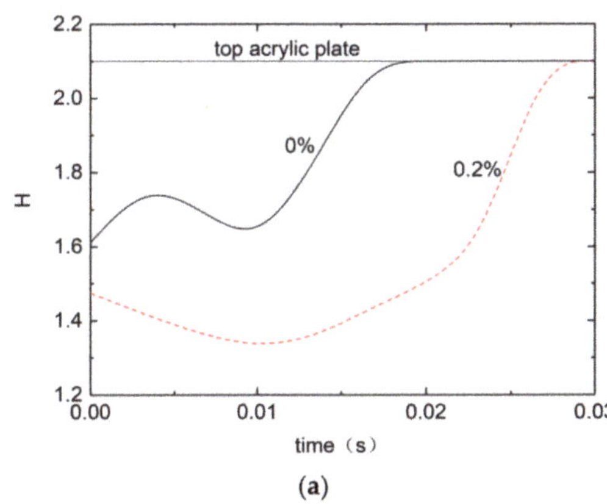

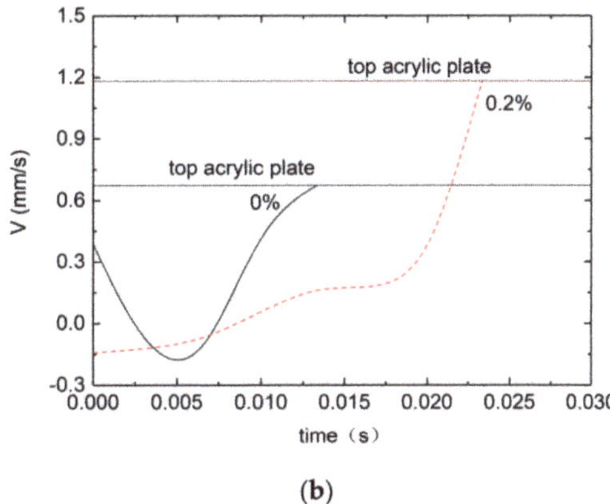

Figure 8. Droplet position and velocity vs. time for Newtonian and viscoelastic droplets. (**a**) Position of the droplet tip. (**b**) Velocity at the droplet tip for water and 0.2% PEO aqueous solution taking off on a superhydrophobic surface (150°).

The critical voltage for the take-off phenomenon of the droplet on the superhydrophobic (contact angle 150°) substrate is further investigated, as show in Figure 9. Three different droplet volumes of 5 μL, 10 μL, and 20 μL are chose to analyze the effect of droplet size along with PEO concentrations ranging from 0 to 1%. The take-off voltage required for water droplet (0%) is significantly higher than that of PEO aqueous solution, and the take-off voltage required by PEO aqueous solution has a smooth linear decay with the increase in concentration. As the concentration of PEO aqueous solution increases, the dielectric constant decreases gradually, and so the required voltage also decreases gradually. The critical take-off voltage increases with the droplet volume, regardless of the PEO concentration. Previous studies reveal that the practical contact area between a hydrophobic surface and water governs the net charge amount of the droplet on the surface [44,45]. A larger droplet thus increases the Coulomb force because it increases the electric charge on the droplet. A larger droplet size also increases the moving resistance because of the increase in the three-phase (solid–liquid–air) contact lines that are known to govern the movement of water droplets on solid surfaces [46]. Droplet size has more of an effect on Coulomb force than on the moving resistance, as the former depends on the contact area while the latter depends on the contact line. Therefore, the electric field required for droplet movement decreases with the increase in droplet size. Under a vertical electric field, Coulomb force on the droplet should overcome the sum of the adhesion force and gravity. Both adhesion force and gravity increase as the droplet size increases. The adhesion force is originally the same as the moving resistance mentioned above, and depends on the three-phase contact lines. However, gravity to a water droplet depends on the droplet mass, namely the droplet volume. The droplet size has more of an effect on gravity than on Coulomb force because the former depends on the droplet volume while the latter depends on the contact area, which relates to the square of the droplet radius [47]. The electric field required for droplet launching thus increases as the droplet size increases.

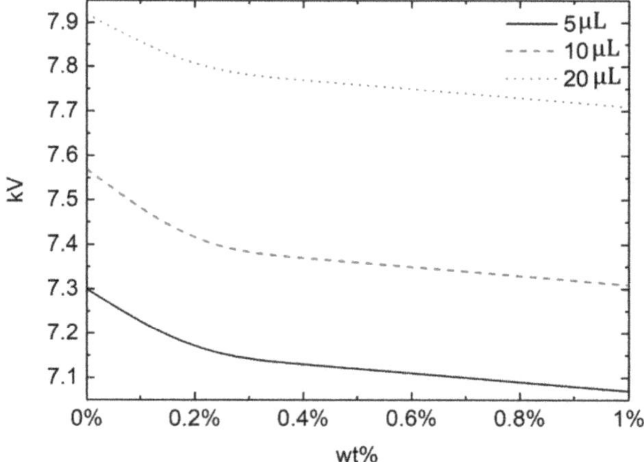

Figure 9. Critical voltage for the take-off phenomenon against PEO concentration for three different droplet volumes.

4. Concluding Remarks

The dynamics of viscoelastic droplets on surfaces with different wettability activated by a pin-plate electrode system with a dielectric plate is investigated for the first time. The viscoelasticity of droplets is controlled by changing the concentration of PEO in the aqueous solution. It is found that on hydrophilic and weakly hydrophobic surfaces (up to contact angle 105°), droplets under high electric field equilibrated to a stationary shape after severe transient oscillation. Regardless of wettability, droplets tend to equilibrate to a stationary shape under sufficiently low electric field. On highly hydrophobic substrates, droplets are stretched by the electric field to reach the top acrylic plate, and a bridge is formed. The liquid bridge formed by viscoelastic droplets show sustained filament in the middle, which thins and spins with the action of the electric field, which is analogous to an EHD electrospinning. Both water and PEO droplets take off from a substrate with even higher hydrophobicity. As in the bridging phenomenon, liquid filaments appear at the bottom of PEO droplets during the take-off process, in contrast to water droplets. The critical voltages for the phenomena reported in this work are also investigated with the viscoelasticity, the wettability, and the droplet size as parameters. We found that the voltage required for water take-off is higher than that of PEO aqueous solution, and the voltage required for take-off decreases linearly with the increase of PEO concentration. The voltage required for the droplet to take off increases with the droplet volume. The pin-plate electrode system is used to control the dielectric polarization to obtain the electric field. The motion of droplets can be controlled precisely by changing the position of the needle along with other parameters reported here. This work can be further extended to develop a new droplet manipulation method of practical importance.

Author Contributions: Conceptualization, B.S.W.; methodology, B.S.W.; investigation, B.S.W.; writing—original draft preparation, B.S.W.; writing—review and editing, S.W.J.; supervision, S.W.J.; project administration, S.W.J.; funding acquisition, S.W.J. All authors have read and agreed to the published version of the manuscript.

Funding: This work is funded by the grant NRF-2022R1A2C2002799 of the National Research Foundation of Korea.

Conflicts of Interest: The authors have declared no conflict of interest.

References

1. Wu, M.C. Optoelectronic tweezers. *Nat. Photonics* **2011**, *5*, 322–324. [CrossRef]
2. Lim, M.B.; Felsted, R.G.; Zhou, X.; Smith, B.E.; Pauzauskie, P.J. Patterning of graphene oxide with optoelectronic tweezers. *Appl. Phys. Lett.* **2018**, *113*, 031106. [CrossRef]
3. Tian, W.C.; Erin, F. Introduction to microfluidics. In *Microfluidics for Biological Applications*; Springer: Boston, MA, USA, 2008; pp. 1–34.
4. Khobaib, K.; Rozynek, Z.; Hornowski, T. Mechanical properties of particle-covered droplets probed by nonuniform electric field. *J. Mol. Liq.* **2022**, *354*, 118834. [CrossRef]
5. Smith, D.P.H. The electrohydrodynamic atomization of liquids. *IEEE Trans. Ind. Appl.* **1986**, *3*, 527–535. [CrossRef]
6. Bhardwaj, N.; Kundu, S.C. Electrospinning: A fascinating fiber fabrication technique. *Biotechnol. Adv.* **2010**, *28*, 325–347. [CrossRef]
7. Wang, J.-C.; Zheng, H.; Chang, M.-W.; Ahmad, Z.; Li, J.-S. Preparation of active 3D film patches via aligned fiber electrohydrodynamic (EHD) printing. *Sci. Rep.* **2017**, *7*, srep43924. [CrossRef]
8. Pethig, R. Dielectrophoresis: Status of the theory, technology, and applications. *Biomicrofluidics* **2010**, *4*, 022811. [CrossRef]
9. Mhatre, S.; Thaokar, R.M. Drop motion, deformation, and cyclic motion in a non-uniform electric field in the viscous limit. *Phys. Fluids* **2013**, *25*, 072105. [CrossRef]
10. Hagedorn, R.; Fuhr, G.; Müller, T.; Gimsa, J. Traveling-wave dielectrophoresis of microparticles. *Electrophoresis* **1992**, *13*, 49–54. [CrossRef]
11. Zaman, M.A.; Padhy, P.; Ren, W.; Wu, M.; Hesselink, L. Microparticle transport along a planar electrode array using moving dielectrophoresis. *J. Appl. Phys.* **2021**, *130*, 034902. [CrossRef]
12. Pollack, M.G.; Fair, R.B.; Shenderov, A.D. Electrowetting-based actuation of liquid droplets for microfluidic applications. *Appl. Phys. Lett.* **2000**, *77*, 1725. [CrossRef]
13. Wang, J.-C.; Chang, M.-W.; Ahmad, Z.; Li, J.-S. Fabrication of patterned polymer-antibiotic composite fibers via electrohydrodynamic (EHD) printing. *J. Drug Deliv. Sci. Technol.* **2016**, *35*, 114–123. [CrossRef]
14. Cloupeau, M.; Prunet-Foch, B. Electrostatic spraying of liquids in cone-jet mode. *J. Electrost.* **1989**, *22*, 135–159. [CrossRef]
15. Duft, D.; Achtzehn, T.; Müller, R.; Huber, B.A.; Leisner, T. Rayleigh jets from levitated microdroplets. *Nature* **2003**, *421*, 128. [CrossRef]
16. Zeleny, J. Instability of Electrified Liquid Surfaces. *Phys. Rev.* **1917**, *10*, 1–6. [CrossRef]
17. Taylor, G.I. Disintegration of water drops in an electric field. *Proc. R. Soc. London. Ser. A Math. Phys. Sci.* **1964**, *280*, 383–397. [CrossRef]
18. Corson, L.T.; Mottram, N.; Duffy, B.; Wilson, S.K.; Tsakonas, C.; Brown, C. Dynamic response of a thin sessile drop of conductive liquid to an abruptly applied or removed electric field. *Phys. Rev. E* **2016**, *94*, 043112. [CrossRef]
19. Collins, R.T.; Jones, J.J.; Harris, M.T.; Basaran, O.A. Electrohydrodynamic tip streaming and emission of charged drops from liquid cones. *Nat. Phys.* **2007**, *4*, 149–154. [CrossRef]

20. Collins, R.T.; Sambath, K.; Harris, M.T.; Basaran, O.A. Universal scaling laws for the disintegration of electrified drops. *Proc. Natl. Acad. Sci. USA* **2013**, *110*, 4905–4910. [CrossRef]
21. Gañán-Calvo, A.M.; José, M.M. Revision of capillary cone-jet physics: Electrospray and flow focusing. *Phys. Rev. E* **2009**, *79*, 066305. [CrossRef]
22. Yu, M.; Ahn, K.H.; Lee, S.J. Design optimization of ink in electrohydrodynamic jet printing: Effect of viscoelasticity on the formation of Taylor cone jet. *Mater. Des.* **2016**, *89*, 109–115. [CrossRef]
23. Swan, J.W. Stress and other effects produced in resin and in a viscid compound of resin and oil by electrification. *Proc. R. Soc. Lond.* **1898**, *62*, 38–46. [CrossRef]
24. Miksis, M.J.; Cheng, K.J. Shape and stability of a drop on a conducting plane in an electric field. *Phys.-Chem. Hydrodyn.* **1989**, *11*, 9–20.
25. Basaran, O.A.; Scriven, L. Axisymmetric shapes and stability of pendant and sessile drops in an electric field. *J. Colloid Interface Sci.* **1990**, *140*, 10–30. [CrossRef]
26. Quilliet, C.; Bruno, B. Electrowetting: A recent outbreak. *Curr. Opin. Colloid Interface Sci.* **2001**, *6*, 34–39. [CrossRef]
27. Reznik, S.N.; Yarin, A.L.; Theron, A.; Zussman, E. Transient and steady shapes of droplets attached to a surface in a strong electric field. *J. Fluid Mech.* **2004**, *516*, 349–377. [CrossRef]
28. Mugele, F.; Baret, J.-C. Electrowetting: From basics to applications. *J. Phys. Condens. Matter* **2005**, *17*, R705–R774. [CrossRef]
29. Corson, L.T.; Tsakonas, C.; Duffy, B.R.; Mottram, N.J.; Sage, I.C.; Brown, C.V.; Wilson, S.K. Deformation of a nearly hemispherical conducting drop due to an electric field: Theory and experiment. *Phys. Fluids* **2014**, *26*, 122106. [CrossRef]
30. Tsakonas, C.; Corson, L.T.; Sage, I.C.; Brown, C.V. Electric field induced deformation of hemispherical sessile droplets of ionic liquid. *J. Electrost.* **2014**, *72*, 437–440. [CrossRef]
31. Glière, A.; Roux, J.-M.; Achard, J.-L. Lift-off of a conducting sessile drop in an electric field. *Microfluid. Nanofluidics* **2013**, *15*, 207–218. [CrossRef]
32. Traipattanakul, B.; Tso, C.Y.; Chao, Y.H.C. Study of jumping water droplets on superhydrophobic surfaces with electric fields. *Int. J. Heat Mass Transf.* **2017**, *115*, 672–681. [CrossRef]
33. Stamatopoulos, C.; Bleuler, P.; Pfeiffer, M.; Hedtke, S.; von Rohr, P.R.; Franck, C.M. Influence of Surface Wettability on Discharges from Water Drops in Electric Fields. *Langmuir* **2019**, *35*, 4876–4885. [CrossRef] [PubMed]
34. Merdasi, A.; Moosavi, A.; Shafii, M.B. Electrowetting-induced droplet jumping over topographically structured surfaces. *Mater. Res. Express* **2019**, *6*, 086333. [CrossRef]
35. Ji, J.; Qian, S.; Liu, Z. Electroosmotic Flow of Viscoelastic Fluid through a Constriction Microchannel. *Micromachines* **2021**, *12*, 417. [CrossRef] [PubMed]
36. Casas, L.; Ortega, J.A.; Gómez, A.; Escandón, J.; Vargas, R.O. Analytical Solution of Mixed Electroosmotic/Pressure Driven Flow of Viscoelastic Fluids between a Parallel Flat Plates Micro-Channel: The Maxwell Model Using the Oldroyd and Jaumann Time Derivatives. *Micromachines* **2020**, *11*, 986. [CrossRef] [PubMed]
37. Escandón, J.; Torres, D.; Hernández, C.; Vargas, R. Start-Up Electroosmotic Flow of Multi-Layer Immiscible Maxwell Fluids in a Slit Microchannel. *Micromachines* **2020**, *11*, 757. [CrossRef] [PubMed]
38. Mei, L.; Zhang, H.; Meng, H.; Qian, S. Electroosmotic Flow of Viscoelastic Fluid in a Nanoslit. *Micromachines* **2018**, *9*, 155. [CrossRef]
39. Li, Y.; Zhang, H.; Li, Y.; Li, X.; Wu, J.; Qian, S.; Li, F. Dynamic control of particle separation in deterministic lateral displacement separator with viscoelastic fluids. *Sci. Rep.* **2018**, *8*, 1–9. [CrossRef]
40. Mei, L.; Qian, S.; Mei. Qian Electroosmotic Flow of Viscoelastic Fluid in a Nanochannel Connecting Two Reservoirs. *Micromachines* **2019**, *10*, 747. [CrossRef]
41. Omori, T.; Ishikawa, T. Swimming of Spermatozoa in a Maxwell Fluid. *Micromachines* **2019**, *10*, 78. [CrossRef]
42. Ferraro, P.; Coppola, S.; Grilli, S.; Paturzo, M.; Vespini, V. Dispensing nano–pico droplets and liquid patterning by pyroelectrodynamic shooting. *Nat. Nanotechnol.* **2010**, *5*, 429–435. [CrossRef] [PubMed]
43. Caputo, D.; de Cesare, G.; Vecchio, N.L.; Nascetti, A.; Parisi, E.; Scipinotti, R. Polydimethylsiloxane material as hydrophobic and insulating layer in electrowetting-on-dielectric systems. *Microelectron. J.* **2014**, *45*, 1684–1690. [CrossRef]
44. Takeda, K.; Nakajima, A.; Murata, Y.; Hashimoto, K.; Watanabe, T. Control of Water Droplets on Super-Hydrophobic Surfaces by Static Electric Field. *Jpn. J. Appl. Phys.* **2002**, *41*, 287–291. [CrossRef]
45. Yatsuzaka, K.; Mizuno, Y.; Asano, K. Electrification phenomena of distilled water dripping and sliding on polymer surface. *J. Instrum. Electrostat. Jpn.* **1992**, *16*, 401–410.
46. Chen, W.; Fadeev, A.Y.; Hsieh, M.C.; Öner, D.; Youngblood, J.; McCarthy, T.J. Ultrahydrophobic and Ultralyophobic Surfaces: Some Comments and Examples. *Langmuir* **1999**, *15*, 3395–3399. [CrossRef]
47. Miwa, M.; Nakajima, A.; Fujishima, A.; Hashimoto, K.; Watanabe, T. Effects of the Surface Roughness on Sliding Angles of Water Droplets on Superhydrophobic Surfaces. *Langmuir* **2000**, *16*, 5754–5760. [CrossRef]

 *micromachines*

Article

Transient Two-Layer Electroosmotic Flow and Heat Transfer of Power-Law Nanofluids in a Microchannel

Shuyan Deng * and Tan Xiao

Institute of Architecture and Civil Engineering, Guangdong University of Petrochemical Technology, Maoming 525011, China; xiaotan@gdupt.edu.cn
* Correspondence: sydeng4-c@my.cityu.edu.hk; Tel.: +86-173-7689-1017

Abstract: To achieve the optimum use and efficient thermal management of two-layer electroosmosis pumping systems in microdevices, this paper studies the transient hydrodynamical features in two-layer electroosmotic flow of power-law nanofluids in a slit microchannel and the corresponding heat transfer characteristics in the presence of viscous dissipation. The governing equations are established based on the Cauchy momentum equation, continuity equation, energy equation, and power-law nanofluid model, which are analytically solved in the limiting case of two-layer Newtonian fluid flow by means of Laplace transform and numerically solved for two-layer power-law nanofluid fluid flow. The transient mechanism of adopting conducting power-law nanofluid as a pumping force and that of pumping nonconducting power-law nanofluid are both discussed by presenting the two-layer velocity, flow rates, temperature, and Nusselt number at different power-law rheology, nanoparticle volume fraction, electrokinetic width and Brinkman number. The results demonstrate that shear thinning conducting nanofluid represents a promising tool to drive nonconducting samples, especially samples with shear thickening features. The increase in nanoparticle volume fraction promotes heat transfer performance, and the shear thickening feature of conducting nanofluid tends to suppress the effects of viscous dissipation and electrokinetic width on heat transfer.

Keywords: transient two-layer flow; electroosmotic flow; power-law nanofluid; heat transfer; Laplace transform; nanoparticle volume fraction

Citation: Deng, S.; Xiao, T. Transient Two-Layer Electroosmotic Flow and Heat Transfer of Power-Law Nanofluids in a Microchannel. *Micromachines* **2022**, *13*, 405. https://doi.org/10.3390/mi13030405

Academic Editors: Jin-yuan Qian and Kwang-Yong Kim

Received: 19 January 2022
Accepted: 26 February 2022
Published: 1 March 2022

Publisher's Note: MDPI stays neutral with regard to jurisdictional claims in published maps and institutional affiliations.

1. Introduction

It is well known that in microchannels the contact between the electrolyte solution and the solid surface of the channel wall leads to the rearrangement of charged ions, inducing an electric double layer (EDL) near the channel wall. In the presence of EDL, a layer of conducting fluid under a tangentially-applied electric field moves forward, forming electroosmotic flow (EOF); this phenomenon is called electroosmosis. Due to such favorable attributes as its ease of integration, plug-like profile, and the independence of its non-mechanical parts, the electroosmosis pumping mechanism has become a common transport phenomena in microfluidic devices [1]. In order to meet the growing demand for electroosmosis-based applications, a large number of works have theoretically studied EOF from different point of view. The transport characteristics of EOF in containers with different geometries, including slit microchannels [2], microtubes [3,4], rectangular microchannels [5], elliptic microchannels [6], and T-shaped microchannels [7] have been investigated. Several working liquids in microdevices, such as biomedical lab-on-a-chip, show nonlinear rheological behaviors, which is where the use of non-Newtonian fluid modeling becomes relevant. The nonlinear relationship between the shear stress and shear rate has been carefully treated using the power-law model [8,9], Casson model [10], Maxwell model [11], Carreau model [12], etc. The power-law model was first proposed by Das and Chakraborty [8] to describe the rheological behavior of blood, which has received great attention due to its wide coverage and the simple rheological relation [2,13]. The various aspects of power-law fluid for EOF have been discussed; the power-law model incorporates the shear thinning rheological behavior encountered in DNA solutions, and shear thickening rheological behavior encountered in cornstarch solution [14]. Recently, external environmental effects on EOF, such as a rotating frame or peristalsis, have been considered. In microchannel flow, a rotating environment induces Coriolis force, which causes a secondary flow; this is applied in biofluid transportation, drug delivery, and DNA analysis [15]. The theoretical analysis of rotating EOF was first studied by Chang and Wang [16], and has subsequently been extended in literature [17,18]. In order to obtain a comprehensive understanding of the intricate mechanism of rotational flow for biofluid flow, Kaushik et al., engaged in a transient analysis of the rotational electrohydrodynamics of power-law fluids under the effect of EDL [19]. Moreover, in the application of EOF to biomedical and biochemical analysis, peristalsis is introduced to assist in the EOF of biofluids, and thus electroosmosis-modulated peristaltic flow has recently become a frequently-studied research topic [20].

Owing to the application of external electric fields in electroosmosis-driven flow systems, the applied electric voltage leads to an inherent byproduct of the Ohmic resistance of electrolytes, namely, the Joule heating effect. Joule heating-induced heat transfer exerts an influence on transport performance by altering the electric properties of working liquids, especially for certain thermally-liable samples, which has been widely discussed in numerous works [21–25]. To optimize the hydrodynamic transport process and minimize the Joule heating effect, combined electric and magnetic fields are applied to working liquids in order to improve the actuation mechanism in microfluidics, which has the advantage of lower voltage operation, convenient manufacture, and the independence of moving parts [26,27]. On the other, to promote heat transfer and reduce entropy generation in the heat exchange equipment of microfluidics, nanofluid is created by adding nanosized metal particles, which possesses boosted thermal conductivity compared to conventional pure fluids, to the sample [28–30]. Nanofluid flow has been extensively applied in different fields, as it has none of the usual drawbacks such as sedimentation, blockage, and pressure drop [31–33]. Al_2O_3–water Nanofluid is used for cooling microprocessors or other microelectronic components due to its enhanced thermal conductivity [34], which exhibits shear thinning rheological behavior in certain ranges of nanoparticle volume fraction [35]. Moreover, in order to provide a better understanding of blood flow and other non-Newtonian biological flows in biomicrofluidic chips (such as pseudoplastic aqueous nanoliquid flow driven by electroosmosis and peristalsis and Cu/CuO–blood microvascular nanoliquid flow under thermal, microrotation, and electromagnetic field effects) were studied in [36,37], respectively. Carboxymethyl cellulose (CMC) water with γ-Al_2O_3, TiO_2 and CuO particles has been experimentally investigated for the achievement of efficient thermal management in microelectronics [38]. A comprehensive literature survey of topics in nanofluid flow indicates that the viscosity of several nanofluids mentioned above shows a nonlinear dependence on the shear rate and volume fraction of nanoparticles; thus, the power-law nanofluid model is proposed to precisely describe the rheological behavior of such nanofluids [39–42]. Regarding the power-law nanofluid, the heat transfer characteristics in magneto-hydrodynamic flow [43], convection flow [44], and EOF [45–47] have been extensively investigated.

The great advent of technologies in microelectrical mechanical systems (MEMS) enables the delivery, mixing, or separation of multi-liquids at microscales. However, certain liquids, for instance oil, blood, and ethanol, have low electrical conductivity ($<10^{-6}\,\mathrm{S\,m^{-1}}$) and are defined as nonconducting fluids [48], which fail to be driven by electroosmosis. Furthermore, for certain liquids the applied electric voltage leads to undesirable problems such as the generation of gases, fluctuation of PH value, or electrochemical decomposition. In this context, Brask et al. developed a two-layer flow system where the conducting fluid driven by electroosmotic force is adopted as driving mechanism to drag the nonconducting fluid; this has gained great attention in recent decades [49]. The steady hydrodynamic behaviors of two-Newtonian fluid EOF in a rectangular microchannel [50], two-power-law fluid combined electroosmotics with pressure driven flow in a microtube [51], and Newtonian–Casson fluid EOF in a microtube [48] have all been theoretically studied in this context. In terms of transient hydrodynamical behaviors, Gao et al. characterized the transient two-layer EOF of Newtonian fluids in a rectangular microchannel by presenting analytical velocities and flow rates at different viscosities and different electroosmotic properties [52]. Su et al., presented semi-analytical velocities for two-layer combined electroosmotic and pressure driven flow of Newtonian fluids in a slit microchannel at different electric and hydrodynamic parameters [53]. Time periodic transport characteristics of two-Newtonian liquid combining electroosmotic and pressure driven flows in a microtube have been studied numerically [54]. In order to improve transport efficiency and reduce the Joule heating effect in a two-layer pumping system, a magnetic field can be applied in addition to the pressure gradient to form a magneto-hydrodynamic EOF; the corresponding entropy generation analysis has been conducted as well [55,56]. Two-layer EOF assisted by peristalsis force was proposed by Ranjit et al., who analyzed entropy generation and heat transfer in such cases [57]. External factors such as a rotating environment [58] or varying wall shapes together with zeta potential [59] have been considered in two-layer electroosmotic systems, and the resulting influence on hydrodynamic behavior has been discussed. Furthermore, because of its desirable thermal conductivity properties, nanofluids can be applied in two-layer mixed convection flows, which are characterized by the power-law nanofluid model; the outcomes can help with the promotion of heat transfer performance [60]. Entropy generation and heat transfer in immiscible EOF of two conducting power-law nanofluid flows through a microtube have been analyzed by computing their temperature, Nusselt number, and entropy generation at different nanoparticle volume fractions and different rheological and electroosmotic properties [61]; the rheological effect of the peripheral fluid plays a dominant role in thermal performance as compared to that of inner fluid.

The application of chemical mixing/separation in thermofluidic micropumps has become increasingly frequent; therefore, the corresponding microscale cooling and heat exchangers need to be carefully designed as working liquids combined with nanoparticles show nonlinear rheological behavior in nature. To the authors' best knowledge, the underlying mechanism of the transient transport process as it develops from an unsteady to a steady state in two-power-law nanofluid EOF in a slit microchannel remains to be discovered. Therefore, this paper studies transient two-layer flow with one layer of conducting power-law nanofluid and one layer of nonconducting power-law nanofluid in a slit microchannel, with consideration of Joule heating and viscous dissipation. The governing equations are established based on the Cauchy momentum equation, continuity equation, energy equation, and power-law nanofluid rheological relation, which are analytically solved for two-Newtonian fluid flow and numerically solved for two-power-law nanofluid flow. For the hydrodynamic aspect, the mechanisms involved in using power-law nanofluid as a pumping force as well as those of pumping power-law nanofluid as the system develops from unsteady to a steady state are carefully discussed by presenting the time evolution of velocity and flow rate at different parameters. For the thermal aspect, in order to guarantee efficient thermal performance, the heat transfer characteristics arising from the interplay of the nanoparticles, power-law rheological behavior, viscous dissipation, and electrokinetic effects in a two-layer system are analyzed theoretically. The results are relevant for assisting in determining the operating parameters for optimal performance of microdevices characterized by multi-fluid delivery, mixing, or cooling.

2. Problem Formulation

2.1. Electric Potential Distribution

A two-layer immiscible EOF of power-law nanofluids in a slit microchannel is considered where the power-law nanofluid in layer I is conducting and the power-law nanofluid in layer II is nonconducting (Figure 1). The Al_2O_3 nanoparticles are suspended and uniformly distributed in a two-layer power-law base fluid system where carboxymethyl cellulose–water (CMC-water) can be represented by shear thinning fluid [38] and, without loss of generality, the base fluids fall into the categories of a shear thinning fluid and a shear thickening one. According to previously published work [50], the zeta potential difference near the two-liquid interface ($y^* = 0$) is negligible. The heights of layer I and layer II are represented by H. In the two-layer system, the EDL forms near the channel wall within the region of layer I, which creates the electric potential distribution ψ^*. When a uniform electric field E_z^* is tangentially exerted across layer I, the conducting power-law nanofluid moves forward under the electroosmotic force due to the existence of EDL near the lower channel wall, which drags the nonconducting power-law nanofluid in layer II via the interfacial viscous stress, eventually forming a two-layer EOF.

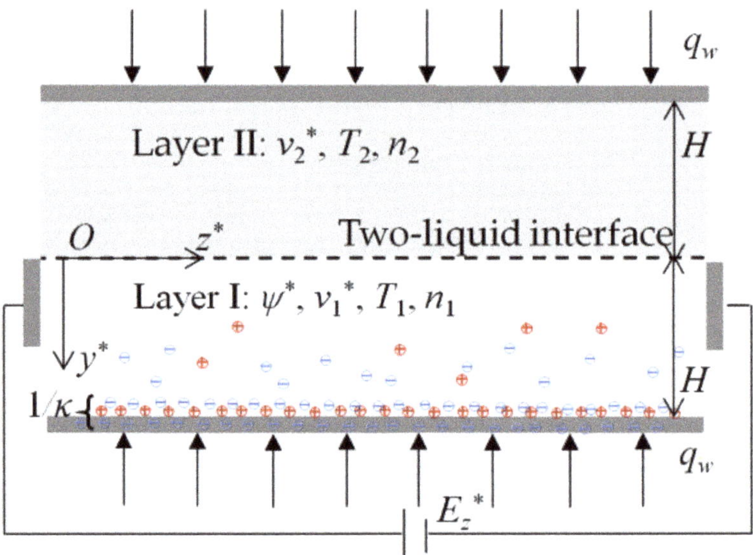

Figure 1. Schematic of two-layer EOF of power-law nanofluids in a slit microchannel.

It is assumed that the zeta potential ζ^* is small and the EDL thickness is far less than the height of microchannel; thus, the EDLs near the channel walls will not overlap. Eventually, based on electrostatic theory, the electric potential distribution ψ^* is governed by the well-known Poisson–Boltzmann (P-B) equations [50]

$$\frac{d^2\psi^*}{dy^{*2}} = -\frac{\rho_e}{\varepsilon} \tag{1}$$

$$\rho_e = -2n_0 z_0 e \sinh\left(\frac{z_0 e \psi^*}{k_B T_0}\right) \tag{2}$$

According to the established model in the scientific literature [32,62], when nanoparticles with an order of nm and with a volume fraction of $\phi \leq 10\%$ are distributed in the channel with μm-sized height, it is physically reasonable to assume that the EDLs around the nanoparticles are rather small compared to the EDLs near the channel walls, and can thus be neglected, as well as that there is no electrophoretic force on the nanoparticles. Correspondingly, the nanoparticles have no influence on the local electric charge density ρ_e, which can be governed by the P-B equations, and their influence on liquid property can be incorporated into the effective viscosity and effective thermal conductivity of the nanofluid [28–30] such that their importance can be attached to the effect of power-law feature and the effect of nanoparticles on the hydrodynamic and thermal characteristics of transient two-layer flow.

The P-B equations are subject to the following boundary conditions:

$$\left.\frac{d\psi^*}{dy^*}\right|_{y^*=0} = 0, \ \psi^*|_{y^*=H} = \zeta^* \tag{3}$$

where ε denotes the dielectric constant, n_0 denotes the ion density, z_0 denotes the valence, e is the electron charge, and k_B and T_0 represent the Boltzmann constant and the absolute temperature, respectively.

Introducing the nondimensional variables $\psi = ez_0\psi^*/(k_BT_0)$, $\zeta = ez_0\zeta^*/(k_BT_0)$, and $y = y^*/H$, $K = \kappa H$ with $\kappa^2 = 2e^2z_0^2n_0/(\varepsilon k_BT_0)$, and applying Debye–Hückel approximation ($\sinh\psi \approx \psi$ [50]) to Equations (1)–(3) when $|\zeta^*| \leq 0.025$ V, the nondimensional versions of the P-B equations can be rewritten as

$$\frac{d^2\psi}{dy^2} = K^2\psi \tag{4}$$

$$\left.\frac{d\psi}{dy}\right|_{y=0} = 0, \ \psi|_{y=1} = \zeta \tag{5}$$

Solving Equations (4) and (5) yields the electric potential distribution, as below:

$$\psi = \frac{\zeta\cosh(Ky)}{\cosh(K)} \tag{6}$$

With Equation (6), the electroosmotic force driving the conducting nanofluid can be obtained.

2.2. Two-Layer Velocity Distribution and Flow Rates

Focusing on the hydrodynamical aspects of transient two-layer EOF of power-law nanofluids in a slit microchannel, the governing equations for velocity distribution are represented by the Cauchy momentum equation and the continuity equation, as below:

$$\nabla \cdot \vec{v}^* = 0 \tag{7}$$

$$\rho\left[\frac{\partial\vec{v}^*}{\partial t^*} + \left(\vec{v}^* \cdot \nabla\right)\vec{v}^*\right] = \nabla \cdot \vec{\tau} + \vec{f} - \nabla p \tag{8}$$

where $\vec{v}$ is the velocity vector, ρ is the liquid density, t^* is the time, τ is the shear stress, $\vec{f}$ denotes the body force vector, and ∇p denotes the pressure gradient. The channel is open-ended and no pressure gradient is induced. The following assumptions are made for the purpose of analysis: (1) the properties of the nanofluids are independent of the external electric field, ion concentration, and temperature [48,50]; (2) the two-layer flow is immiscible, laminar, and incompressible, and the two-liquid interface remains distinguishable [48,50]; and (3) the gravity force and buoyancy force of the nanofluids are neglected [32]. As a result, there is only velocity component along z^* direction v^*_i (y^*,t^*), with $i = 1,2$ and the body force equal to the electroosmotic force, $f_z = E_z\rho_e$. In a system with two power-law nanofluids flowing through a slit microchannel, the shear stress of a power-law nanofluid is $\tau_i = \eta_{effi}.\partial v^*_i/\partial y^*$, where η_{effi} implies the effective dynamic viscosity of the power-law nanofluid, which nonlinearly depends on the shear rate $\partial v^*_i/\partial y^*$ and the nanoparticle volume fraction ϕ, namely, $\eta_{effi} = m_0^*/(1+\phi)^{5/2}.\left|\partial v_i^*/\partial y^*\right|^{n_i-1}$ [13,28,43,51], where m_0^* is the consistency viscosity coefficient, n_i is the flow behavior index, the subscript $i = 1$ represents the conducting nanofluid in layer I, and $i = 2$ represents the nonconducting nanofluid in layer II. Note that $n_i < 1$ corresponds to shear thinning base fluid, $n_i = 1$ corresponds to Newtonian base fluid and $n_i > 1$ corresponds to shear thickening base fluid.

Accordingly, the modified Cauchy momentum equation for the conducting power-law nanofluid in layer I under the electroosmotic force is expressed as

$$\frac{\partial v_1^*}{\partial t^*} = \frac{m_0^*}{(1+\phi)^{5/2}\rho_1}\frac{\partial}{\partial y^*}\left(\left|\frac{\partial v_1^*}{\partial y^*}\right|^{n_1-1}\frac{\partial v_1^*}{\partial y^*}\right) + \frac{1}{\rho_1}\rho_e E_z^* \text{ for } 0 \le y^* \le H \tag{9}$$

The modified Cauchy momentum equation of nonconducting power-law nanofluid in layer II is expressed as

$$\frac{\partial v_2^*}{\partial t^*} = \frac{m_0^*}{(1+\phi)^{5/2}\rho_2}\frac{\partial}{\partial y^*}\left(\left|\frac{\partial v_2^*}{\partial y^*}\right|^{n_2-1}\frac{\partial v_2^*}{\partial y^*}\right) \text{ for } -H \le y^* \le 0 \tag{10}$$

The velocities and shear stresses of the two-layer liquid satisfy the matching conditions, the velocities at the channel walls satisfy the no-slip condition, and the two-layer flow is initially set as motionless [55], thusly:

$$v_1^*\big|_{y^*=0} = v_2^*\big|_{y^*=0}, \ \eta_{eff1}\cdot\frac{\partial v_1^*}{\partial y^*}\bigg|_{y^*=0} = \eta_{eff2}\cdot\frac{\partial v_2^*}{\partial y^*}\bigg|_{y^*=0}, \ v_1^*\big|_{y^*=H}=0, \ v_1^*\big|_{y^*=-H}=0, \ v_i^*\big|_{t^*=0}=0 \tag{11}$$

With the introduction of the nondimensional variables $t = t^* m_0/(\rho_1 H^2)$, $v_i = v_i^*/U$, and $y = y^*/H$, by replacing Equations (2) and (6) with Equations (9)–(11), the nondimensional versions of the governing equations can be obtained as follow

$$\frac{\partial v_1}{\partial t} = \frac{m_1}{m_0}\frac{\partial}{\partial y}\left(\left|\frac{\partial v_1}{\partial y}\right|^{n_1-1}\frac{\partial v_1}{\partial y}\right) - \frac{GE_z\zeta}{\cosh(K)}\cosh(Ky) \text{ for } 0 \le y \le 1 \tag{12}$$

$$\frac{\partial v_2}{\partial t} = \rho_r\frac{m_2}{m_0}\frac{\partial}{\partial y}\left(\left|\frac{\partial v_2}{\partial y}\right|^{n_2-1}\frac{\partial v_2}{\partial y}\right) \text{ for } -1 \le y \le 0 \tag{13}$$

$$v_1\big|_{y=1}=v_2\big|_{y=-1}=0, v_1\big|_{y=0}=v_2\big|_{y=0}, m_1\left|\frac{\partial v_1}{\partial y}\right|^{n_1-1}\cdot\frac{\partial v_1}{\partial y}\bigg|_{y=0}=m_2\left|\frac{\partial v_2}{\partial y}\right|^{n_2-1}\cdot\frac{\partial v_2}{\partial y}\bigg|_{y=0}, v_i\big|_{t=0}=0 \tag{14}$$

where $m_i = (U/H)^{n_i-1}m_0/(1-\phi)^{5/2}$, $\rho_r = \rho_2/\rho_1$, $G = 2zen_0\zeta^*/(\rho_1 U^2)$, and $E_z = E_z^* HRe/\zeta^*$, with $Re = \zeta^* UH/m_0$, U the reference velocity.

With the nondimensional transient velocities v_1 and v_2 solved as in Equations (12)–(14), the transient flow rates are defined as follows:

$$Q_1 = \int_0^1 v_1 dy, \ Q_2 = \int_{-1}^0 v_2 dy \tag{15}$$

As time elapses, the transient velocities for layer I and layer II, namely, v_1 and v_2, reach steady status, and are then expressed as $v_{s1}(y) = \lim_{t\to\infty} v_1(y,t)$ and $v_{s2}(y) = \lim_{t\to\infty} v_2(y,t)$. To compare the flow rate of the conducting nanofluid in layer I (flow rate I) and the nonconducting nanofluid in layer II (flow rate II) with different parameters, the steady flow rate ratio is defined as

$$Q_r = \frac{\int_0^1 v_{s1} dy}{\int_{-1}^0 v_{s2} dy} \tag{16}$$

2.3. Two-Layer Temperature Distribution and Heat Transfer Rate

With the steady velocity distribution obtained from Equations (12)–(14), the temperature distribution for the thermally fully developed two-layer flow can be determined from the following energy equation:

$$(\rho c_p)_{eff}\left(\frac{\partial T}{\partial t^*} + \vec{v}_s^* \cdot \nabla T\right) = k_{eff}\nabla^2 T + \lambda E_z^{*2} + \eta_{eff}\Phi^* \tag{17}$$

where T denotes the temperature distribution, $\vec{v}_s$ means the steady velocity vector, c_p means the specific heat at constant pressure, k is the thermal conductivity, λ is the electric conductivity of base fluid, Φ^* measures the viscous dissipation effect, and the subscript *eff* means the nanofluid.

The assumption that the two-layer flow is fully thermally developed leads to the vanishing of the unsteady part of Equation (17), $\partial T/\partial t^*$, hence producing the following energy equations for the conducting nanofluid and nonconducting nanofluid, respectively, along with their corresponding boundary conditions [26,55]:

$$(\rho c_p)_{eff1}v_{s1}^*\frac{\partial T_1}{\partial z^*} = k_{eff1}\frac{d^2 T_1}{dy^{*2}} + \lambda E_z^2 + \eta_{eff1}\left(\frac{dv_{s1}^*}{dy^*}\right)^2 \text{ for } 0 \leq y^* \leq H \tag{18}$$

$$(\rho c_p)_{eff2}v_{s2}^*\frac{\partial T_2}{\partial z^*} = k_{eff2}\frac{d^2 T_2}{dy^{*2}} + \eta_{eff2}\left(\frac{dv_{s2}^*}{dy^*}\right)^2 \text{ for } -H \leq y^* \leq 0 \tag{19}$$

$$T_1\big|_{y^*=0} = T_2\big|_{y^*=0}, \; k_{eff1}\frac{dT_1}{dy^*}\bigg|_{y^*=0} = k_{eff2}\frac{dT_2}{dy^*}\bigg|_{y^*=0}, \; T_1\big|_{y^*=H} = T_w, \; T_2\big|_{y^*=-H} = T_w \tag{20}$$

where T_w means the temperature at the channel wall, subscript $i = 1$ implies the conducting nanofluid, and $i = 2$ implies the nonconducting nanofluid. Regarding the thermal properties of power-law nanofluids, the model of Choi and Yu has been applied [30,63] as it is capable of predicting the thermal conductivity of nanoliquids suspended with various kind of nonspherical nanoparticles, namely, $(\rho c_p)_{effi}= \phi(\rho c_p)_p + (1 - \phi)(\rho c_p)_b$ and $k_{effi} = \frac{k_p+2k_{bi}+2(k_p-k_{bi})(1+\omega)^3\phi}{k_p+2k_{bi}-2(k_p-k_{bi})(1+\omega)^3\phi}k_{bi}$, where ω represents the ratio of nanolayer thickness to the original particle radius and the subscripts p and b mean nanoparticles and base fluid, respectively. The left-hand side of Equation (18) measures the heat generation due to axial conduction, while the right-hand sides of Equation (18) measure the heat generation caused by heat diffusion, heat generation from Joule heating, and heat generation caused by viscous dissipation. Imposing the constant heat flux boundary condition $q_w\equiv const$ for the fully thermally developed flow above, namely, $\partial[(T_w - T_i)/(T_w - T_m)]/\partial y^* = 0$, leads to $\partial T_1/\partial y^* = \partial T_2/\partial y^* = dT_w/dy^* = dT_m/dy^*\equiv const$, in which T_m implies the mean temperature [26,55]. Furthermore, the overall energy balance condition over an elemental control volume results in

$$\frac{dT_m}{dz^*} = \frac{2q_w + \lambda HE_z^2 + \frac{m_0}{(1-\phi)^{5/2}}\left(\int_0^H \left|\frac{\partial v_{s1}^*}{\partial y^*}\right|^{n_1-1}\frac{\partial v_{s1}^*}{\partial y^*}dy^* + \int_{-H}^0 \left|\frac{\partial v_{s2}^*}{\partial y^*}\right|^{n_2-1}\frac{\partial v_{s2}^*}{\partial y^*}dy^*\right)}{H(\rho c_p)_{eff1}v_{ms1}^* + H(\rho c_p)_{eff2}v_{ms2}^*} \tag{21}$$

where $v_{ms1}^* = \int_0^H v_{s1}^* dy^*/H$ and $v_{ms2}^* = \int_{-H}^0 v_{s2}^* dy^*/H$ are the dimensional average velocities in layer I and layer II, respectively. Introducing the nondimensional temperature variable $\theta_i = k_{f1}(T_i - T_w)/(q_w H)$ and placing Equation (21) into Equations (18)–(20) yields

$$\frac{d^2\theta_1}{dy^2} + \frac{k_{f1}}{k_{eff1}}(S + Br\Phi_1) = \frac{k_{f1}}{k_{eff1}}\frac{(2 + S + Br\Gamma_1 + m_r Br\Gamma_2)}{v_{ms1} + (\rho c_p)_r v_{ms2}}v_{s1} \text{ for } 0 \leq y \leq 1 \tag{22}$$

$$\frac{d^2\theta_2}{dy^2} + \frac{k_{f1}}{k_{eff2}} m_r Br\Phi_2 = \frac{k_{f1}}{k_{eff2}} \frac{(2 + S + Br\Gamma_1 + m_r Br\Gamma_2)}{v_{ms1}/(\rho c_p)_r + v_{ms2}} v_{s2} \text{ for } -1 \leq y \leq 0 \qquad (23)$$

$$\theta_1\big|_{y=1} = \theta_2\big|_{y=-1} = 0, \ \theta_1\big|_{y=0} = \theta_2\big|_{y=0}, \ k_{eff1}\frac{d\theta_1}{dy}\bigg|_{y=0} = k_{eff2}\frac{d\theta_2}{dy}\bigg|_{y=0} \qquad (24)$$

where $v_{ms1} = \int_0^1 v_{s1}dy$, $v_{ms2} = \int_{-1}^0 v_{s2}dy$, $\Phi_i = \left|dv_{si}/dy\right|^{n_i-1}(dv_{si}/dy)^2$, $\Gamma_1 = \int_0^1 \Phi_1 dy$, $\Gamma_2 = \int_{-1}^0 \Phi_2 dy$, $(\rho c_p)_r = (\rho c_p)_{eff2}/(\rho c_p)_{eff1}$, $m_r = m_2/m_1$, $Br = m_1 U^2/(q_w H)$ is the Brinkman number, and $S = \lambda H E_z^2/q_w$ is the Joule heating parameter.

With the two-layer temperature distributions solved from Equations (22)–(24), the heat transfer performance can be examined by evaluating heat transfer rate as represented by the Nusselt number, $Nu = hD_h/k_{eff}$. With the convective heat transfer coefficient at the channel surface $h = q_w/(T_w - T_m)$ and the characteristic height $D_h = 2H$ [55,57], the further rearrangement produces

$$Nu = -\frac{2k_{f1}}{k_{eff f1}\theta_m} \qquad (25)$$

where $\theta_m = \left(\int_0^1 v_{s1}\theta_1 dy + \int_{-1}^0 v_{s2}\theta_2 dy\right)/\left(\int_0^1 v_{s1}dy + \int_{-1}^0 v_{s2}dy\right)$ is the mean temperature.

2.4. Entropy Generation Analysis

Based on the second law of thermodynamics, a certain amount of energy is inevitably destroyed during the heat transfer process, that is, thermal irreversibility inherently accompanies the thermal behaviors and reduces system efficiency. This irreversibility is represented by the entropy generation rate, and the thermal performance of system is thereby assessed. The local entropy generation over a given cross-section of the microchannel can be given for two-layer flow as

$$S_{l1}^*(y^*) = \frac{k_{eff1}}{T_1^2}\left(\frac{dT_1}{dy^*}\right)^2 + \frac{\sigma E_z^2}{|T_1|} + \frac{m_{eff1}}{|T_1|}\left|\frac{dv_{s1}^*}{dy^*}\right|^{n_i-1}\left(\frac{dv_{s1}^*}{dy^*}\right)^2 \text{ for } 0 \leq y \leq 1 \qquad (26)$$

$$S_{l2}^*(y^*) = \frac{k_{eff2}}{T_2^2}\left(\frac{dT_2}{dy^*}\right)^2 + \frac{m_{eff2}}{|T_2|}\left|\frac{dv_{s2}^*}{dy^*}\right|^{n_2-1}\left(\frac{dv_{s2}^*}{dy^*}\right)^2 \text{ for } -1 \leq y \leq 0 \qquad (27)$$

in which the right-hand side of Equation (26) implies entropy generation caused by heat transfer, Joule heating and viscous dissipation, respectively. With the introduction of the nondimensional groups $S_{li} = H^2 S_{li}^*/k_{f1}$ and $\Theta = k_{f1}T_w/(q_w H)$, the respective nondimensional local entropy generation is obtained as follows

$$S_{l1}(y) = \frac{k_{eff1}}{k_{f1}(\theta_1 + \Theta)^2}\left(\frac{d\theta_1}{dy}\right)^2 + \frac{S}{|\theta_1 + \Theta|} + \frac{Br}{|\theta_1 + \Theta|}\left|\frac{dv_{s1}}{dy}\right|^{n_1-1}\left(\frac{dv_{s1}}{dy}\right)^2 \text{ for } 0 \leq y \leq 1 \qquad (28)$$

$$S_{l2}(y) = \frac{k_{eff2}}{k_{f1}(\theta_2 + \Theta)^2}\left(\frac{d\theta_2}{dy}\right)^2 + \frac{Br}{|\theta_2 + \Theta|}\left|\frac{dv_{s2}}{dy}\right|^{n_2-1}\left(\frac{dv_{s2}}{dy}\right)^2 \text{ for } -1 \leq y \leq 0 \qquad (29)$$

The total entropy generation can be computed by integrating Equations (28) and (29) over the relevant cross-section area of microchannel:

$$S_t = \int_0^1 S_{l1}dy + \int_{-1}^0 S_{l2}dy \qquad (30)$$

2.5. Solutions of Modelling and Validation

2.5.1. In the Case of Newtonian Fluids

The analytical velocities and analytical temperatures of a two-layer transient EOF of Newtonian fluids are first solved, and can then be employed to validate the numerical algorithm proposed for power-law nanofluid flow. Specifically, when $n_i = 1$ and $\phi = 0$, Equations (12)–(14) reduce to

$$\frac{\partial v_1^N}{\partial t} = \frac{\partial^2 v_1^N}{\partial y^2} - \frac{GE_z\zeta}{\cosh(K)} \cosh(Ky) \text{ for } 0 \leq y \leq 1 \tag{31}$$

$$\frac{\partial v_2^N}{\partial t} = \frac{\partial^2 v_2^N}{\partial y^2} \text{ for } -1 \leq y \leq 0 \tag{32}$$

$$v_1^N\big|_{y=0} = v_2^N\big|_{y=0}, \ \frac{\partial v_1^N}{\partial y}\bigg|_{y=0} = \frac{\partial v_2^N}{\partial y}\bigg|_{y=0}, \ v_1^N\big|_{y=1} = 0, \ v_1^N\big|_{y=-1} = 0, \ v_1^N\big|_{t=0} = 0 \tag{33}$$

where the superscript N denotes the special case of Newtonian fluid. Using the method of Laplace transformation, the analytical expressions of the transient velocities are obtained as follows:

$$\begin{aligned}
v_1^N &= \frac{GE_z\zeta}{K^2}\left[\frac{1-\cosh(K)}{2\cosh(K)}(y-1) + \frac{\cosh(Ky)}{\cosh(K)} - 1\right] \\
&+ 8GE_z\zeta \sum_{P=1}^{\infty} \frac{(-1)^{P+1}e^{-(2P-1)^2\pi^2t/4}}{(2P-1)\pi[4K^2+(2P-1)^2\pi^2]} \cos\left[\frac{(2P-1)\pi y}{2}\right] \\
&+ GE_z\zeta \sum_{P=1}^{\infty} \frac{[1/\cosh(K)-(-1)^{P+1}]e^{-(P\pi)^2t}}{P\pi[K^2+(P\pi^2)]} \sin(P\pi y) \text{ for } 0 \leq y \leq 1
\end{aligned} \tag{34}$$

$$\begin{aligned}
v_2^N &= \frac{GE_z\zeta}{K^2}\left[\frac{1-\cosh(K)}{2\cosh(K)}(y-1) + \frac{1}{\cosh(K)} - 1\right] \\
&+ 8GE_z\zeta \sum_{P=1}^{\infty} \frac{(-1)^{P+1}e^{-(2P-1)^2\pi^2t/4}}{(2P-1)\pi[4K^2+(2P-1)^2\pi^2]} \cos\left[\frac{(2P-1)\pi y}{2}\right] \\
&+ GE_z\zeta \sum_{P=1}^{\infty} \frac{[1/\cosh(K)-(-1)^{P+1}]e^{-(P\pi)^2t}}{P\pi[K^2+(P\pi^2)]} \sin(P\pi y) \text{ for } -1 \leq y \leq 0
\end{aligned} \tag{35}$$

for which the solving procedure is presented in the Appendix A in the interest of conciseness and readability.

As time tends to infinity, the two-fluid velocities reach a steady status as follows:

$$v_{s1}^N(y) = \lim_{t\to\infty} v_1^N(y,t) = \frac{GE_z\zeta}{K^2}\left[\frac{1-\cosh(K)}{2\cosh(K)}(y-1) + \frac{\cosh(Ky)}{\cosh(K)} - 1\right] \text{ for } 0 \leq y \leq 1 \tag{36}$$

$$v_{s2}^N(y) = \lim_{t\to\infty} v_2^N(y,t) = \frac{GE_z\zeta}{K^2}\left[\frac{1-\cosh(K)}{2\cosh(K)}(y-1) + \frac{1}{\cosh(K)} - 1\right] \text{ for } -1 \leq y \leq 0 \tag{37}$$

which are exactly the solutions to Equations (31)-(33) when the temporal term $\partial v_i/\partial t$ vanishes.

Replacing Equations (36) and (37) with Equations (22)–(24), the analytical temperature distributions are solved by integrating Equations (22) and (23) twice and combining them with Equation (24):

$$\theta_1^N = A_1 y^3 + A_2\cosh(Ky) + A_3 y^2 + A_4\sinh(Ky) + A_5\left[\frac{\cosh(Ky)^2}{K^2} - y^2\right] + D_1 y + D_2 \tag{38}$$

$$\theta_2^N = B_1 y^3 + B_2 y^2 + D_3 y + D_4 \tag{39}$$

with the coefficients A, B, and D presented in the Appendix A for conciseness.

2.5.2. In the Case of Power-Law Nanofluids

Due to the nonlinearity of a two-layer EOF of power-law nanofluids, Equations (12)–(14) and (22)–(24) are numerically solved based on the explicit finite difference method. At first, the following discretization is introduced: $t_l = l\Delta t$, $y_j = j\Delta y$, $v_{i,j}^l = v_i(j\Delta y, l\Delta t)$, and $\theta_{i,j}^l = \theta_i(j\Delta y, l\Delta t)$, $l = 1, 2, \ldots, L$ and $j = 1, 2, \ldots, J$.

At first, the numerical velocities at the channel walls are easily determined by discretizing the no slip conditions in Equation (14):

$$v_{1,J}^l = v_{2,1}^l = 0 \tag{40}$$

The numerical velocities at the two-liquid interface are computed using the bisection method from the discretized version of the two-liquid interface matching conditions in Equation (14)

$$v_{1,1}^l = v_{2,J}^l, \; m_1 \left(\frac{-3v_{1,1}^l + 4v_{1,2}^l - v_{1,3}^l}{2\Delta y} \right)^{n_1} = m_2 \left(\frac{v_{2,J-2}^l - 4v_{2,J-1}^l + 3v_{2,J}^l}{2\Delta y} \right)^{n_2} \tag{41}$$

Note that for the proposed numerical algorithm, when $n_i \geq 1$, let the initial two-layer velocities be $v_1^0 = v_2^0 = 0$, and when $n_i < 1$, to avoid the singularity caused by the zero denominator, let the initial two-layer velocities be $v_1^0 = v_2^0 = nonzero$.

Then, the velocities of the bulk liquid can be computed by means of the following numerical algorithm:

$$v_{1,j}^{l+1} = v_{1,j}^l + [\Lambda_{1,j}^l + GE_z\zeta \cosh(Ky_j) / \cosh(K)] \cdot \Delta t \tag{42}$$

$$v_{2,j}^{l+1} = v_{2,j}^l + \Lambda_{2,j}^l \cdot \Delta t \tag{43}$$

where $\Lambda_{i,j}^l = \frac{m_i}{m_0} \left[(g_{i,j}^l)^{n_i-1} \frac{v_{i,j+1}^l - 2v_{i,j}^l + v_{i,j-1}^l}{\Delta y^2} + (n_i - 1)(g_{i,j}^l)^{n_i-2} \frac{g_{i,j+1}^l - g_{i,j-1}^l}{2\Delta y} \frac{v_{i,j+1}^l - v_{i,j-1}^l}{2\Delta y} \right]$, $g_{i,j}^l = \left| \frac{v_{i,j+1}^l - v_{i,j-1}^l}{2\Delta y} \right|$, $i = 1, 2$, and $j = 2, 3, \ldots, J - 1$. When t is great enough, the transient velocities reach steady status, i.e., they are examined by $\|v^l - v^{l+1}\| < Er$, with Er being a specified criterion, and the steady velocities v_{s1} and v_{s2} are solved numerically. Consequently, the flow rates can be computed from Equations (15) and (16) by means of the Simpson composite integration method [61].

The temporal term $\partial\theta_i/\partial t$ is introduced to Equations (22) and (23) to allow the same numerical algorithm to be used to solve both velocity distribution and temperature distribution. As time approaches to infinity, the fully thermally developed temperature distribution can be obtained. More specifically, at first the numerical temperatures at the boundaries and initial condition can be easily determined from

$$\theta_{1,J}^l = \theta_{2,1}^l = 0, \; \theta_1^0 = \theta_2^0 = 0 \tag{44}$$

$$\theta_{1,1}^l = \theta_{2,J}^l, \; k_{eff1} \frac{-3\theta_{1,1}^l + 4\theta_{1,2}^l - \theta_{1,3}^l}{2\Delta y} = k_{eff2} \frac{\theta_{2,J-2}^l - 4\theta_{2,J-1}^l + 3\theta_{2,J}^l}{2\Delta y} \tag{45}$$

Then, the temperature distributions of the bulk liquid are computed based on the following numerical algorithm:

$$\theta_{1,j}^{l+1} = \theta_{1,j}^{l+1} + (\Pi_{1,j}^l + k_{f1}S/k_{eff1}) \cdot \Delta t \tag{46}$$

$$\theta_{2,j}^{l+1} = \theta_{2,j}^l + (\Pi_{2,j}^l) \cdot \Delta t \tag{47}$$

where $\Pi_{i,j}^l = \left[\frac{\theta_{i,j+1}^l - 2\theta_{i,j}^l + \theta_{i,j-1}^l}{\Delta y^2} + a_i \frac{k_{f1}}{k_{effi}} \Phi_{i,j}^l - \frac{k_{f1}}{k_{effi}} \frac{2+S+Br\Gamma_1^l+m_r Br\Gamma_2^l}{b_i v_{ms1}^l + c_i v_{ms2}^l} v_{si}^l \right]$, $\Phi_{i,j}^l = (g_{i,j}^l)^{n_i-1} \left(\frac{v_{si,j+1}^l - v_{si,j-1}^l}{2\Delta y} \right)^2$, $a_1 = b_1 = c_2 = 1$, $a_2 = m_r$, $b_2 = 1/(\rho c_p)_r$, $c_1 = (\rho c_p)_r$, $i = 1, 2, j = 2, 3, \ldots, J - 1$, and the integrals $\Gamma_1^l = \int_0^1 \Phi_1^l dy$, $\Gamma_2^l = \int_{-1}^0 \Phi_2^l dy$, $v_{ms1}^l = \int_0^1 v_{s1}^l dy$, $v_{ms2}^l = \int_{-1}^0 v_{s2}^l dy$ are computed using the Simpson composite integration method. As time grows, if $\|\theta^l - \theta^{l+1}\| < Er$, the thermally fully developed numerical temperature distribution is obtained, based on which the Nusselt number in Equation (25) and total entropy generation in Equation (30) can be computed and the heat transfer analysis is conducted.

The physical parameters regarding the hydrodynamic properties of two power-law nanofluids are: $m_0^* = 9 \times 10^{-4}$ Nm$^{-2}s^n$, $H = 1 \times 10^{-4}$ m, and $U = 1 \times 10^{-4}$ m·s^{-1}, which is of the same order as the Helmholtz–Smoluchowski velocity [52,53,61]. The physical parameters regarding the electric properties of two power-law nanofluids are : $\varepsilon_2 = 7.08 \times 10^{-10}$ F/m, $e = 1.6 \times 10^{-19}$ C, $z_0 = 1$, $E_z^* = 1 \times 10^4$ V·m^{-1}, $k_B = 1.38 \times 10^{-23}$ J·K^{-1}, and $\zeta^* = -0.025$ V [61]. To facilitate discussion, let the thermophysical properties of the fluids remain constant; the thermal conductivity of two base fluids are $k_{b1} = k_{b2} = 0.618$ Wm^{-1}K^{-1}, the thermal conductivity of Al$_2$O$_3$ is $k_p = 40$ Wm^{-1}K^{-1}, $T_0 = 293$ K, and the Joule heating parameter $S = 1$ [61]. Furthermore, it is essential to provide the ranges of important nondimensional governing parameters based on practical physical uses. From the well-established electroosmotic theory of power-law modeling, $n_i = 0.6{\sim}1.4$ [23] and the width of EDL is far less than the channel height; thus, $K = 10{\sim}100$ [23,52]. Based on the given order of reference (velocity, viscosity, and channel height), the Brinkman number ranges from $Br = 0{\sim}0.1$ [31,55], and the nanoparticle volume fraction $\phi = 0{\sim}0.09$ is suitable for the chosen model of effective viscosity and effective thermal conductivity [44].

Concerning the numerical schemes, let $Er = 1 \times 10^{-8}$ in order to assure that the velocity and temperature reach steady status; then, a test grid dependence is conducted, with a grid system of 1×10^3 chosen. The good agreement between the analytical solutions and numerical solutions depicted in Figure 2 shows that the numerical algorithm proposed here is feasible for computing the two-layer velocity distribution and two-layer temperature distribution and for carrying out the heat transfer analysis.

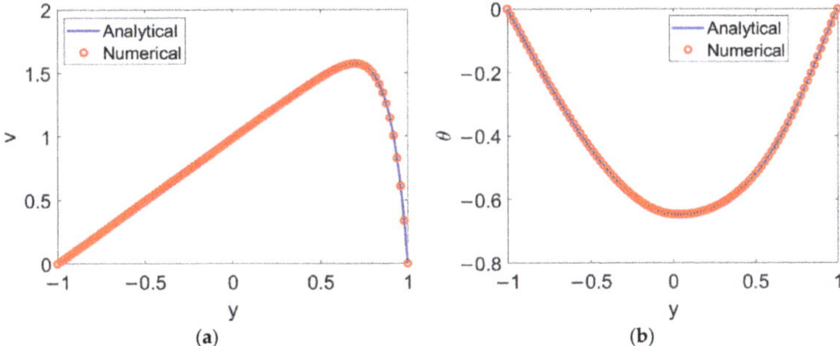

Figure 2. Comparison of analytical solutions and numerical solutions for (**a**) velocity and (**b**) temperature.

3. Results and Discussion

The transient hydrodynamic behavior of two-layer power-law nanofluid EOF is discussed by evaluating the transient two-layer velocities at different times and the flow rates for different governing nondimensional parameters. Then, with the steady velocities computed, the heat transfer analysis and entropy generation analysis are conducted by presenting the two-layer temperature profiles, Nusselt number and entropy generation rate at different governing nondimensional parameters. Specifically, $(\rho C_p)_r = 1$ and $\rho_r = 1$, such that their importance can be attached to the effects of power-law rheology (represented by n_i), the electroosmotic property (represented by electrokinetic width K), the nanoparticle volume fraction, ϕ, and the viscous dissipation effect (represented by the Brinkman number, Br) on hydrodynamic and thermal behaviors.

3.1. Flow Characteristics in Two-Layer

The time evolutions of two-layer velocities with different types of conducting nanofluid, i.e., flow behavior index n_1 at different electrokinetic width K, are presented in Figure 3. It can be seen that at first the conducting nanofluid near the channel wall is set in motion due to the electroosmotic force, and as time elapses the bulk conducting nanofluid attains velocity. As opposed to the single-layer EOF, the flow of bulk conducting fluid fails to form a plug-like profile, as the delivery of momentum via the bulk conducting nanofluid is dissipated through interfacial viscous stress, causing a deviated parabolic profile in layer I. Accordingly, the nonconducting nanofluid is driven by interfacial shear stress; thus, in layer II, the closer to the two-liquid interface the flow, the higher the velocity. No matter what value K or n_1 takes, the conducting nanofluid near the wall at first moves forward and drags the nonconducting fluid through the interfacial hydrodynamical shear stress, and as t increases to 5, the transient two-layer velocity reaches its steady state. The comparison among Figure 3a–c describes the way in which, when the conducting nanofluid is shear thinning, the transient two-layer velocity is augmented with the increase of K, while as shown in Figure 3d–f, when the conducting nanofluid is shear thickening, the transient two-layer velocity shows abatement with the increase of K. The profiles of the steady velocities in Figure 3b,c show that the remarkable increase in the velocity of the conducting nanofluid with K results in a subtle turning of the velocity near the two-liquid interface. Moreover, when pumped by shear thinning nanofluid, the two-layer fluid is more sensitive to changes in the electrokinetic width K than when pumped by shear-thickening nanofluid.

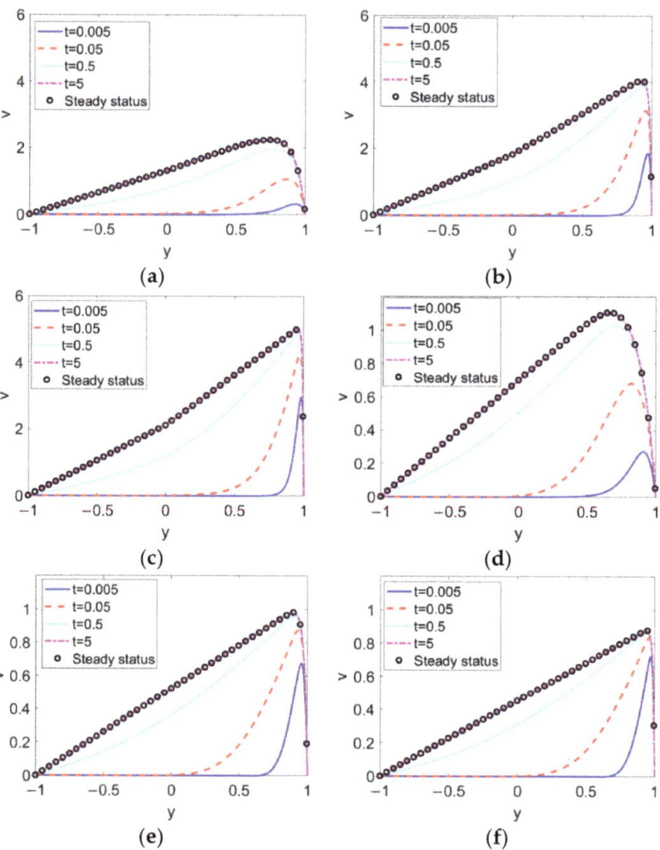

Figure 3. Transient two-layer velocities at different electrokinetic widths K and different flow behavior indexes of conducting fluid n_1 when $Br = 0.02$, $S = 1$, $\zeta = -1$, $n_2 = 1.2$, and $\phi = 0.03$. (**a**) $K = 10$, $n_1 = 0.8$; (**b**) $K = 50$, $n_1 = 0.8$; (**c**) $K = 100$, $n_1 = 0.8$; (**d**) $K = 10$, $n_1 = 1.2$; (**e**) $K = 50$, $n_1 = 1.2$; (**f**) $K = 100$, $n_1 = 1.2$.

When the conducting nanofluid and nonconducting nanofluid change from shear thinning to shear thickening, the time evolutions of the two-layer velocities are as delineated in Figure 4. No matter what type of pumped nonconducting nanofluid is considered, the decrease in the flow behavior index n_1, namely, the shear thinning feature of conducting fluid, accelerates the flow and consequently enhances the two-layer velocity. From the comparison between Figures 4a–e, the change in nonconducting nanofluid type from shear thinning to shear thickening exerts a slight influence on the two-layer flow near the two-liquid interface, and shows little influence on that far away from the two-liquid interface. The influence of n_1 on two-layer flow far outweighs that of n_2 on two-layer flow.

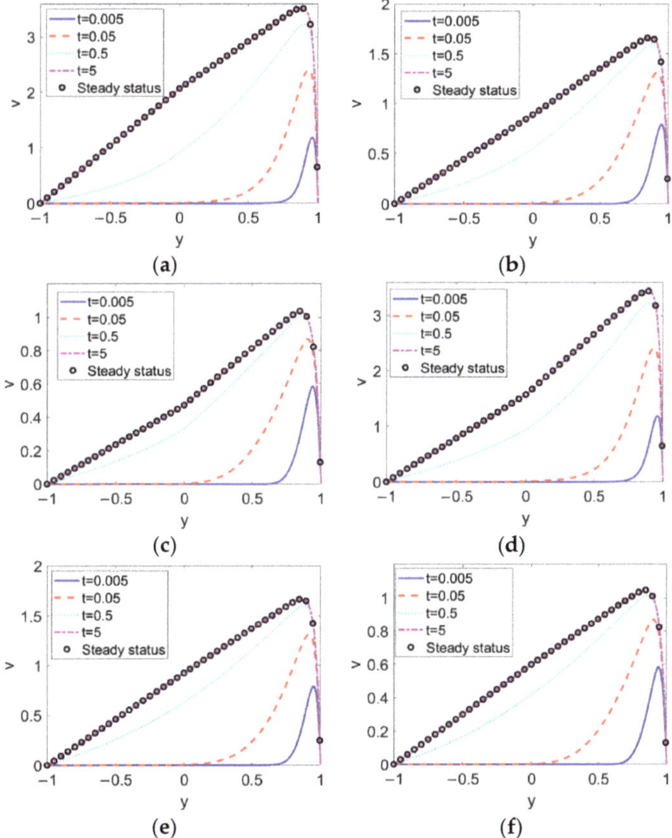

Figure 4. Transient two-layer velocities at different flow behavior indexes n_1 and different flow behavior indexes n_2 when $Br = 0.02$, $S = 1$, $\zeta = -1$, $K = 30$ and $\phi = 0.03$. (**a**) $n_1 = 0.8$, $n_2 = 0.6$; (**b**) $n_1 = 1$, $n_2 = 0.6$; (**c**) $n_1 = 1.2$, $n_2 = 0.6$; (**d**) $n_1 = 0.8$, $n_2 = 1.4$; (**e**) $n_1 = 1$, $n_2 = 1.4$; (**f**) $n_1 = 1.2$, $n_2 = 1.4$.

Because the flow behavior index, n_1, plays a crucial role in two-layer velocity, the time evolutions of flow rates at different flow behavior indexes of conducting nanofluid n_1 are presented in Figure 5 when (a) $\phi = 0$ and (b) $\phi = 0.03$. It is obvious that the unsteady flow rate of conducting nanofluid takes a longer time to reach steady status than that of nonconducting nanofluid; as the conducting nanofluid near the wall is set in motion first, the bulk conducting nanofluid moves forward, and eventually the nonconducting nanofluid is dragged by the bulk conducting fluid. Irrespective of the value of the nanoparticle volume fraction ϕ, the shear thinning feature of the conducting fluid enhances the flow rates of both the conducting and nonconducting nanofluids. A comparison between Figure 5a,b shows that the addition of the nanoparticle to the bulk nanofluid reduces the two-layer flow rate by improving the viscosity of the two fluids, as per Equations (9) and (10).

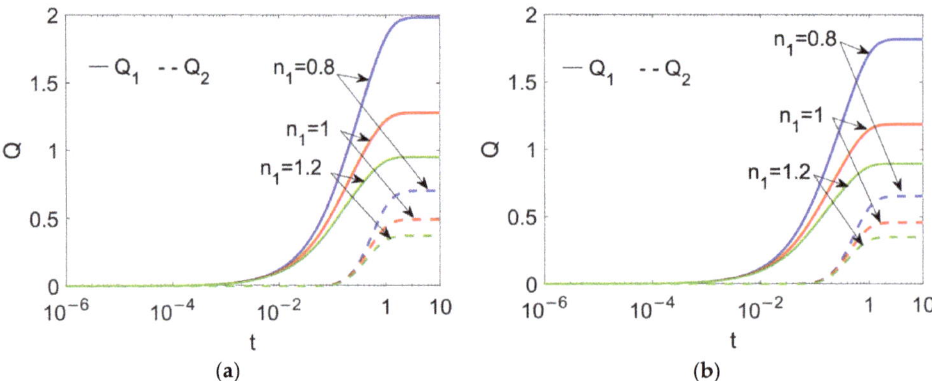

Figure 5. Time evolutions of flow rates of two power-law nanofluids at different flow behavior indexes n_1 and different volume fractions of nanoparticle ϕ when $Br = 0.02$, $S = 1$, $\zeta = -1$, $K = 10$ and $n_2 = 1.2$. (**a**) $\phi = 0$; (**b**) $\phi = 0.03$.

In Figure 6, the time evolutions of flow rates at different electrokinetic widths K are presented when (a) $n_1 = 0.8$ and (b) $n_2 = 1.2$. From Figure 6a, when the pumping conducting nanofluid is shear thinning, the flow rates of the two power-law nanofluids show augmentation with the electrokinetic width K, being consistent with Figure 3; in contrast, when it is shear thickening, the flow rates of both fluids show abatement with K, meaning that the change in fluid type of the pumping nanofluid from shear thinning to shear thickening can reverse the effect of the electrokinetic width. In addition, the flow rates show a noticeable reduction when n_1 increases, irrespective of what value K takes.

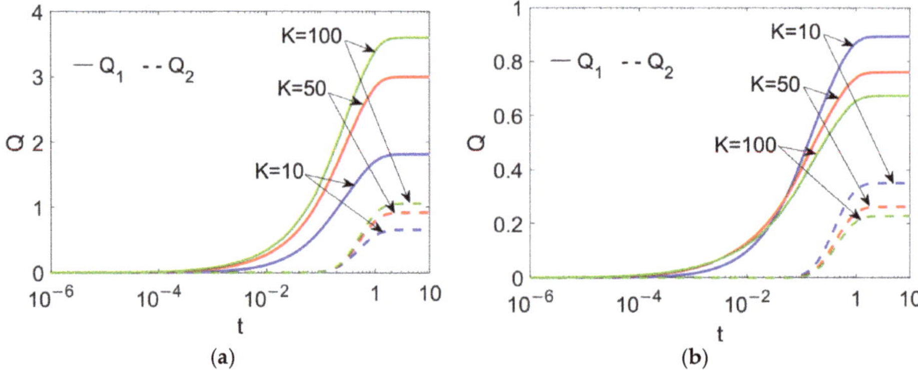

Figure 6. Time evolutions in flow rates of two power-law nanofluids at different flow behavior indexes n_1 and different electrokinetic widths K when $Br = 0.02$, $S = 1$, $\zeta = -1$, $n_2 = 1.2$ and $\phi = 0.03$. (**a**) $n_1 = 0.8$; (**b**) $n_1 = 1.2$.

In Figure 7, the variation in the ratio of flow rate I to flow rate II is plotted for a flow behavior index of conducting nanofluid n_1 at different nanoparticle volume fractions ϕ. When the pumping conducting nanofluid is shear thinning, that is, $n_1 < 1$, flow rate I is more than three times higher than flow rate II and two-layer flow is evidently affected by the change in the nanoparticle volume fraction, ϕ. In contrast, when the pumping conducting nanofluid is shear thickening (i.e., $n_1 > 1$), the two-layer flow is insensitive to the change in flow behavior index n_1 and nanoparticle volume fraction ϕ. Compared to the effect of the nanoparticle volume fraction, the effect of n_1 dominates in the two-layer flow.

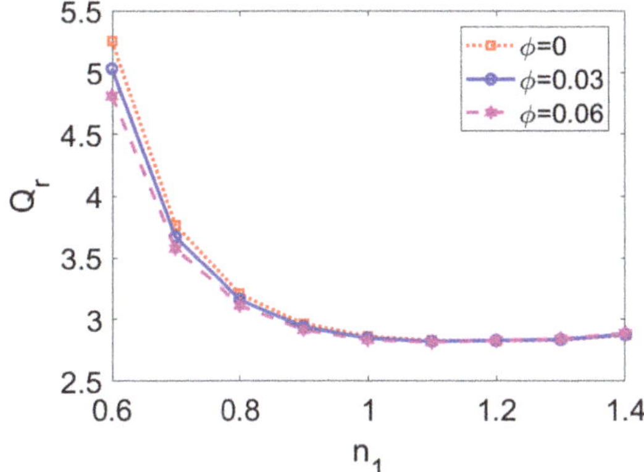

Figure 7. Variation of the ratio of flow rates with flow behavior index n_1 at different nanoparticle volume fractions ϕ when $Br = 0.02$, $S = 1$, $\zeta = -1$, $n_2 = 1.2$, and $K = 30$.

In Figure 8, the variation in the ratio of flow rate I to flow rate II is presented for the flow behavior index of nonconducting nanofluid n_2 at different electrokinetic widths K. A higher flow behavior index of nonconducting nanofluid triggers stronger viscosity, and thus smaller flow rate of nonconducting nanofluid flow, finally leading to an increase in the flow rate ratio. When the type of conducting nanofluid is fixed, although a thinner EDL length improves velocities of both fluids (as shown in Figure 3a–c), the increase in flow rate I is much greater than that of flow rate II, and thus the increasing trend of flow rate ratio with K is observed, which is especially obvious for a shear thickening nonconducting nanofluid. Therefore, when dragging a shear thickening fluid, increasing electrokinetic width exerts more influence on flow rate I than on flow rate II. Furthermore, in comparison with Figure 7, the effect of the flow behavior index, n_2, on the flow rate ratio is almost linear, and the increasing rate is enhanced for greater values of the electrokinetic width, K.

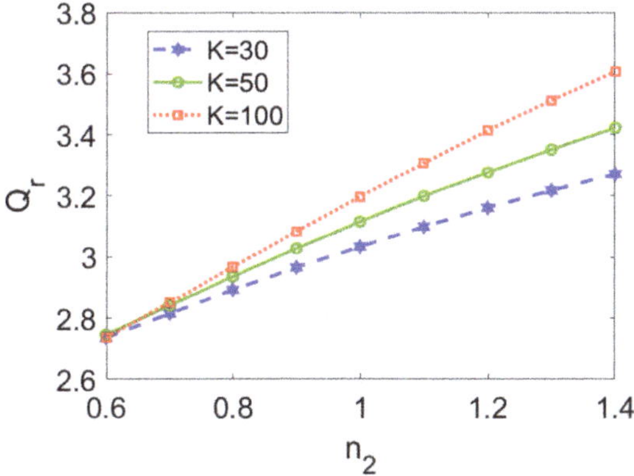

Figure 8. Variation of the ratio of volumetric flow rates with flow behavior index n_2 at different nanoparticle volume fractions when $Br = 0.02$, $S = 1$, $\zeta = -1$, $n_1 = 0.8$, and $\phi = 0.03$.

3.2. Heat Transfer Characteristics in Two-Layer Flow

In Figure 9, the temperature profiles at different flow behavior indexes n_1 are described when (a) $\phi = 0$, (b) $\phi = 0.03$, and (c) $\phi = 0.06$. It is noted that the temperature profiles of the two-layer flow are asymmetric around the two-liquid interface, in which the minimum value occurs within layer I. The temperature difference between the channel wall and bulk fluid is reduced, with the conducting nanofluid changing from a shear thinning fluid to a shear thickening one; therefore, it can be seen that the smaller the velocity gradient of two-layer flow is, the better the heat transfer performance becomes. On the other hand, the increase in the nanoparticle volume fraction, ϕ, leads to an overall increase in the temperature profiles for all types of conducting nanofluid, meaning that the introduction of nanoparticles truly improves the heat transfer in two-layer flow.

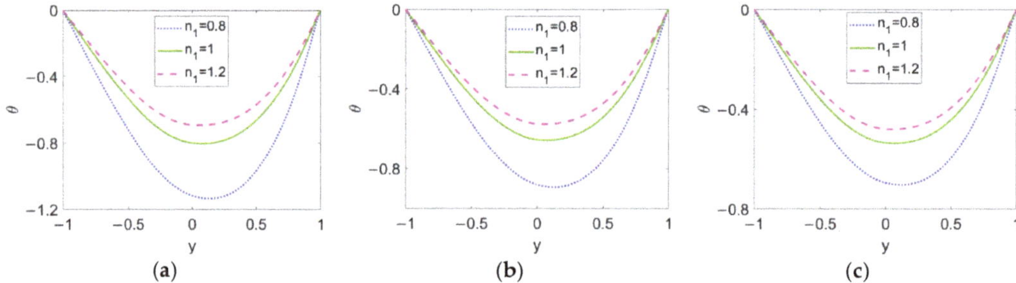

Figure 9. Variation of temperature at different flow behavior indexes n_1 and different nanoparticle volume fractions ϕ when $K = 30$, $n_2 = 1.2$, $\zeta = -1$, $S = 1$, and $Br = 0.02$. (**a**) $\phi = 0$; (**b**) $\phi = 0.03$; (**c**) $\phi = 0.06$.

As shown in Figure 10, the temperature profiles at different electrokinetic widths K are plotted when (a) $n_1 = 0.8$, (b) $n_1 = 1$, and (c) $n_1 = 1.2$ in order to demonstrate the influence of n_1 on the effect of K. The increase in K, namely a thinner EDL thickness, enlarges the temperature difference between the channel wall and bulk liquid and shifts the minimum value of the temperature to the right. This can be explained by the fact that the increase in K enhances the two-layer velocity and causes a drastic change in velocity near the channel wall, as described in Figure 3. The comparison among Figure 10a–c indicates that the shear thickening feature of the conducting nanofluid tends to suppress the effect of the electrokinetic width, K, on temperature profile.

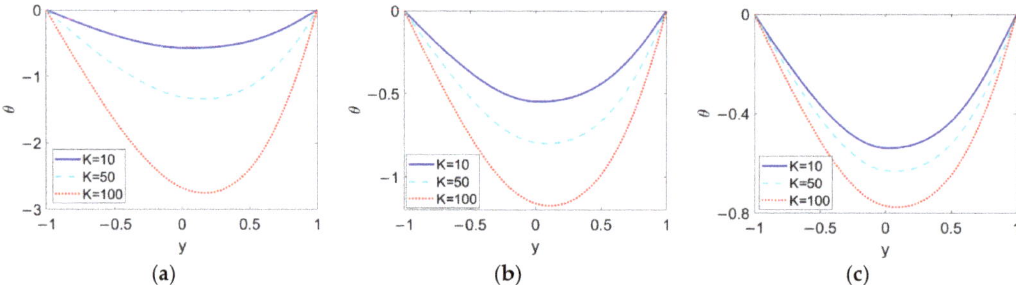

Figure 10. Variation of temperature at different flow behavior indexes n_1 and different electrokinetic widths K when $\phi = 0.03$, $n_2 = 1.2$, $\zeta = -1$, $S = 1$, and $Br = 0.02$. (**a**) $n_1 = 0.8$; (**b**) $n_1 = 1$; (**c**) $n_1 = 1.2$.

As shown in Figure 11, the temperature profiles with different Brinkman numbers Br are plotted when (a) $n_1 = 0.8$, (b) $n_1 = 1$, and (c) $n_1 = 1.2$ in order to show the combined effect of flow behavior index n_1 and Brinkman number Br. As the Brinkman number Br increases, viscous dissipation in the two-layer flow is improved, which further enlarges the temperature difference between the channel wall and bulk liquid, meaning that the consideration of viscous dissipation evidently hinders the heat transfer in two-layer flow. Moreover, as obvious changes in temperature are observed in Figure 11a–c, the Brinkman number Br has an important effect on the temperature distribution for all types of conducting nanofluid; in other word, no matter what type of fluid the two-layer flow is driven by, the effect of Br on temperature cannot be neglected.

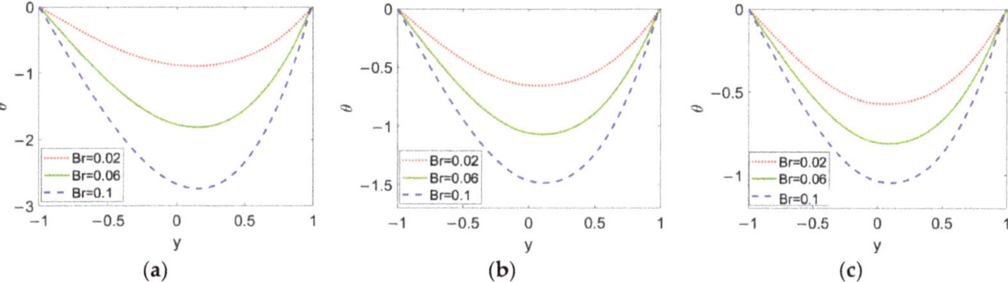

Figure 11. Variation of temperature at different flow behavior indicese n_1 and different Brinkman numbers Br when $\phi = 0.03$, $n_2 = 1.2$, $\zeta = -1$, $S = 1$ and, $K = 30$. (**a**) $n_1 = 0.8$; (**b**) $n_1 = 1$; (**c**) $n_1 = 1.2$.

The temperature profiles at different flow behavior indices of nonconducting nanofluid n_2 are described in Figure 12. When the pumped nonconducting nanofluid changes from a shear thinning fluid to a shear thickening one, the temperature in the vicinity of the two-liquid interface slightly increases, while far away from the two-liquid interface it shows little change. This is because the increase in the flow behavior index, n_2, causes a slight change in the two-layer velocity near the two-liquid interface. Therefore, compared to the influence of the fluid type of the pumping conducting nanofluid, that of the fluid type of the pumped nonconducting nanofluid (n_2) on temperature profile is weak.

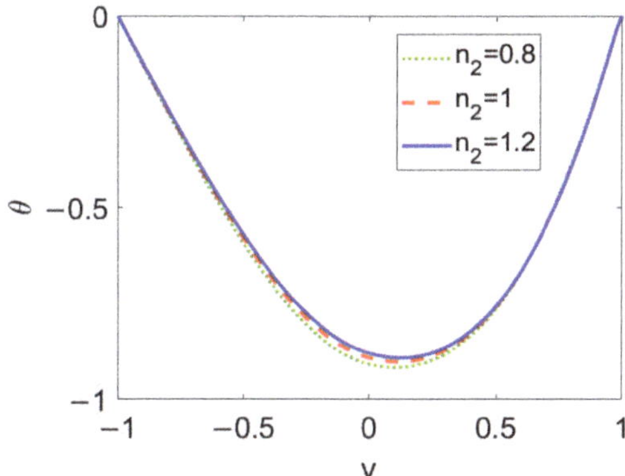

Figure 12. Variation of temperature at different flow behavior indexes n_2 when $\phi = 0.03$, $n_1 = 0.8$, $\zeta = -1$, $S = 1$, and $Br = 0.02$.

The variation in the Nusselt number Nu with flow behavior index n_2 at different electrokinetic widths K is presented in Figure 13. The Nusselt number, Nu, shows a slightly ascending trend with flow behavior index n_2 no matter the value of the electrokinetic width. On the other hand, the increase in electrokinetic width, K, leads to the decrease in Nusselt number, Nu. This is because the rapid change in velocity profile caused by a lower EDL thickness triggers a widened temperature difference by suppressing the heat transfer performance of the two-layer flow. It should be noted that the effect of electroosmotic property of the fluid and the effect of the type of pumped nonconducting nanofluid do not interact with each other.

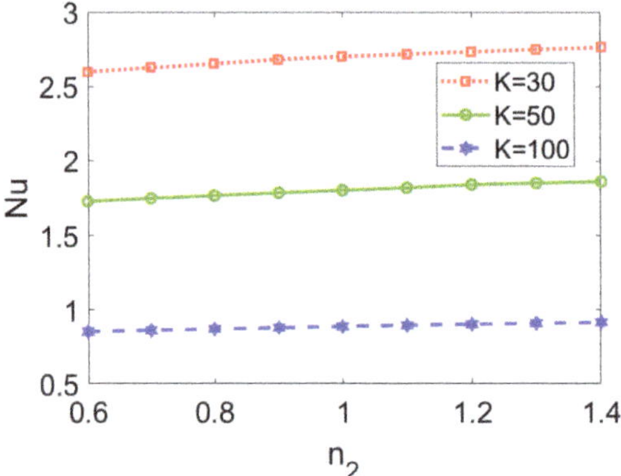

Figure 13. Variation of Nusselt number with flow behavior index n_2 at different electrokinetic widths K when $\phi = 0.03$, $n_1 = 0.8$, $\zeta = -1$, $S = 1$, and $Br = 0.02$.

In Figure 14, the variation of the Nusselt number Nu with (a) electrokinetic width K and (b) Brinkman number Br is presented when choosing different types of conducting nanofluid. As shown in Figure 14a, a greater value of the electrokinetic width K reduces the heat transfer in two-layer flow, causing the dramatic change in velocity and widening temperature difference shown in Figures 3 and 10. Figure 14b shows that the Nusselt number Nu decreases with the Brinkman number Br, as the stronger viscous dissipation effect represented by greater values of Br enlarges the temperature difference between the channel wall and the bulk liquid, impeding heat transfer in two-layer flow. As opposed to the interaction between the effects of K and n_2 predicted in Figure 13, the descending trend of the Nusselt number Nu with K or Br becomes smaller when the conducting nanofluid changes from a shear thinning fluid to a shear thickening one, meaning that when two-layer flow is driven by a shear thinning fluid the heat transfer performance is much more susceptible to changes in electrokinetic width, K, or the Brinkman number, Br.

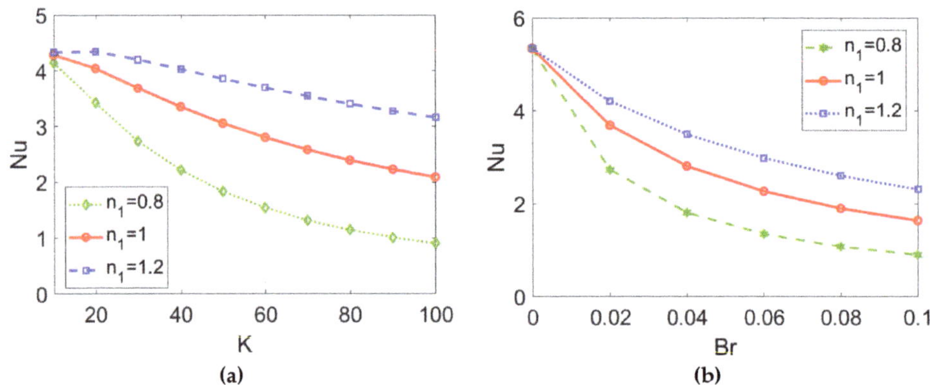

Figure 14. Variation of the Nusselt number Nu with (a) electrokinetic width K and (b) Brinkman number Br at different flow behavior indexes n_1 when $\phi = 0.03$, $\zeta = -1$, $n_2 = 1.2$, and $S = 1$. (**a**) $Br = 0.02$; (**b**) $K = 30$.

Variation of the Nusselt number Nu with flow behavior index n_1 at different nanoparticle volume fractions ϕ is shown in Figure 15, revealing the interaction between the effect of the conducting nanofluid and that of the nanoparticles. As the conducting nanofluid changes from a shear thinning fluid to a shear thickening one, the Nusselt number Nu is augmented, hence the heat transfer performance of two-layer flow is enhanced. Furthermore, the ascending trend of Nu with n_1 becomes more slight. The increment in the nanoparticle volume fraction ϕ tends to improve the variation of Nu with n_1 as a whole, thus promoting the heat transfer of two-layer flow for all types of conducting nanofluid. This means that the addition of nanoparticles has little influence on the effect of the conducting nanofluid type on heat transfer (represented by Nu). According to Equation (25), growth in mean temperature with ϕ leads to the increase in Nu, implying that the temperature difference between the bulk fluid and channel walls is narrowed. Therefore, in practical engineering terms, when a wide temperature difference between the channel wall and the centerline occurs and might cause undesirable results (such as electrochemical decomposition or fluctuation of PH value in working liquids), the introduction of nanoparticles can intensify heat transfer performance and help to avoid the problems mentioned above.

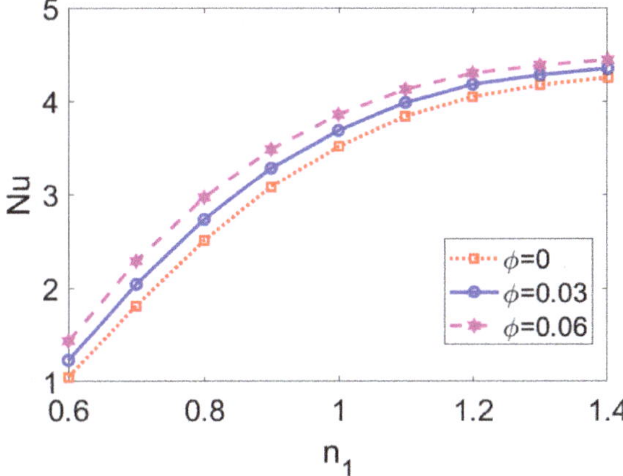

Figure 15. Variation of Nusselt number Nu with flow behavior index n_1 at different volume fractions of nanoparticles ϕ when $K = 30$, $n_2 = 1.2$, $\zeta = -1$, $S = 1$, and $Br = 0.02$.

The variation of total entropy generation S_t with (a) electrokinetic width K and (b) Brinkman number Br is presented in Figure 16 for different types of conducting nanofluids. An almost linear increase of total entropy generation S_t with electrokinetic width K and Brinkman number Br can be observed in Figure 16. From Figure 16a, it can be inferred that a thinner EDL length (represented by a greater value of K) results in a wider velocity gap between the channel wall and the bulk fluid; therefore, the heat transfer is retarded, and entropy generation improves accordingly. Figure 16b implies that the viscous dissipation effect grows stronger with the Brinkman number, leading to an increase in entropy generation. In addition, the increasing trend with K and Br becomes smaller when nanofluid I changes from shear thinning to shear thickening.

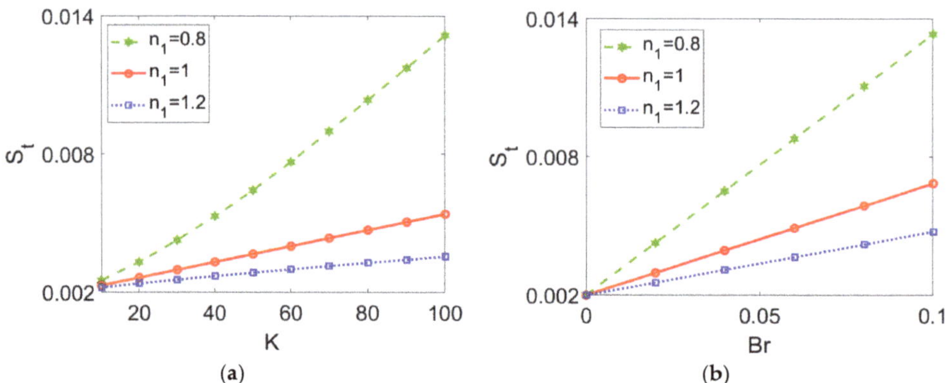

Figure 16. Variation of total entropy generation with (**a**) electrokinetic width K and (**b**) Brinkman number Br at different flow behavior indexes n_1 when $\phi = 0.03$, $n_2 = 1.2$, and $S = 1$. (**a**) $Br = 0.02$; (**b**) $K = 30$.

The variation of total entropy generation with flow behavior index n_1 at different nanoparticle volume fractions ϕ is presented in Figure 17. The entropy generation decreases when the conducting nanofluid changes from a shear thinning type to a shear thickening one. The shear thinning feature of conducting fluid accelerates the two-layer flow and enhances velocity distribution, while the increased velocity gradient and temperature gradient result in greater entropy generation as per Equations (28) and (29). Furthermore, entropy generation in two-layer flow driven by shear thinning nanofluid is more sensitive to changes in nanoparticle volume fraction compared to that driven by shear thickening nanofluid. In practical terms, when the two-layer flow is pumped by shear thinning fluid, it is more likely to utilize nanoparticles to adjust the thermal performance of the system.

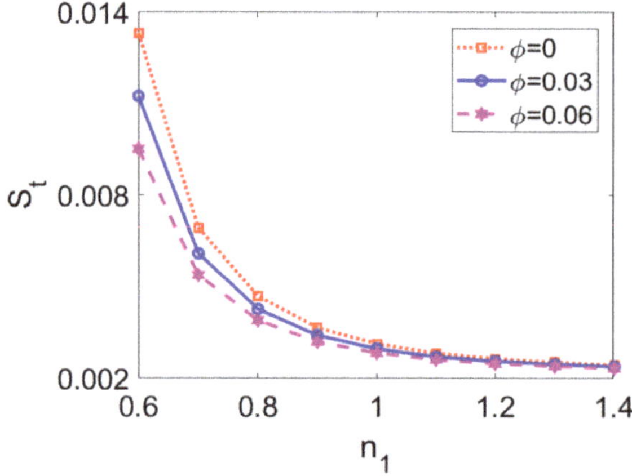

Figure 17. Variation of total entropy generation with flow behavior index n_1 at different nanoparticle volume fractions ϕ when $K = 30$, $n_2 = 1.2$, $S = 1$, and $Br = 0.02$.

4. Conclusions

(1) The hydrodynamic behavior of transient two-layer EOFs of power-law nanofluids in a slit microchannel were investigated by evaluating the transient two-layer velocity distribution at different times and with different two-layer flow rates.

- When driven by shear thinning nanofluid, the two-layer flow accelerates for thinner EDL thicknesses and decelerates when driven by shear thickening nanofluid.

The change in fluid type of pumped nonconducting nanofluid exerts only a slight influence on velocity near the two-liquid interface. It is concluded that compared to the fluid type of pumped nonconducting nanofluid, the fluid type of the pumping conducting nanofluid plays a dominant role in two-layer flow and alters the effect of the electrokinetic width, K.

- In practical terms, the selection of a conducting nanofluid is crucial, as is the use of electrokinetic width to adjust two-layer flow for different types of conducting nanofluid.
- As opposed to the variation of the flow rate ratio with n_2, the variation with n_1 is nonlinear, and the flow rate of two-layer flow driven by shear thinning nanofluid is more sensitive to changes in the nanoparticle volume fraction.

(2) With steady two-layer velocity obtained, the thermally developed heat transfer characteristics were discussed by presenting the temperature distribution, Nusselt number, and total entropy generation at different parameters.

- The fluid type of the pumping conducting nanofluid, Brinkman number, nanoparticle volume fraction, and electrokinetic width all play important roles in the temperature profile, Nusselt number, and total entropy generation; in contrast, the influence of the type of pumped nonconducting nanofluid is weak.
- In terms of the interactive influence of the governing parameters, shear thickening feature of the conducting nanofluid tends to suppress the effects of the Brinkman number and electrokinetic width on heat transfer and entropy generation.
- No matter what type of conducting nanofluid is considered, increasing the nanoparticle volume fraction within a specified range truly enhances the heat transfer performance of two-layer flow.
- Entropy generation in two-layer flow driven by shear thinning nanofluid is more sensitive to changes in electrokinetic width, Brinkman number, and nanoparticle volume fraction.

Author Contributions: Conceptualization, S.D.; Data curation, S.D.; Formal analysis, S.D.; Funding acquisition, S.D. and T.X.; Investigation, S.D.; Methodology, S.D.; Project administration, S.D.; Resources, S.D.; Software, S.D.; Supervision, S.D.; Validation, S.D.; Visualization, S.D.; Writing—original draft, S.D.; Writing—review & editing, S.D. All authors have read and agreed to the published version of the manuscript.

Funding: This work was supported by the Guangdong Basic and Applied Basic Research Foundation (Grant numbers 2021A1515012371 and 2020A1515011241), the National Natural Science Foundation of China (Grant number 11902082), and the Scientific Research Foundation of Universities in Guangdong Province for Young Talents (Grant number 2018KQNCX165). The authors are grateful to the Reviewers and Editors for their valuable comments and suggestions.

Data Availability Statement: Not applicable.

Conflicts of Interest: The authors declare no conflict of interest.

Appendix A

To obtain the analytical solutions for two-layer EOF in the case of Newtonian fluid, Laplace transform to two-layer velocity distributions v_i is introduced:

$$V_i(s) = \mathcal{L}[v_i(t)] = \int_0^\infty v_i^N(y,t)e^{-st}dt \tag{A1}$$

Transforming Equations (31)–(33), and letting $\rho_r = 1$, the ordinary differential equations (ODEs) with respect to s can be obtained as follows:

$$\frac{d^2 V_1}{dy^2} - sV_1 = \frac{1}{s}GE_z\zeta\frac{\cosh(Ky)}{\cosh(K)} \text{ for } 0 \le y \le 1 \tag{A2}$$

$$\frac{d^2 V_2}{dy^2} - s V_2 = 0 \text{ for } -1 \le y \le 0 \tag{A3}$$

$$V_1\big|_{y=0} = V_2\big|_{y=0}, \ \frac{dV_1}{dy}\Big|_{y=0} = \frac{dV_2}{dy}\Big|_{y=0}, \ V_2\big|_{y=-1} = 0, \ V_1\big|_{y=1} = 0 \tag{A4}$$

The solution to the above ODEs (A2)–(A3) takes the following forms:

$$V_1 = C\cosh(Ky) + E\cosh(\sqrt{s}y) + F\sinh(\sqrt{s}y) \text{ for } 0 \le y \le 1 \tag{A5}$$

$$V_2 = M\cosh(\sqrt{s}y) + F\sinh(\sqrt{s}y) \text{ for } -1 \le y \le 0 \tag{A6}$$

Combining the boundary Equation (A4),

$$C = \frac{GE_z\zeta}{s(K^2-s)\cosh(K)}, E = \frac{-GE_z\zeta}{2s(K^2-s)\cosh\sqrt{s}}\left(1 + \frac{\cosh\sqrt{s}}{\cosh(K)}\right)$$

$$F = \frac{GE_z\zeta}{2s(K^2-s)\sinh\sqrt{s}}\left(\frac{\cosh\sqrt{s}}{\cosh(K)} - 1\right), M = \frac{GE_z\zeta}{s(K^2-s)\cosh(K)} + E \tag{A7}$$

Applying the Laplace inverse transform to Equations (A5) and (A6) yields

$$v_i^N = \mathcal{L}^{-1}[V_i] = \frac{1}{2\pi I}\int_{\sigma-I\infty}^{\sigma+I\infty} V_i e^{st} ds \tag{A8}$$

in which I is the imaginary unit, σ is a real number satisfying $\sigma > 0$, and the integral path is presented in Figure A1. Based on the residue theorem, Equation (A8) equals

$$v_1^N = \sum_k \text{Res}[V_1(s)e^{st}, s_k], \text{ for } 0 \le y \le 1 \tag{A9}$$

$$v_2^N = \sum_k \text{Res}[V_2(s)e^{st}, s_k], \text{ for } -1 \le y \le 0 \tag{A10}$$

where $\text{Res}[V_i(s)e^{st}, s_k]$ implies the residue of $V_i(s)e^{st}$ at the isolated singularities, s_k. The isolated singularities s_k are enclosed in the simple closed curve $C_R + L$, as shown in Figure A1. According to Equations (A5)–(A7), $s_k = 0$, K^2, $-(2P-1)^2\pi^2/4$, and $-(P\pi)^2$, with $k = 1, 2, 3, 4$ and $P = 1, 2, \ldots$. Consequently, the residues of $V_i(s)e^{st}$ are evaluated as follows:

$$\text{Res}[V_1(s)e^{st}, 0] = \lim_{s_1 \to 0} \frac{d[s^2 \cdot V_1(s) \cdot e^{st}]}{ds} = \frac{GE_z\zeta}{K^2}\left[\frac{1-\cosh(K)}{2\cosh(K)}(y-1) + \frac{\cosh(Ky)}{\cosh(K)} - 1\right] \tag{A11}$$

$$\text{Res}[V_2(s)e^{st}, 0] = \lim_{s_1 \to 0} \frac{d[s^2 \cdot V_2(s) \cdot e^{st}]}{ds} = \frac{GE_z\zeta}{K^2}\left[\frac{1-\cosh(K)}{2\cosh(K)}(y-1) + \frac{1}{\cosh(K)} - 1\right] \tag{A12}$$

$$\text{Res}[V_i(s)e^{st}, K^2] = \lim_{s_2 \to K^2}(s - K^2) \cdot V_i(s) \cdot e^{st} = 0 \tag{A13}$$

$$\text{Res}\left[V_i(s)e^{st}, -\frac{[(2P-1)\pi]^2}{4}\right] = \lim_{s_{3P} \to -\frac{[(2P-1)\pi]^2}{4}}\left[s + \frac{[(2P-1)\pi]^2}{4}\right] \cdot V_i(s) \cdot e^{st} \tag{A14}$$

$$= \frac{8GE_z\zeta(-1)^{P+1}e^{-(2P-1)^2\pi^2 t/4}}{(2P-1)\pi[4K^2+(2P-1)^2\pi^2]}\cos\left[\frac{(2P-1)\pi y}{2}\right]$$

$$\text{Res}\left[V_i(s)e^{st}, -(P\pi)^2\right] = \lim_{s_{4P} \to -(P\pi)^2}\left[s + (P\pi)^2\right] \cdot V_i(s) \cdot e^{st} \tag{A15}$$

$$= \frac{GE_z\zeta[1/\cosh(K)-(-1)^{P+1}]e^{-(P\pi)^2 t}}{P\pi[K^2+(P\pi^2)]}\sin(P\pi y)$$

with $i = 1, 2$. According to Equations (A9) and (A10), summarizing Equations (A11)–(A15) produces the analytical velocities for two-layer Newtonian fluid EOF, which are exactly expressed by Equations (34) and (35).

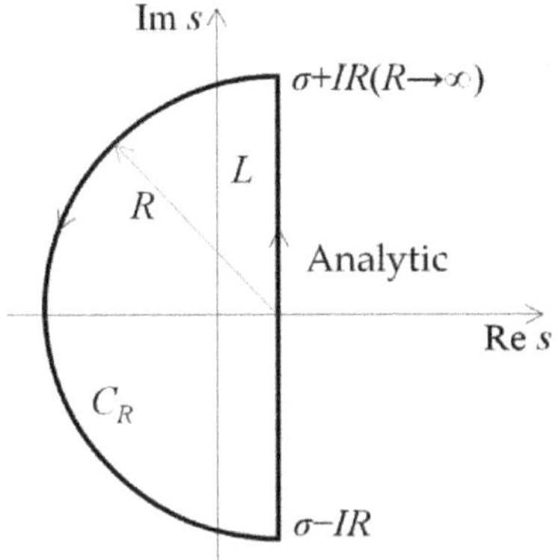

Figure A1. Schematic of the simple closed curve $C_R + L$ in the s-plane.

In addition, the coefficients in Equations (38) and (39) are presented as follows:

$$A_1 = \frac{k_{f1}GE_z\zeta[1-\cosh(K)]}{12K^2\cosh(K)k_{eff1}} \cdot \frac{(2+S+Br\Gamma_1+m_rBr\Gamma_2)}{v_{ms1}+(\rho c_p)_r v_{ms2}},$$

$$A_2 = \frac{k_{f1}GE_z\zeta}{K^4\cosh(K)k_{eff1}} \cdot \frac{(2+S+Br\Gamma_1+m_rBr\Gamma_2)}{v_{ms1}+(\rho c_p)_r v_{ms2}},$$

$$A_3 = -\frac{k_{f1}}{2k_{eff1}}\left[\frac{GE_z\zeta[1+\cosh(K)]}{2K^2\cosh(K)} \cdot \frac{(2+S+Br\Gamma_1+m_rBr\Gamma_2)}{v_{ms1}+(\rho c_p)_r v_{ms2}} + S + \frac{Br(GE_z\zeta)^2[1-\cosh(K)]^2}{4K^4\cosh(K)^2}\right]$$

$$A_4 = -\frac{k_{f1}Br(GE_z\zeta)^2[1-\cosh(K)]}{K^5\cosh(K)^2 k_{eff1}}, \quad A_5 = -\frac{k_{f1}Br(GE_z\zeta)^2}{4K^2\cosh(K)^2 k_{eff1}},$$

$$B_1 = \frac{k_{f1}GE_z\zeta[1-\cosh(K)]}{12K^2\cosh(K)k_{eff1}} \cdot \frac{(2+S+Br\Gamma_1+m_rBr\Gamma_2)}{v_{ms1}/(\rho c_p)_r+v_{ms2}},$$

$$B_2 = \frac{k_{f1}}{2k_{eff1}}\left[\frac{GE_z\zeta[1-\cosh(K)]}{2K^2\cosh(K)} \cdot \frac{(2+S+Br\Gamma_1+m_rBr\Gamma_2)}{v_{ms1}+v_{ms2}(\rho c_p)_r} - m_rBr\Phi_2\right],$$

$$D_1 = -\frac{k_{eff2}(d_1-d_2)}{k_{eff1}+k_{eff2}},$$

$$D_2 = -\frac{k_{eff2}}{k_{eff1}+k_{eff2}}\left(\frac{k_{eff1}}{k_{eff2}}d_1 + d_2\right),$$

$$D_3 = \frac{k_{eff1}}{k_{eff2}}(KA_4 + D_1),$$

$$D_4 = (A_2 + A_5/K^2 + D_2),$$

$$d_1 = A_1 + A_2\cosh(K) + A_3 + A_4\sinh(K) + A_5\left[\frac{\cosh(K)^2}{K^2} - 1\right]$$

$$d_2 = \left(-B_1 + B_2 - \frac{k_{eff1}}{k_{eff2}}KA_4 + A_2 + \frac{A_5}{K^2}\right).$$

References

1. Bruus, H. *Theoretical Microfluidics*; Oxford University Press: Oxford, UK; New York, NY, USA; New York, NY, USA, 2008.
2. Vasu, N.; De, S. Electroosmotic flow of power-law fluids at high zeta potentials. *Colloids Surf. A Physicochem. Eng. Asp.* **2010**, *368*, 44–52. [CrossRef]
3. Kang, Y.; Yang, C.; Huang, X. Dynamic aspects of electroosmotic flow in a cylindrical microcapillary. *Int. J. Eng. Sci.* **2002**, *40*, 2203–2221. [CrossRef]
4. Moghadam, A.J. Electrokinetic-Driven Flow and Heat Transfer of a Non-Newtonian Fluid in a Circular Microchannel. *J. Heat Transf.* **2013**, *135*, 021705. [CrossRef]
5. Marcos; Yang, C.; Wong, T.N.; Ooi, K.T. Dynamic aspects of electroosmotic flow in rectangular microchannels. *Int. J. Eng. Sci.* **2004**, *42*, 1459–1481. [CrossRef]
6. Srinivas, B. Electroosmotic flow of a power law fluid in an elliptic microchannel. *Colloids Surf. A Physicochem. Eng. Asp.* **2016**, *492*, 144–151. [CrossRef]
7. Bianchi, F.; Ferrigno, R.; Girault, H.H. Finite Element Simulation of an Electroosmotic-Driven Flow Division at a T-Junction of Microscale Dimensions. *Anal. Chem.* **2000**, *72*, 1987–1993. [CrossRef]
8. Das, S.; Chakraborty, S. Analytical solutions for velocity, temperature and concentration distribution in electroosmotic microchannel flows of a non-Newtonian bio-fluid. *Anal. Chim. Acta* **2006**, *559*, 15–24. [CrossRef]
9. Tang, G.H.; Li, X.F.; He, Y.L.; Tao, W.Q. Electroosmotic flow of non-Newtonian fluid in microchannels. *J. Non-Newtonian Fluid Mech.* **2009**, *157*, 133–137. [CrossRef]
10. Ng, C.-O. Combined pressure-driven and electroosmotic flow of Casson fluid through a slit microchannel. *J. Non-Newtonian Fluid Mech.* **2013**, *198*, 1–9. [CrossRef]
11. Li, H.; Jian, Y. Dispersion for periodic electro-osmotic flow of Maxwell fluid through a microtube. *Int. J. Heat Mass Transf.* **2017**, *115*, 703–713. [CrossRef]
12. Mehta, S.K.; Pati, S.; Mondal, P.K. Numerical study of the vortex-induced electroosmotic mixing of non-Newtonian biofluids in a nonuniformly charged wavy microchannel: Effect of finite ion size. *Electrophoresis* **2021**, *42*, 2498–2510. [CrossRef] [PubMed]
13. Bharti, R.P.; Harvie, D.J.E.; Davidson, M.R. Electroviscous effects in steady fully developed flow of a power-law liquid through a cylindrical microchannel. *Int. J. Heat Fluid Flow* **2009**, *30*, 804–811. [CrossRef]
14. Shit, G.C.; Mondal, A.; Sinha, A.; Kundu, P.K. Electro-osmotic flow of power-law fluid and heat transfer in a micro-channel with effects of Joule heating and thermal radiation. *Phys. A Stat. Mech. Its Appl.* **2016**, *462*, 1040–1057. [CrossRef]
15. Siva, T.; Kumbhakar, B.; Jangili, S.; Mondal, P.K. Unsteady electro-osmotic flow of couple stress fluid in a rotating microchannel an analytical solution. *Phys. Fluids* **2020**, *32*, 102013. [CrossRef]
16. Chang, C.-C.; Wang, C.-Y. Rotating electro-osmotic flow over a plate or between two plates. *Phys. Rev. E* **2011**, *84*, 056320. [CrossRef]
17. Si, D.-Q.; Jian, Y.-J.; Chang, L.; Liu, Q.-S. Unsteady Rotating Electroosmotic Flow Through a Slit Microchannel. *J. Mech.* **2016**, *32*, 603–611. [CrossRef]
18. Qi, C.; Ng, C.-O. Rotating electroosmotic flow of viscoplastic material between two parallel plates. *Colloids Surf. A Physicochem. Eng. Asp.* **2017**, *513*, 355–366. [CrossRef]
19. Kaushik, P.; Mondal, P.K.; Chakraborty, S. Rotational electrohydrodynamics of a non-Newtonian fluid under electrical double-layer phenomenon: The role of lateral confinement. *Microfluid. Nanofluidics* **2017**, *21*, 122. [CrossRef]
20. Ranjit, N.K.; Shit, G.C. Entropy generation on electro-osmotic flow pumping by a uniform peristaltic wave under magnetic environment. *Energy* **2017**, *128*, 649–660. [CrossRef]
21. Escandón, J.; Bautista, O.; Méndez, F. Entropy generation in purely electroosmotic flows of non-Newtonian fluids in a microchannel. *Energy* **2013**, *55*, 486–496. [CrossRef]
22. Babaie, A.; Saidi, M.H.; Sadeghi, A. Heat transfer characteristics of mixed electroosmotic and pressure driven flow of power-law fluids in a slit microchannel. *Int. J. Therm. Sci.* **2012**, *53*, 71–79. [CrossRef]
23. Zhao, C.L.; Yang, C. Joule heating induced heat transfer for electroosmotic flow of power-law fluids in a microcapillary. *Int. J. Heat Mass Transf.* **2012**, *55*, 2044–2051. [CrossRef]
24. Zhao, C.; Zhang, W.; Yang, C. Dynamic Electroosmotic Flows of Power-Law Fluids in Rectangular Microchannels. *Micromachines* **2017**, *8*, 34. [CrossRef]
25. Goswami, P.; Mondal, P.K.; Datta, A.; Chakraborty, S. Entropy Generation Minimization in an Electroosmotic Flow of Non-Newtonian Fluid: Effect of Conjugate Heat Transfer. *J. Heat Transf.* **2016**, *138*, 051704. [CrossRef]
26. Jian, Y.; Si, D.; Chang, L.; Liu, Q. Transient rotating electromagnetohydrodynamic micropumps between two infinite microparallel plates. *Chem. Eng. Sci.* **2015**, *134*, 12–22. [CrossRef]
27. Jian, Y. Transient MHD heat transfer and entropy generation in a microparallel channel combined with pressure and electroosmotic effects. *Int. J. Heat Mass Transf.* **2015**, *89*, 193–205. [CrossRef]
28. Brinkman, H.C. The Viscosity of Concentrated Suspensions and Solutions. *J. Chem. Phys.* **1952**, *20*, 571. [CrossRef]
29. Choi, U.S. Enhancing thermal conductivity of fluids with nanoparticles. *ASME FED* **1995**, *231*, 99–103.
30. Yu, W.; Choi, S.U.S. The Role of Interfacial Layers in the Enhanced Thermal Conductivity of Nanofluids: A Renovated Maxwell Model. *J. Nanoparticle Res.* **2003**, *5*, 167–171. [CrossRef]

31. Zhao, G.; Wang, Z.; Jian, Y. Heat transfer of the MHD nanofluid in porous microtubes under the electrokinetic effects. *Int. J. Heat Mass Transf.* **2019**, *130*, 821–830. [CrossRef]

32. Zhao, G.P.; Jian, Y.J.; Li, F.Q. Heat transfer of nanofluids in microtubes under the effects of streaming potential. *Appl. Therm. Eng.* **2016**, *100*, 1299–1307. [CrossRef]

33. Kalteh, M.; Abbassi, A.; Saffar-Avval, M.; Harting, J. Eulerian–Eulerian two-phase numerical simulation of nanofluid laminar forced convection in a microchannel. *Int. J. Heat Fluid Flow* **2011**, *32*, 107–116. [CrossRef]

34. Nguyen, C.T.; Roy, G.; Gauthier, C.; Galanis, N. Heat transfer enhancement using Al_2O_3–water nanofluid for an electronic liquid cooling system. *Appl. Therm. Eng.* **2007**, *27*, 1501–1506. [CrossRef]

35. Lee, J.-H.; Hwang, K.S.; Jang, S.P.; Lee, B.H.; Kim, J.H.; Choi, S.U.S.; Choi, C.J. Effective viscosities and thermal conductivities of aqueous nanofluids containing low volume concentrations of Al_2O_3 nanoparticles. *Int. J. Heat Mass Transf.* **2008**, *51*, 2651–2656. [CrossRef]

36. Jayavel, P.; Jhorar, R.; Tripathi, D.; Azese, M.N. Electroosmotic flow of pseudoplastic nanoliquids via peristaltic pumping. *J. Braz. Soc. Mech. Sci. Eng.* **2019**, *41*, 61. [CrossRef]

37. Tripathi, D.; Prakash, J.; Tiwari, A.K.; Ellahi, R. Thermal, microrotation, electromagnetic field and nanoparticle shape effects on cu-cuo/blood flow in microvascular vessels. *Microvasc. Res.* **2020**, *132*, 104065. [CrossRef] [PubMed]

38. Hojjat, M.; Etemad, S.G.; Bagheri, R.; Thibault, J. Rheological characteristics of non-Newtonian nanofluids: Experimental investigation. *Int. Commun. Heat Mass Transf.* **2011**, *38*, 144–148. [CrossRef]

39. Chang, H.; Jwo, C.S.; Lo, C.H.; Tsung, T.T.; Kao, M.J.; Lin, H.M. Rheology of CuO nanoparticle suspension prepared by ASNSS. *Rev. Adv. Mater. Sci.* **2005**, *10*, 128–132.

40. Pak, B.C.; Cho, Y.I. Hydrodynamic and heat transfer study of dispersed fluids with submicron metallic oxide particles. *Exp. Heat Transf.* **1998**, *11*, 151–170. [CrossRef]

41. Hojjat, M.; Etemad, S.G.; Bagheri, R.; Thibault, J. Turbulent forced convection heat transfer of non-Newtonian nanofluids. *Exp. Therm. Fluid Sci.* **2011**, *35*, 1351–1356. [CrossRef]

42. Zhou, S.-Q.; Ni, R.; Funfschilling, D. Effects of shear rate and temperature on viscosity of alumina polyalphaolefins nanofluids. *J. Appl. Phys.* **2010**, *107*, 054317. [CrossRef]

43. Lin, Y.; Zheng, L.; Zhang, X.; Ma, L.; Chen, G. MHD pseudo-plastic nanofluid unsteady flow and heat transfer in a finite thin film over stretching surface with internal heat generation. *Int. J. Heat Mass Transf.* **2015**, *84*, 903–911. [CrossRef]

44. Si, X.; Li, H.; Zheng, L.; Shen, Y.; Zhang, X. A mixed convection flow and heat transfer of pseudo-plastic power law nanofluids past a stretching vertical plate. *Int. J. Heat Mass Transf.* **2017**, *105*, 350–358. [CrossRef]

45. Shehzad, N.; Zeeshan, A.; Ellahi, R. Electroosmotic Flow of MHD Power Law Al_2O_3-PVC Nanouid in a Horizontal Channel: Couette-Poiseuille Flow Model. *Commun. Theor. Phys.* **2018**, *69*, 655. [CrossRef]

46. Deng, S. Thermally Fully Developed Electroosmotic Flow of Power-Law Nanofluid in a Rectangular Microchannel. *Micromachines* **2019**, *10*, 363. [CrossRef]

47. Ganguly, S.; Sarkar, S.; Hota, T.K.; Mishra, M. Thermally developing combined electroosmotic and pressure-driven flow of nanofluids in a microchannel under the effect of magnetic field. *Chem. Eng. Sci.* **2015**, *126*, 10–21. [CrossRef]

48. Liu, M.; Liu, Y.; Guo, Q.; Yang, J. Modeling of electroosmotic pumping of nonconducting liquids and biofluids by a two-phase flow method. *J. Electroanal. Chem.* **2009**, *636*, 86–92. [CrossRef]

49. Brask, A.; Goranovic, G.; Bruus, H. Electroosmotic pumping of nonconducting liquids by viscous drag from a secondary conducting liquid. In Proceedings of the 2003 Nanotechnology Conference and Trade Show, San Francisco, CA, USA, 23–27 February 2003; pp. 190–193.

50. Gao, Y.; Wong, T.N.; Yang, C.; Ooi, K.T. Two-fluid electroosmotic flow in microchannels. *J. Colloid Interface Sci.* **2005**, *284*, 306–314. [CrossRef]

51. Deng, S.; Xiao, T.; Wu, S. Two-layer combined electroosmotic and pressure-driven flow of power-law fluids in a circular microcapillary. *Colloids Surf. A Physicochem. Eng. Asp.* **2021**, *610*, 125727. [CrossRef]

52. Gao, Y.D.; Wong, T.N.; Yang, C.; Ooi, K.T. Transient two-liquid electroosmotic flow with electric charges at the interface. *Colloids Surf. A* **2005**, *266*, 117–128. [CrossRef]

53. Su, J.; Jian, Y.-J.; Chang, L.; Li, Q.-S. Transient electro-osmotic and pressure driven flows of two-layer fluids through a slit microchannel. *Acta Mech. Sin.* **2013**, *29*, 534–542. [CrossRef]

54. Moghadam, A.J.; Akbarzadeh, P. AC two-immiscible-fluid EOF in a microcapillary. *J. Braz. Soc. Mech. Sci. Eng.* **2019**, *41*, 194. [CrossRef]

55. Xie, Z.-Y.; Jian, Y.-J. Entropy generation of two-layer magnetohydrodynamic electroosmotic flow through microparallel channels. *Energy* **2017**, *139*, 1080–1093. [CrossRef]

56. Xie, Z.; Jian, Y. Entropy generation of magnetohydrodynamic electroosmotic flow in two-layer systems with a layer of non-conducting viscoelastic fluid. *Int. J. Heat Mass Transf.* **2018**, *127*, 600–615. [CrossRef]

57. Ranjit, N.K.; Shit, G.C.; Tripathi, D. Entropy generation and joule heating of two layered elctroosmotic flow in the peristaltically induced micro-channel. *Int. J. Mech. Sci.* **2019**, *153–154*, 430–444. [CrossRef]

58. Zheng, J.; Jian, Y. Rotating electroosmotic flow of two-layer fluids through a microparallel channel. *Int. J. Mech. Sci.* **2018**, *136*, 293–302. [CrossRef]

59. Qi, C.; Ng, C.-O. Electroosmotic flow of a two-layer fluid in a slit channel with gradually varying wall shape and zeta potential. *Int. J. Heat Mass Transf.* **2018**, *119*, 52–64. [CrossRef]
60. Li, B.; Zhang, W.; Zhu, L.; Zheng, L. On mixed convection of two immiscible layers with a layer of non-Newtonian nanofluid in a vertical channel. *Powder Technol.* **2017**, *310*, 351–358. [CrossRef]
61. Deng, S.; Li, M.; Yang, Y.; Xiao, T. Heat transfer and entropy generation in two layered electroosmotic flow of power-law nanofluids through a microtube. *Appl. Therm. Eng.* **2021**, *196*, 117314. [CrossRef]
62. Prakash, J.; Tripathi, D.; Bég, O.A. Comparative study of hybrid nanofluids in microchannel slip flow induced by electroosmosis and peristalsis. *Appl. Nanosci.* **2020**, *10*, 1693–1706. [CrossRef]
63. Hemmat Esfe, M.; Saedodin, S.; Mahian, O.; Wongwises, S. Thermal conductivity of Al_2O_3/water nanofluids. *J. Therm. Anal. Calorim.* **2014**, *117*, 675–681. [CrossRef]

 *micromachines*

Article

Dynamics of Tri-Hybrid Nanoparticles in the Rheology of Pseudo-Plastic Liquid with Dufour and Soret Effects

Enran Hou [1], Fuzhang Wang [2,3], Umar Nazir [4], Muhammad Sohail [4,*], Noman Jabbar [4] and Phatiphat Thounthong [5]

[1] College of Mathematics, Huaibei Normal University, Huaibei 235000, China; houenran@163.com
[2] Department of Mathematics, Nanchang Institute of Technology, Nanchang 330044, China; wangfuzhang1984@163.com
[3] College of Mathematics and Statistics, Xuzhou University of Technology, Xuzhou 221018, China
[4] Department of Applied Mathematics and Statistics, Institute of Space Technology, P.O. Box 2750, Islamabad 44000, Pakistan; nazir_u2563@yahoo.com (U.N.); noman.jabbar777@gmail.com (N.J.)
[5] Renewable Energy Research Centre, Department of Teacher Training in Electrical Engineering, Faculty of Technical Education, King Mongkut's University of Technology North Bangkok, 1518 Pracharat 1 Road, Bangsue, Bangkok 10800, Thailand; phatiphat.t@fte.kmutnb.ac.th
* Correspondence: muhammad_sohail111@yahoo.com

Abstract: The rheology of different materials at the micro and macro levels is an area of great interest to many researchers, due to its important physical significance. Past experimental studies have proved the efficiency of the utilization of nanoparticles in different mechanisms for the purpose of boosting the heat transportation rate. The purpose of this study is to investigate heat and mass transport in a pseudo-plastic model past over a stretched porous surface in the presence of the Soret and Dufour effects. The involvement of tri-hybrid nanoparticles was incorporated into the pseudo-plastic model to enhance the heat transfer rate, and the transport problem of thermal energy and solute mechanisms was modelled considering the heat generation/absorption and the chemical reaction. Furthermore, traditional Fourier and Fick's laws were engaged in the thermal and solute transportation. The physical model was developed upon Cartesian coordinates, and boundary layer theory was utilized in the simplification of the modelled problem, which appears in the form of coupled partial differential equations systems (PDEs). The modelled PDEs were transformed into corresponding ordinary differential equations systems (ODEs) by engaging the appropriate similarity transformation, and the converted ODEs were solved numerically via a Finite Element Procedure (FEP). The obtained solution was plotted against numerous emerging parameters. In addition, a grid independent survey is presented. We recorded that the temperature of the tri-hybrid nanoparticles was significantly higher than the fluid temperature. Augmenting the values of the Dufour number had a similar comportment on the fluid temperature and concentration. The fluid temperature increased against a higher estimation of the heat generation parameter and the Eckert numbers. The impacts of the buoyancy force parameter and the porosity parameter were quite opposite on the fluid velocity.

Keywords: tri-hybrid nanoparticles; Soret and Dufour effect; boundary layer analysis; finite element scheme; heat generation; constructive and destructive chemical reaction

Citation: Hou, E.; Wang, F.; Nazir, U.; Sohail, M.; Jabbar, N.; Thounthong, P. Dynamics of Tri-Hybrid Nanoparticles in the Rheology of Pseudo-Plastic Liquid with Dufour and Soret Effects. *Micromachines* **2022**, *13*, 201. https://doi.org/10.3390/mi13020201

Academic Editors: Lanju Mei and Shizhi Qian

Received: 15 January 2022
Accepted: 25 January 2022
Published: 27 January 2022

Publisher's Note: MDPI stays neutral with regard to jurisdictional claims in published maps and institutional affiliations.

1. Introduction

In the last few decades, many researchers have become interested in the study of shear thinning fluids, due to the many fascinating industrial and everyday applications [1]. A small number of these include wall paint, printing ink, nail polish, whipped cream, ketchup, and engine oil. Shear thinning fluid can also be called pseudo-plastic fluid and is considered to display the behaviors of both Newtonian fluid and of plastic fluid. In shear thinning fluid, the more stress is applied, the more freely the fluid flows. This property is a useful characteristic for its use in materials such as paint, oils, and cream. Eberhard et al. [2] computed the effective viscosity for Newtonian and non-Newtonian materials past through a porous surface. They first considered the rheology of the power law model to estimate an effective shear rate. In their investigation, they assumed the constant permeability. Rosti and Takagi [3] studied the shear thinning and shear thickening behaviors of materials by studying different important aspects. Moreover, they recorded several important features through changing the phase and volume fractions. Gul et al. [4] computed the exact solution in the case of lifting and drainage with slip conditions for the power law model of thin film. They approximated the flow rate and the skin friction coefficient and plotted numerous sketches against different involved parameters for fluid velocity. They recorded the decline in the velocity field for the escalating values of the slip parameter. Pseudo-plastic nanofluid obeying Brownian motion and thermophoresis past over a vertical cylinder was examined by Hussein et al. [5]. They solved the converted modelled equations numerically and monitored the decline in velocity against the curvature parameter and fluid parameter. Abdelsalam and Sohail [6] studied the involvement of the bio-convection phenomenon in viscous nanofluid comprising variable properties past over a bidirectional stretched surface engaging an optimal homotopy scheme. They established an error analysis and performed a comparative study, noting the depreciation in the motile density profile against the Peclet and Lewis numbers. Sohail and Naz [7] investigated the stretched Sutterby fluid flow in a cylinder and presented the dynamical survey while considering thermophoresis, Brownian motion; thermal and concentration relaxation times. They used the polar coordinates to derive the physical model in the form of coupled partial differential equations systems (PDEs) and then transformed these into ordinary differential equations systems (ODEs) by engaging the appropriate similarity transformation while utilizing the approach of boundary layer theory. Afterwards, converted ODEs were tackled analytically. They monitored the decline in fluid temperature against the Prandtl number and the concentration was controlled for the higher values of the Schmidt number. Chu et al. [8] examined the involvement of chemical reaction and activation energy in the nanofluid flow problem. They solved the resulting equations numerically via a finite element procedure, recording the decline in the fluid velocity against the magnetic parameter. Hina et al. [9] used a long wavelength approach to model the pseudo-plastic fluid problem with wall and slip properties in a curved channel under the peristaltic transport phenomenon. They solved the boundary value problem numerically and the solution was plotted for different parametric values in Mathematica 15.0 software; they also examined the symmetric pattern for fluid velocity against a larger curvature parameter. Salahuddin et al. [10] numerically stimulated the magnetohydrodynamic (MHD) flow for a pseudo-plastic fluid over a stretching cylinder with a MHD effect and temperature dependent thermal conductivity. They derived the physical model into a mathematical form by engaging boundary layer theory. The derived problem was highly nonlinear in nature and was presented in the form of PDEs. Similarity transformation was used to simplify the problem and to convert PDEs into ODEs. A numerical solution was obtained via the Keller box scheme, and they recorded the decline in fluid velocity for higher values of the Weissenberg number. Alam et al. [11] presented a study on the drainage and lifting MHD pseudo-plastic problem and addressed the exact solution, finding the decline in velocity against the Stokes number. The phenomenon of tapering is observed in the peristaltic flow of the pseudo-plastic model with variable viscosity studied by Hayat et al. [12]. They engaged a perturbation approach to solving the boundary value problem in the symmetric channel and observed an increase in velocity against the higher values of the magnetic parameter. Moreover, they plotted streamlines for different parameters. Further important contributions are reported in [13–15].

The principle behind synthesizing nanofluid composites is to enhance the properties of a single nanoparticle that possesses either improved thermal conductivity or improved rheological properties. By framing, nanofluids are made using single nanoparticles which are useful in developing a greater ability to absorb heat energy and the rheological properties are improved in various fluids. In a similar way, a composite of nanofluids is most significant in view of an improved transfer of thermal energy and rheological properties in liquids. Composites of nanoparticles are known as nanofluids, while composites of two or more nanofluids are called hybrid nanoparticles. Composites of three kinds of nanofluids are known as tri-hybrid nanoparticles. Tri-hybrid nanoparticles are recognized as the most significant to the betterment of thermal conductivity. Applications of such composite particles are relevant in the making of electronic heaters, the production of solar energy, nuclear safety, the pharmaceutical industry, etc.

Tri-hybrid nanoparticles have been studied by various scholars. For example, Manjunatha et al. [16] discussed the effect of tri-hybrid nanoparticles in energy transfer phenomena considering convective conditions. Nazir et al. [17] studied significant thermal growth for ternary hybrid nanoparticles as compared to the thermal growth for hybrid nanofluid and nanoparticles in complex fluid over a heated two-dimensional frame. They implemented a finite element approach to achieve numerical results. Chen et al. [18] scrutinized the thermal properties along with the Ternary hybrid nanostructures in graphene oxide/graphene and MoS_2/zirconia while they found an improvement in the tribiological and mechanical properties. Zayan, Mohammed et al. [19] investigated novel ternary hybrid nanostructures in view of the thermal additives. Shafiq et al. [20] studied Walters' B liquid in nanoparticles in view of dual stratification including the stagnation point using a Riga plate. Swain et al. [21] discussed features related to hybrid nanoparticles in the presence of a chemical reaction towards a stretching surface including slip conditions. Mebarek-Oudina et al. [22] performed a useful model study regarding hybrid nanoparticles under magnetic parameter in view of convection heat energy in a trapezoidal cavity. Warke et al. [23] numerically investigated the stagnation point behavior in the presence of thermal radiation impact over a heated surface. Dadheech et al. [24] captured the impacts related to heat energy transfer in the presence of hybrid nanoparticles using the role of hybrid nanoparticles under a magnetic parameter. Marzougui et al. [25] studied entropy generation and heat transfer including nanoparticles in a lid-driven cavity under a magnetic field. Oudina [26] simulated convective heat transfer using a category of nanoparticles based on titanium nanofluids using heat source terms. Dhif et al. [27] analyzed the role of hybrid nanofluid in solar collectors. Zamzari et al. [28] discussed the influences of mixed convection in a vertical heated channel whereby they determined aspects of entropy generation. Li et al. [29] determined the flow and the thermal characterizations in non-Newtonian liquid, adding nanoparticles using a non-Fourier approach alongside a Prandtl approach in the presence of a Darcy–Forchheimer motion. Mehrez et al. [30] investigated the impact of a magnetic field in ferro-fluid in a heated channel. Khashi'ie et al. [31] discussed the influences of hybrid nanofluid along with the shape factor while considering thermal radiation effect. Esfe et al. [32] analyzed the role of non-Newtonian liquid, adding hybrid nanoparticles to non-Newtonian material using variable viscosity. Mehrez and Cafsi [33] captured the role of hybrid nanofluid in a heated cavity via a pulsating inlet condition.

A surveying of the available literature shows that no adequate study has been performed involving ternary hybrid nanoparticles. This contribution aims to fill this gap in the research. A literature survey is covered in Section 1; the modeling is included in Section 2 with attention given to several important physical effects; the computational strategy is explained in Section 3; the results are analyzed in Sections 4 and 5.

The preparation approach associated with ternary hybrid nanoparticles is captured in Figure 1.

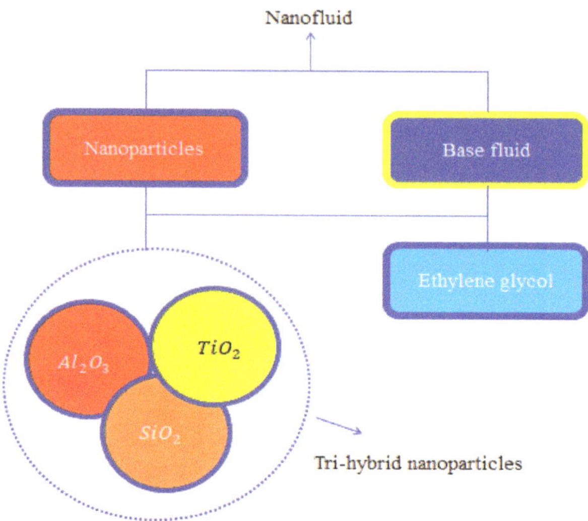

Figure 1. A description of the tri-hybrid approach in nanofluids.

2. Description of Constructing Model

The rheology of a two-dimensional heat and mass diffusion transfer model in a pseudo-plastic liquid past a vertical surface was considered in the presence of a Darcy–Forchheimer model. A phenomenon associated with Dufour and Soret impacts was analyzed. The base fluid is assumed as ethylene glycol in a pseudo-plastic liquid inserting three kinds of nanoparticles (Al_2O_3, SiO_2, and TiO_2). The thermal properties of silicon dioxide, ethylene glycol, aluminum oxide, and tritium dioxide are considered in Table 1. The following assumptions are detailed below.

- Two-dimensional flow of the pseudo-plastic material is considered;
- Fourier's law and Fick's law are assumed;
- Chemical reaction and heat generation are addressed;
- Transfer of heat is characterized in the presence of the Dufour and Soret effects;
- Viscous dissipation is removed;
- Darcy–Forchheimer porous theory is analyzed, and the vertical surface is plotted in Figure 2.

Table 1. Thermal properties [34] of density, electrical conductivity, and thermal conductivity.

	K (Thermal Conductivity)	σ (Electrical Conductivity)	ρ (Density)
$C_2H_6O_2$	0.253	4.3×10^{-5}	1113.5
Al_2O_3	32.9	5.96×10^{7}	6310
TiO_2	8.953	2.4×10^{6}	4250
SiO_2	1.4013	3.5×10^{6}	2270

The power law model associated with shear stress is defined as:

$$\tau_{xy} = -n\left(\left|\frac{\partial u}{\partial y}\right|^{m-1}\right)\frac{\partial u}{\partial y}. \tag{1}$$

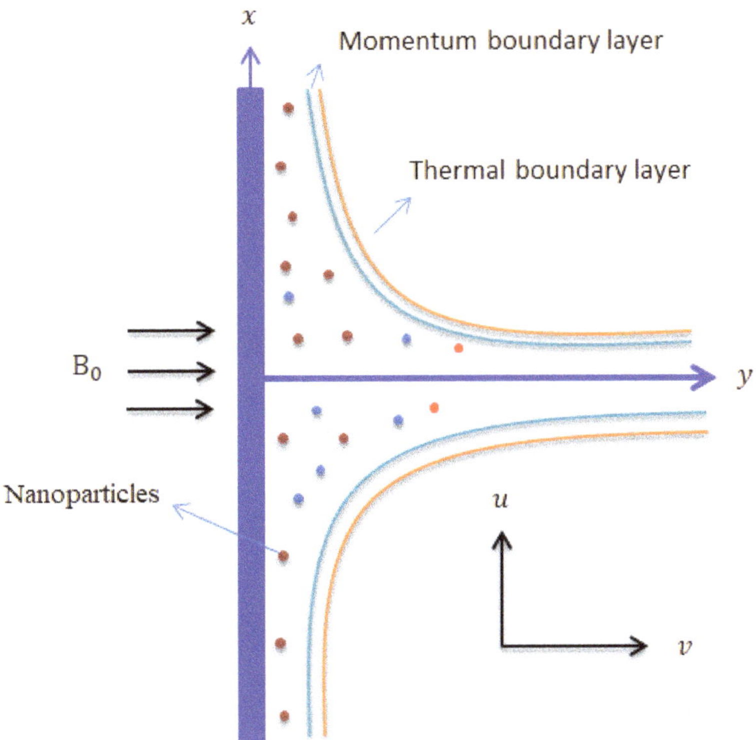

Figure 2. Illustration of the geometry of the current analysis.

Equation (1) is known as the power law model regarding the shear stress of pseudo-plastic liquid. The fluids category is based on the numerical values of m. The present model becomes a Newtonian fluid model when $m = 1$ while the present model can be converted into a dilatant fluid mode when $m > 1$, and the present model becomes a pseudo-plastic liquid model when $0 < m < 1$.

Boundary layer approximations are used to derive a system of PDEs on conservations laws regarding momentum, thermal energy, and mass diffusion. The present model is considered in terms of two-dimensional flow as well as steady and incompressible flow. The modeled PDEs (partial differential equations) are deduced Refs. [35–37] as:

$$\frac{\partial u}{\partial x} + \frac{\partial v}{\partial y} = 0, \tag{2}$$

$$u\frac{\partial u}{\partial x} + v\frac{\partial v}{\partial y} = \nu_{tehnf}\frac{\partial}{\partial x}\left(\left|\frac{\partial u}{\partial y}\right|^{m-1}\frac{\partial u}{\partial y}\right) - \frac{\nu_{tehnf}}{k^s}F_D u - \frac{F_D}{(k^s)^{\frac{1}{2}}}u^2 + g\alpha(T - T_\infty) + g\beta(C - C_\infty), \tag{3}$$

$$u\frac{\partial T}{\partial x} + v\frac{\partial T}{\partial y} = \frac{K_{tehnf}}{(\rho C_P)_{tehnf}}\frac{\partial^2 T}{\partial y^2} + \frac{Q(T - T_\infty)}{(\rho C_P)_{tehnf}} + \frac{k_t D_{tehnf}}{(C_P)_f C_s}\frac{\partial^2 C}{\partial^2 y} + \frac{\mu_{tehnf}}{(\rho C_P)_{tehnf}}\left|\frac{\partial u}{\partial y}\right|^{m+1}, \tag{4}$$

$$u\frac{\partial C}{\partial x} + v\frac{\partial C}{\partial y} = D_{tehnf}\frac{\partial^2 C}{\partial y^2} + \frac{k_t D_{tehnf}}{T_M}\frac{\partial^2 T}{\partial y^2} - K(C - C_\infty). \tag{5}$$

Equation (2) is a continuity equation for two-dimensional flows as well as steady and incompressible flows. Equation (3) is termed as a momentum equation in the presence of pseudo-plastic liquid inserting correlations of tri-hybrid nanoparticles using bouncy forces, while Equations (4) and (5) are concentration and thermal energy equations including the effects of the chemical reaction; viscous dissipation; heat source; the Dufour and Soret influences. In Equation (3), the terms on the left-hand side are known as the inertial force, the first term on the right-hand side is the viscous force in the presence of pseudo-plastic liquid, whereas the last two terms on the right-hand side of Equation (3) are due bouncy forces and the two middle terms on the right-hand side of Equation (3) are modeled on Darcy–Forchheimer law. Terms on the left-hand side of Equation (4) are convection terms in heat transfer phenomena, the first term on the right-hand side is a conduction term in heat transfer phenomena, the second term on the right-hand side of Equation (4) occurs due to the heat source, while the third and last terms on the right-hand side of Equation (4) are formulated based on the effects of Soret and viscous dissipation. The second term on the right-hand side of Equation (5) is the Dufour effect, the last term on right-hand side signifies a chemical reaction, the first term on the right-hand side and the terms on the left-hand side occur due to the diffusion of mass species in view of convection and of conduction, respectively.

The desired (boundary conditions) BCs are:

$$u = u_w,\ v = -v_w,\ C = C_w,\ T = T_w \text{ at } y = 0, u \to u_\infty,\ C \to C_\infty,\ T \to T_\infty \text{ when } y \to \infty. \tag{6}$$

Transformations are defined as:

$$\theta = \frac{T - T_\infty}{T_w - T_\infty},\ \theta = \frac{C - C_\infty}{C_w - C_\infty},\ \zeta = y\left(\frac{U^{2-m}}{x v_f}\right)^{\frac{1}{m+1}},\ \Psi = F\left(x v_f U^{2m-1}\right)^{\frac{1}{m+1}}. \tag{7}$$

Dimensionless ODEs are formulated via defined transformations:

$$\left(|F''|^{m-1}F''\right)' + \frac{1}{m+1}F''F - \epsilon F' - \frac{v_f}{v_{tehnf}}F_R\left(F'^2\right) + \frac{v_f}{v_{tehnf}}[\lambda_n\theta + \lambda_M\phi] = 0, \tag{8}$$

$$\theta'' + \frac{Pr}{m+1}F\theta' + \frac{k_f(\rho C_p)_{tehnf}}{k_{tehnf}(\rho C_p)_f}PrEc|F''|^{m+1} + \frac{k_f(\rho C_p)_{tehnf}}{k_{tehnf}(\rho C_p)_f}PrD_f\phi'' + \frac{k_f}{k_{tehnf}}H_hPr\theta = 0, \tag{9}$$

$$\phi'' + \frac{(1-\varphi_b)^{-2.5}Sc}{(1-\varphi_c)^{2.5}m+1(1-\varphi_a)^{2.5}}F\phi' - \frac{(1-\varphi_b)^{-2.5}Sc}{(1-\varphi_a)^{2.5}(1-\varphi_c)^{2.5}}K_c\phi + ScS_r\theta'' = 0. \tag{10}$$

Defined correlations in the motion of tri-hybrid nanoparticles are [34]:

$$\rho_{tehnf} = (1-\varphi_a)\left\{(1-\varphi_b)\left[(1-\varphi_c)\rho_f + \varphi_c\rho_3\right] + \varphi_b\rho_2\right\} + \varphi_c\rho_1, \tag{11}$$

$$\frac{\mu_f}{(1-\varphi_a)^{2.5}(1-\varphi_b)^{2.5}(1-\varphi_c)^{2.5}},\ \frac{K_{tenf}}{K_{nf}} = \frac{K_2 + 2K_{nf} - 2\varphi_a\left(K_{nf} - K_2\right)}{K_2 + 2K_{nf} + \varphi_b\left(K_{nf} - K_2\right)}, \tag{12}$$

$$\frac{K_{tehnf}}{K_{hnf}} = \frac{K_1 + 2K_{hnf} - 2\varphi_a\left(K_{hnf} - K_1\right)}{K_1 + 2K_{hnf} + \varphi_a\left(K_{hnf} - K_1\right)},\ \frac{K_{nf}}{K_f} = \frac{K_3 + 2K_f - 2\varphi_c\left(K_f - K_3\right)}{K_3 + 2K_f + \varphi_c\left(K_f - K_3\right)}, \tag{13}$$

$$\frac{\sigma_{tenf}}{\sigma_{hnf}} = \frac{\sigma_1(1+2\varphi_a) - \varphi_{hnf}(1-2\varphi_a)}{\sigma_1(1-\varphi_a) + \sigma_{hnf}(1+\varphi_a)},\ \frac{\sigma_{hnf}}{\sigma_{nf}} = \frac{\sigma_2(1+2\varphi_b) + \varphi_{nf}(1-2\varphi_b)}{\sigma_2(1-\varphi_b) + \sigma_{nf}(1+\varphi_b)}, \tag{14}$$

$$\frac{\sigma_{nf}}{\sigma_f} = \frac{\sigma_3(1+2\varphi_c) + \varphi_f(1-2\varphi_c)}{\sigma_3(1-\varphi_c) + \sigma_f(1+\varphi_c)}. \tag{15}$$

$$\rho_{tehnf} = (1-\varphi_a)\left\{(1-\varphi_b)\left[(1-\varphi_c)\rho_f + \varphi_c\rho_3\right] + \varphi_b\rho_2\right\} + \varphi_c\rho_1, \tag{16}$$

$$\frac{\mu_f}{(1-\varphi_a)^{2.5}(1-\varphi_b)^{2.5}(1-\varphi_c)^{2.5}}, \quad \frac{K_{tenf}}{K_{nf}} = \frac{K_2 + 2K_{nf} - 2\varphi_a\left(K_{nf} - K_2\right)}{K_2 + 2K_{nf} + \varphi_b\left(K_{nf} - K_2\right)}, \tag{17}$$

$$\frac{K_{tehnf}}{K_{hnf}} = \frac{K_1 + 2K_{hnf} - 2\varphi_a\left(K_{hnf} - K_1\right)}{K_1 + 2K_{hnf} + \varphi_a\left(K_{hnf} - K_1\right)}, \quad \frac{K_{nf}}{K_f} = \frac{K_3 + 2K_f - 2\varphi_c\left(K_f - K_3\right)}{K_3 + 2K_f + \varphi_c\left(K_f - K_3\right)}, \tag{18}$$

$$\frac{\sigma_{tenf}}{\sigma_{hnf}} = \frac{\sigma_1(1+2\varphi_a) - \varphi_{hnf}(1-2\varphi_a)}{\sigma_1(1-\varphi_a) + \sigma_{hnf}(1+\varphi_a)}, \quad \frac{\sigma_{hnf}}{\sigma_{nf}} = \frac{\sigma_2(1+2\varphi_b) + \varphi_{nf}(1-2\varphi_b)}{\sigma_2(1-\varphi_b) + \sigma_{nf}(1+\varphi_b)}, \tag{19}$$

$$\frac{\sigma_{nf}}{\sigma_f} = \frac{\sigma_3(1+2\varphi_c) + \varphi_f(1-2\varphi_c)}{\sigma_3(1-\varphi_c) + \sigma_f(1+\varphi_c)}. \tag{20}$$

The surface force at the surface of the wall is:

$$C_f = -\frac{\tau_w}{U^2 \rho_f}, \quad \tau_w = \left(\frac{\partial u}{\partial y}\left|\frac{\partial u}{\partial y}\right|^{m-1}\right), \tag{21}$$

Using the value of τ_w in Equation (21) and:

$$C_f = -\frac{\left(\frac{\partial u}{\partial y}\left|\frac{\partial u}{\partial y}\right|^{m-1}\right)_{y=0}}{U^2 \rho_f}, \tag{22}$$

Implementing the value of $\frac{\partial u}{\partial y}$ in Equations (1) and (22) becomes:

$$(Re)^{\frac{1}{m+1}} C_f = -\frac{(1-\varphi_b)^{-2.5}}{(1-\varphi_a)^{2.5}(1-\varphi_c)^{2.5}}\left[F''(0)|F''(0)|^{m-1}\right]. \tag{23}$$

Nusselt number (Nu) is modeled as:

$$Nu = \frac{xQ_w}{(T_w - T_\infty)k_f}, \quad Q_w = -K_{Thnf}\left(\frac{\partial T}{\partial y}\right), \tag{24}$$

Using the value of Q_w, we get

$$Nu = \frac{-xK_{Thnf}\left(\frac{\partial T}{\partial y}\right)_{y=0}}{(T_w - T_\infty)k_f}, \quad (Re)^{\frac{-1}{m+1}} Nu = -\frac{K_{Thnf}}{k_f}\theta'(0). \tag{25}$$

The rate of mass diffusion is:

$$Sc = \frac{xM_w}{(C_w - C_\infty)D_f}, \quad M_w = -D_{Thnf}\left(\frac{\partial C}{\partial y}\right), \tag{26}$$

Now, Equation (26) is reduced as

$$Sc = \frac{-xD_{Thnf}\left(\frac{\partial C}{\partial y}\right)_{y=0}}{(C_w - C_\infty)D_f}, \tag{27}$$

$$(Re)^{\frac{-1}{m+1}} Sc = -\frac{(1-\phi_b)^{-2.5}}{(1-\phi_a)^{2.5}(1-\phi_b)^{2.5}}\phi'(0). \tag{28}$$

3. Numerical Scheme

The current model's associated boundary conditions were numerically simulated with the help of a finite element algorithm. The concept upon which the finite element method rests is the division of the required domain into elements (finite). FES (finite element scheme) [8,17] is discussed here. The flow chart of the finite element algorithm is mentioned in Figure 3. This approach has been used in several computational fluid dynamics (CFD) problems; the advantages of using this kind of approach are mentioned below.

➤ Complex type geometries are easily tackled using the finite element method;
➤ Physical problems in applied science are numerically solved by finite element formulation (FEM);
➤ FEM needs a low level of investment in view of time and resources;
➤ An important role of FEM is to simulate various types of boundary conditions;
➤ It has the ability to perform discretization regarding derivatives.

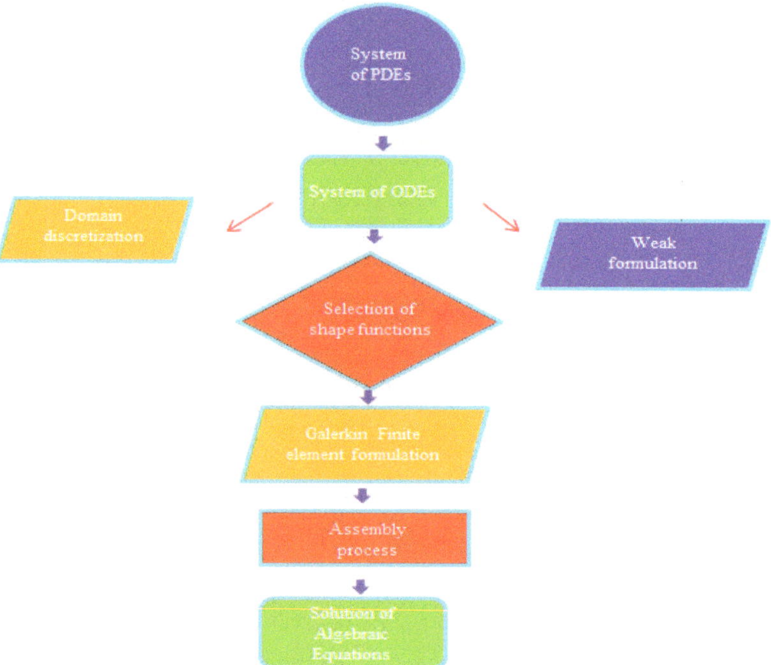

Figure 3. Flow chart regarding the finite element scheme.

3.1. Domain Discretization

Firstly, the domain was discretized into small numbers of elements, whereas the approximation solution was made using the concept of the division of elements. This obtained approximation solution is assumed as a linear polynomial.

3.2. Choice of Shape Function

The shape function plays a vital role in developing the approximation solution along with the nodal value. The nodal value and the shape function for the solution of the current model are defined as:

$$F = \sum_{j=1}^{ll} \Psi_n F_n, \ \Psi_n = (-1)^{n-1} \frac{\xi_{n+1} - \xi}{\xi_{n+1} - \xi_p}, \ \text{here } n = 1, 2. \tag{29}$$

3.3. Residuals

Observe that the present model is known as a strong form model, whereas weak form models are made via the approach related to GFE (Galerkin finite element). The residual is:

$$\int_{\Omega} (\Psi_a R)\, d\Omega. \tag{30}$$

3.4. Assembly Approach

The concept of an assembly approach is implemented in the development of a global stiffness matrix and stiffness matrices. The linearization of algebraic equations is accomplished via the Picard linearization approach.

3.5. Testing of Error Analysis and Mesh Free Analysis

The error analysis of the current investigation is addressed as:

$$Max\left|\Omega_i^r - \Omega_i^{r-1}\right| < 10^{-8}. \tag{31}$$

The convergence of problem is ensured within 300 elements. Table 2 captures the convergence of problem.

Table 2. Grid-independent analyses of concentration, temperature, and velocity at mid of 270 elements.

Number of Elements	$F'\left(\frac{\xi_{max}}{2}\right)$	$\theta\left(\frac{\xi_{max}}{2}\right)$	$\phi\left(\frac{\xi_{max}}{2}\right)$
30	0.01780117292	0.2638819583	0.1043559742
60	0.02008198265	0.2370957263	0.09636478785
90	0.02028786930	0.2277953868	0.09382548703
120	0.02030867335	0.2231793011	0.09257835690
150	0.02029805160	0.2204312762	0.09183716567
180	0.02028205279	0.2186104527	0.09134595367
210	0.02026649287	0.2173159996	0.09099653209
240	0.02025264615	0.2163487639	0.09073525408
270	0.02024062141	0.2155986751	0.09053255564

3.6. Validation of Numerical Results

Table 3 depicts the validation of simulations against already published numerical values [35–37]. Observe that the present flow model is reduced into flow models [35–37] by implementing the values of $\varphi_a = \varphi_b = \varphi_c = 0$, $F_r = \epsilon = \lambda_n = \lambda_m = 0$ into the current model. So, we have found that there is good agreement between the simulations and previously published works.

Table 3. Validation of the numerical results for skin friction coefficient by considering: $\varphi_a = \varphi_b = \varphi_c = 0$, $F_r = \epsilon = \lambda_n = \lambda_m = 0$.

	Skin Friction Coefficient	Present Work Skin Friction Coefficient
Sakiadis [35]	−0.44375	−0.442735
Fox et al. [36]	−0.4437	−0.443639
Chen [37]	−0.4438	−0.442837

4. Results and Discussion

The development of the desired model was carried out under the effects of the Soret and Dufour models in a pseudo-plastic material over a vertical frame in the presence of heat energy and mass species transport. A chemical reaction occurred and dual behavior was addressed relating to heat generation and heat absorption. A mixture of silicon dioxide, aluminum oxide, and titanium oxide in ethylene glycol was considered for analysis of the heat energy and mass species characteristics versus the physical parameters. A detailed discussion of the various parameters is illustrated below.

4.1. Analysis Related to Motion into Particles

The effect of bouncy force on the flow analysis, an effect of (F_r) Forchheimer (m) power law, is addressed in Figures 4–7. Figure 4 addresses the motion into particles by applying the influence of F_r. We observed that the motion of particles became reduced versus the impact of F_r. A mixture of silicon dioxide, aluminum oxide, and titanium oxide in ethylene glycol was inserted during the flow of particles. The frictional force became higher when F_r was increased. In view of layers, layers regarding momentum at the boundary were decreased versus the impact of F_r. Therefore, the fluid became thinner in the case of increasing values for F_r. A Forchheimer number was also used for the declination into motion of the fluid particles. It was the most significant in reducing the momentum layers. Physically, it was the ratio of the pressure reduction into fluid particles which was based on inertia and resistance. So, a higher Forchheimer number created a resistance force among the boundary layers of the fluid particles. The role of power law number is visually represented in the motion of particles considered in Figure 6. The behavior of the fluid was based on the values of power law number. It can be observed that the motion was slowed by applying a variation of power law number. Shear thinning, shear thickening, and fluids category were based on values of power law number. For $m = 0$, fluid became Newtonian. So, fluid motion in the case of Newtonian fluid was dominated when compared to non-Newtonian fluid. MBLs associated within momentum were decreased versus the investigation of power law number. The layers became thick versus the impact of power law number. A power law parameter was used significantly to produce a frictional force among the fluid layers. So, frictional force caused a declination in the flow regarding nanoparticles. Further, layers associated with momentum decreased in function against the values of power law number. The effect of buoyancy forces was created due to a vertical heated sheet. The effects related to buoyancy forces on the flow analysis are observed in Figures 5 and 7. Figure 4 details the relation between motion particles and λ_N. This occurred due to the effect of temperature gradient on the flow analysis. Argumentation related to the motion of particles was boosted when λ_N was increased. In the case of λ_N, MBLs were also inclined versus an effect of λ_N. An impact of λ_M was produced due to the effect of a concentration gradient on the flow. However, in this case, the motion regarding particles slowed. A bouncy parameter was generated due to the use of a vertical surface, which was the reason for producing a bouncy parameter. In this case, a gravitational force was placed on the surfaces via perpendicular direction. The direction of the gravitational force and the flow direction were observed as being opposite. Therefore, flow is slowed versus the impact of a bouncy number.

4.2. Analysis Related to Thermal Energy into Particles

The effects of Eckert number, the heat generation/heat absorption numbers, and of Dufour number on thermal energy are observed in Figures 8–10. A mixture of silicon dioxide, aluminum oxide, and titanium oxide was inserted into ethylene glycol. Figure 8 illustrates the impact of the Eckert number on thermal energy curves inserting ternary hybrid nanoparticles. In this case, heat energy is inclined versus the Eckert number. This inclined impact on temperature curves occurred due to the existence of viscous dissipation. Note that the term regarding viscous dissipation has also been recognized in previously completed work on particles. So, the increase in the viscous dissipation was based on previously completed work. Hence, the work done was increased for the enhancement of particles. Therefore, particles absorb more heat energy when the Eckert number is increased. Thermal layers were based on the viscous dissipation. Higher viscous dissipation created more thermal layers. Hence the increasing function was investigated among the thermal layers and the Eckert number. Figure 9 predicts a dual role regarding heat phenomena. The roles of heat phenomena are known as heat generation and heat absorption, while these behaviors are based on the numerical values of h_s. Negative numerical values are due to heat absorption, whereas positive numerical values are due to heat generation. Heat energy is boosted when h_s is increased because the external source is adjusted at the boundary of the surface. So, due to an external source, heat energy is augmented. The description of the effects of the Dufour number on the thermal layers is illustrated in Figure 10. Heat energy is observed as increasing the function against the variation of the Dufour number. Fluid particles absorb more heat energy in the case of the Dufour number. The comparative investigation between the effects of hybrid nanoparticles, nanostructures, and tri-hybrid nanoparticles on heat energy is considered in Figure 10. A mixture of silicon dioxide, aluminum oxide, and titanium oxide is known as tri-hybrid nanoparticles and a mixture of silicon dioxide and aluminum oxide is known as hybrid nanoparticles, whereas ethylene glycol is the base liquid. In Figure 10, a solid line indicates a tri-hybrid, a dotted line is used for plotting hybrid nanostructures whereas a dashed line is used for nanofluid and a dash-dot line is generated to indicate fluid. We observed that tri-hybrid nanoparticles absorbed maximum heat energy rather than hybrid nanoparticles, fluid, or nanofluid. Hence, tri-hybrid nanoparticles were observed to have more significance for the development and the maximization of heat energy.

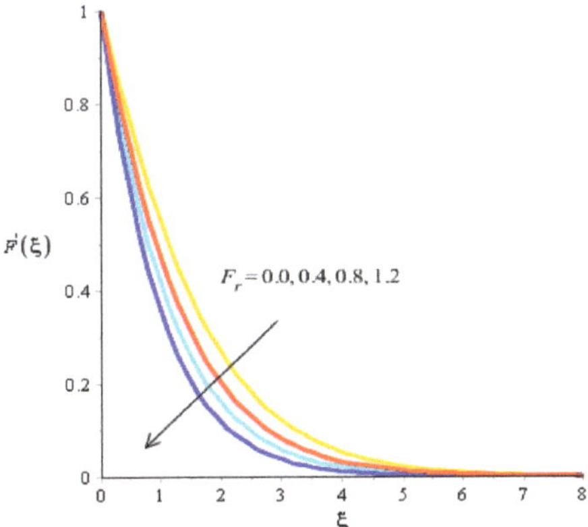

Figure 4. Analysis of velocity curves versus F_r.

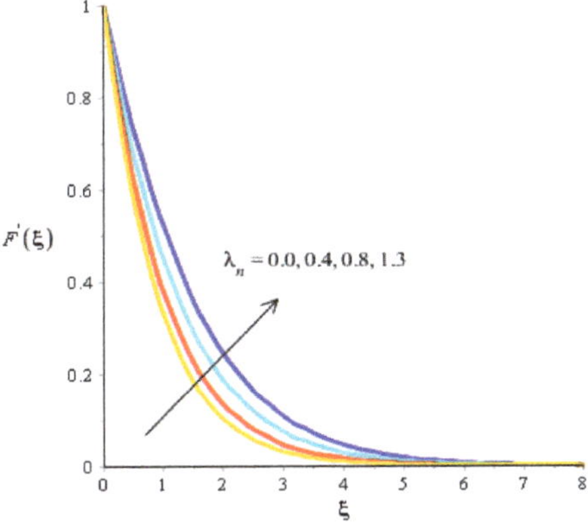

Figure 5. Analysis of velocity curves versus λ_n.

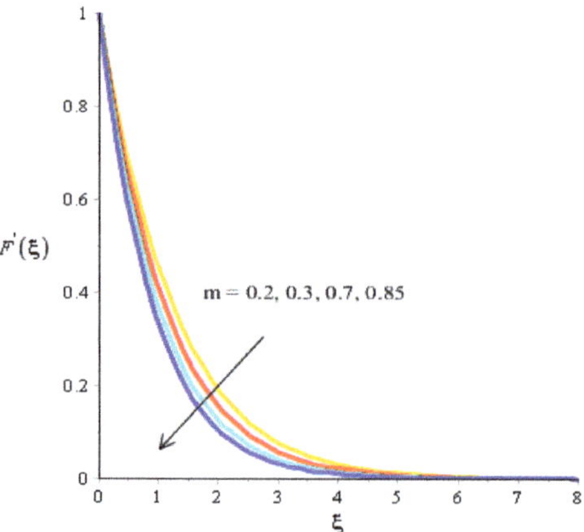

Figure 6. Analysis of velocity curves versus *m*.

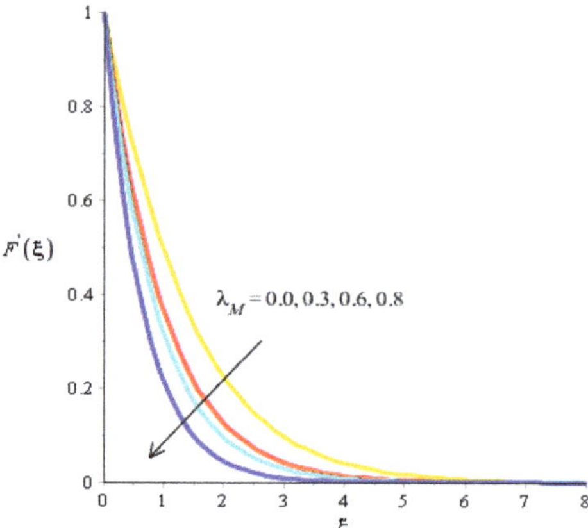

Figure 7. Analysis of velocity curves versus λ_M.

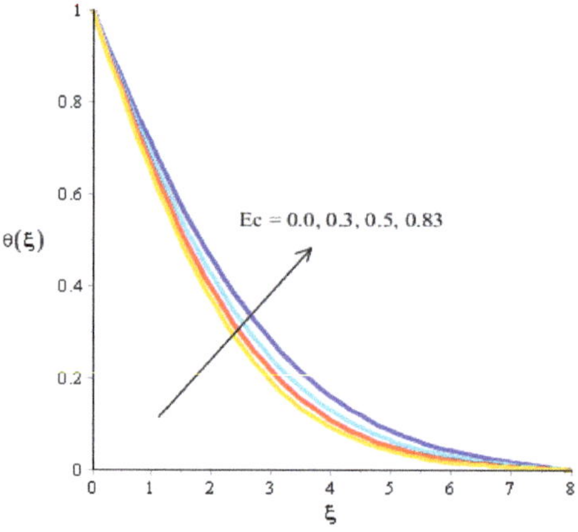

Figure 8. Analysis of the temperature curves versus *Ec*.

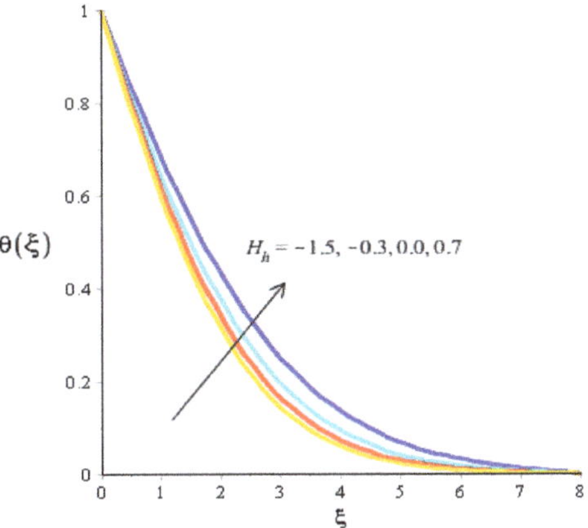

Figure 9. Analysis of the temperature curves versus H_h.

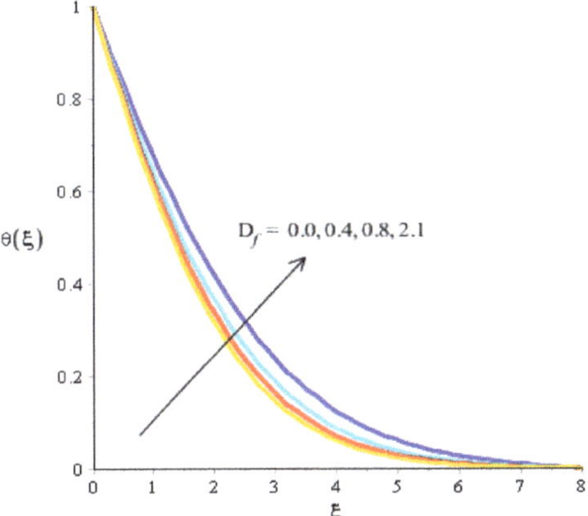

Figure 10. Analysis of the temperature curves versus D_f.

4.3. Analysis Related to Mass Species

The variations related to the Soret number, Schmidt number, chemical reaction and bouncy force in mass species are captured in Figures 11–14. The effect of a Soret number on curves related to the concentration is visualized in Figure 11. Mass species are increased against the distribution in the Soret number. Figure 12 depicts the impact of Bouncy force on mass species. A reduction in mass species versus the distribution in bouncy force was noticed. Figure 13 captures the behavior of the Schmidt number in mass species. The concentration of particles was decreased when Sc was inclined. Physically, Sc is the ratio for mass diffusivity and kinematic viscosity. In view of the physical properties, kinematic viscosity of fluid particles is inclined versus the impact of Sc. However, mass diffusivity is declined against the variation in Sc. Concentration layers have a decreasing function versus the role of Sc. The role of K_c on the concentration is estimated in Figure 14, including ternary hybrid nanostructures. The dual character of the chemical reaction on the concentration is observed. Two types of reactions based on destructive and generative reactions were addressed for this case. These reactions were based on the values of K_c while the negative values were due to a destructive reaction and the positive values were due to a generative chemical reaction. For both types of reactions, the concentration into particles was declined. Further, the concentration for a destructive reaction was higher than the concentration for a generative reaction.

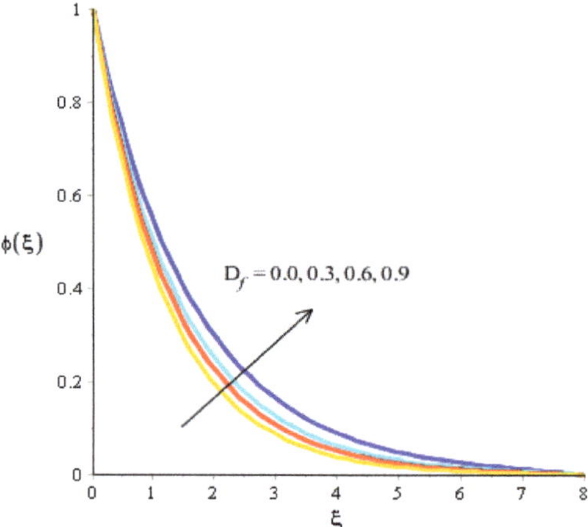

Figure 11. Analysis of concentration curves versus D_f.

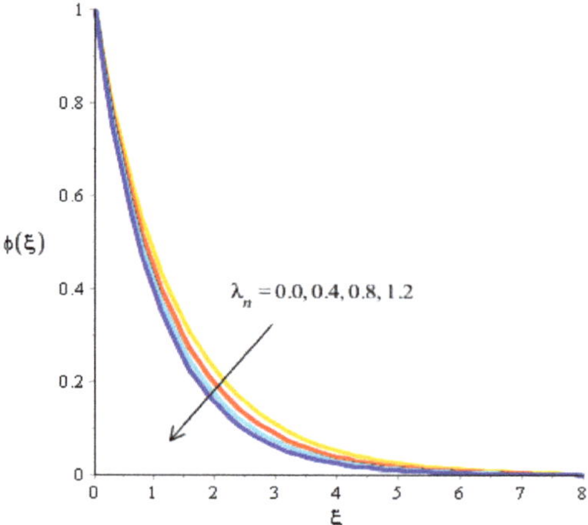

Figure 12. Analysis of concentration curves versus λ_n.

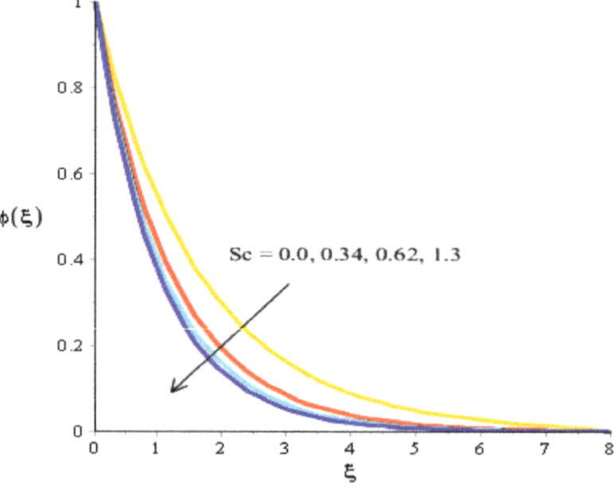

Figure 13. Analysis of concentration curves versus S_c.

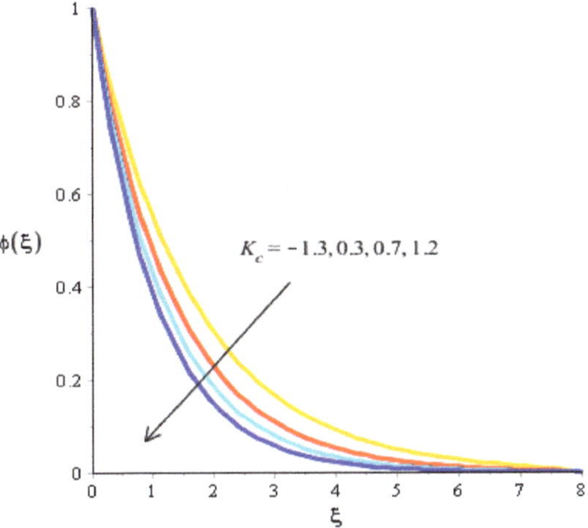

Figure 14. Analysis of concentration curves versus K_c.

5. Conclusions

A finite element scheme was engaged for handling the modelled physical problem which was past over a stretching sheet containing a mixture of nanoparticles in a pseudo-plastic material under the Soret and Dufour effects. Several important plots were displayed to capture the features of different parameters involved in the fluid velocity, temperature, and concentration fields. Important findings are listed below:

- Confirmation of convergence analysis occurred at 270 elements.
- Forchheimer number, power law number, and λ_M caused a decline in the thickness of the momentum boundary layer. However, an increment was investigated in the flow versus argument values of the bouncy parameter (λ_n).
- Temperature distribution was maximized versus higher impacts of the Eckert number, heat generation, and the Dufour number while the thickness associated with thermal layers was increased.
- The concentration field was decreased against the argument values of the Schmidt number, chemical reaction number, and bouncy number, whereas the concentration field was enhanced against higher values of the Dufour number.
- The approach of utilizing ternary hybrid nanoparticles was found to be a significant factor in obtaining maximum thermal energy.

Author Contributions: Conceptualization, E.H. and F.W.; methodology, N.J.; software, U.N.; validation, M.S., E.H. and P.T.; formal analysis, P.T.; investigation, N.J.; resources, U.N.; data curation, M.S.; writing—original draft preparation, U.N.; writing—review and editing, E.H.; visualization, F.W.; supervision, M.S.; project administration, M.S.; funding acquisition, E.H. All authors have read and agreed to the published version of the manuscript.

Funding: This work was supported by the University Natural Science Research Project of Anhui Province (Project nos. KJ2020B06 and KJ2020ZD008).

Data Availability Statement: All supporting data is included in the paper.

Conflicts of Interest: The authors declare no conflict of interest.

Nomenclature

Symbols	Used for	Symbols	Used for
(v, u)	Velocity components (ms^{-1})	ν	Kinematic viscously $(\text{m}^2\text{s}^{-1})$
y, x	Space coordinates (m)	T	Temperature (K)
m	Power law number	ρ	Fluid density (Kgm^{-3})
g	Gravitational acceleration (ms^{-2})	K	Thermal conductivity (Wm^{-1})
C	Mass concentration (Kgm^{-3})	C_∞	Ambient concentration (Kgm^{-3})
T_∞	Ambient temperature (K)	μ	Fluid viscosity $(\text{Kgm}^{-1}\text{s}^{-1})$
Q	Heat source $(\text{JKs}^{-1}\text{m}^{-3})$	C_p	Specific heat capacitance $(\text{JKg}^{-1}\text{m}^{-3})$
D	Mass diffusion $(\text{m}^2\text{s}^{-1})$	BCs	Boundary conditions
T_w	Wall temperature (K)	C_w	Wall concentration (Kgm^{-3})
ξ	Independent variable	ODEs	Ordinary differential equations
F	Dimensionless velocity	θ	Dimensionless temperature
ϕ	Dimensionless concentration	φ	Volume fraction
ϵ	Porosity number	λ_n	Bouncy number
λ_m	Bouncy number	Pr	Prandtl number
Ec	Eckert number	D_f	Dufour number
H_h	Heat generation number	Sc	Schmidt number
K_c	Chemical reaction number	S_r	Soret number
C_f	Skin friction coefficient	Nu	Nusselt number
$tehnf$	Tri-hybrid nanoparticles	Re	Reynolds number
CFD	Computational fluid dynamics	FES	Finite element approach
∞	Infinity	σ	Electrical conductivity (sm^{-1})
$\varphi_a, \varphi_b, \varphi_c$	Volume fractions	R	Residual function
Ψ_a	Shape function	Q_w	Heat flux
m	Consistency coefficient	τ_{xxy}	Shear stress
F	Dimensionless velocity field	hnf	Hybrid nanofluid
nf	Nanofluid	$tehnf$	Tri-hybrid nanoparticles
f	Fluid	$a, b, c, 1, 2, 3$	Subscripts regarding nanoparticles

References

1. Barnes, H.A.; Hutton, J.F.; Walters, K. *An Introduction to Rheology, Rheology Series 3*; Elsevier: Amsterdam, The Netherlands, 1989.
2. Eberhard, U.; Seybold, H.J.; Floriancic, M.; Bertsch, P.; Jiménez-Martínez, J.; Andrade, J.S., Jr.; Holzner, M. Determination of the effective viscosity of non-Newtonian fluids flowing through porous media. *Front. Phys.* **2019**, *7*, 71. [CrossRef]
3. Rosti, M.E.; Takagi, S. Shear-thinning and shear-thickening emulsions in shear flows. *Phys. Fluids* **2021**, *33*, 083319. [CrossRef]
4. Gul, T.; Shah, R.A.; Islam, S.; Ullah, M.; Khan, M.A.; Zaman, A.; Haq, Z. Exact Solution of Two Thin Film Non-Newtonian Immiscible Fluids on a Vertical Belt. *J. Basic Appl. Sci. Res.* **2014**, *4*, 283–288.
5. Hussain, F.; Hussain, A.; Nadeem, S. Thermophoresis and Brownian model of pseudo-plastic nanofluid flow over a vertical slender cylinder. *Math. Probl. Eng.* **2020**, *2020*, 8428762. [CrossRef]
6. Abdelsalam, S.I.; Sohail, M. Numerical approach of variable thermophysical features of dissipated viscous nanofluid comprising gyrotactic micro-organisms. *Pramana* **2020**, *94*, 1–12. [CrossRef]
7. Sohail, M.; Naz, R. Modified heat and mass transmission models in the magnetohydrodynamic flow of Sutterby nanofluid in stretching cylinder. *Phys. A Stat. Mech. Its Appl.* **2020**, *549*, 124088. [CrossRef]
8. Chu, Y.M.; Nazir, U.; Sohail, M.; Selim, M.M.; Lee, J.R. Enhancement in thermal energy and solute particles using hybrid nanoparticles by engaging activation energy and chemical reaction over a parabolic surface via finite element approach. *Fractal Fract.* **2021**, *5*, 119. [CrossRef]
9. Hina, S.; Hayat, T.; Mustafa, M.; Alsaedi, A. Peristaltic transport of pseudoplastic fluid in a curved channel with wall properties and slip conditions. *Int. J. Biomath.* **2014**, *7*, 1450015. [CrossRef]
10. Salahuddin, T.; Malik, M.Y.; Hussain, A.; Bilal, S. Combined effects of variable thermal conductivity and MHD flow on pseudoplastic fluid over a stretching cylinder by using Keller box method. *Inf. Sci. Lett.* **2016**, *5*, 2. [CrossRef]
11. Alam, M.K.; Siddiqui, A.M.; Rahim, M.T.; Islam, S.; Avital, E.J.; Williams, J.J.R. Thin film flow of magnetohydrodynamic (MHD) pseudo-plastic fluid on vertical wall. *Appl. Math. Comput.* **2014**, *245*, 544–556. [CrossRef]
12. Hayat, T.; Iqbal, R.; Tanveer, A.; Alsaedi, A. Variable viscosity effect on MHD peristaltic flow of pseudoplastic fluid in a tapered asymmetric channel. *J. Mech.* **2018**, *34*, 363–374. [CrossRef]

13. Sohail, M.; Ali, U.; Zohra, F.T.; Al-Kouz, W.; Chu, Y.M.; Thounthong, P. Utilization of updated version of heat flux model for the radiative flow of a non-Newtonian material under Joule heating: OHAM application. *Open Phys.* **2021**, *19*, 100–110. [CrossRef]

14. Sohail, M.; Raza, R. Analysis of radiative magneto nano pseudo-plastic material over three dimensional nonlinear stretched surface with passive control of mass flux and chemically responsive species. *Multidiscip. Modeling Mater. Struct.* **2020**, *16*, 1061–1083. [CrossRef]

15. Shafiq, A.; Khan, I.; Rasool, G.; Sherif, E.S.M.; Sheikh, A.H. Influence of single-and multi-wall carbon nanotubes on magnetohydrodynamic stagnation point nanofluid flow over variable thicker surface with concave and convex effects. *Mathematics* **2020**, *8*, 104. [CrossRef]

16. Manjunatha, S.; Puneeth, V.; Gireesha, B.J.; Chamkha, A. Theoretical Study of Convective Heat Transfer in Ternary Nanofluid Flowing past a Stretching Sheet. *J. Appl. Comput. Mech.* **2021**, 1–8. [CrossRef]

17. Nazir, U.; Sohail, M.; Hafeez, M.B.; Krawczuk, M. Significant Production of Thermal Energy in Partially Ionized Hyperbolic Tangent Material Based on Ternary Hybrid Nanomaterials. *Energies* **2021**, *14*, 6911. [CrossRef]

18. Chen, Z.; Yan, H.; Lyu, Q.; Niu, S.; Tang, C. Ternary hybrid nanoparticles of reduced graphene oxide/graphene-like MoS2/zirconia as lubricant additives for bismaleimide composites with improved mechanical and tribological properties. *Compos. Part A Appl. Sci. Manuf.* **2017**, *101*, 98–107. [CrossRef]

19. Zayan, M.; Rasheed, A.K.; John, A.; Muniandi, S.; Faris, A. Synthesis and Characterization of Novel Ternary Hybrid Nanoparticles as Thermal Additives in H$_2$O. *ChemRxiv* **2021**. [CrossRef]

20. Shafiq, A.; Mebarek-Oudina, F.; Sindhu, T.N.; Abidi, A. A study of dual stratification on stagnation point Walters' B nanofluid flow via radiative Riga plate: A statistical approach. *Eur. Phys. J. Plus* **2021**, *136*, 1–24. [CrossRef]

21. Swain, K.; Mebarek-Oudina, F.; Abo-Dahab, S.M. Influence of MWCNT/Fe$_3$O$_4$ hybrid nanoparticles on an exponentially porous shrinking sheet with chemical reaction and slip boundary conditions. *J. Therm. Anal. Calorim.* **2021**, *147*, 1561–1570. [CrossRef]

22. Mebarek-Oudina, F.; Fares, R.; Aissa, A.; Lewis, R.W.; Abu-Hamdeh, N.H. Entropy and convection effect on magnetized hybrid nano-liquid flow inside a trapezoidal cavity with zigzagged wall. *Int. Commun. Heat Mass Transf.* **2021**, *125*, 105279. [CrossRef]

23. Warke, A.S.; Ramesh, K.; Mebarek-Oudina, F.; Abidi, A. Numerical investigation of the stagnation point flow of radiative magnetomicropolar liquid past a heated porous stretching sheet. *J. Therm. Anal. Calorim.* **2021**, 1–12. [CrossRef]

24. Dadheech, P.K.; Agrawal, P.; Mebarek-Oudina, F.; Abu-Hamdeh, N.H.; Sharma, A. Comparative heat transfer analysis of MoS$_2$/C$_2$H$_6$O$_2$ and SiO$_2$-MoS$_2$/C$_2$H$_6$O$_2$ nanofluids with natural convection and inclined magnetic field. *J. Nanofluids* **2020**, *9*, 161–167. [CrossRef]

25. Marzougui, S.; Mebarek-Oudina, F.; Magherbi, M.; Mchirgui, A. Entropy generation and heat transport of Cu-water nanoliquid in porous lid-driven cavity through magnetic field. *Int. J. Numer. Methods Heat Fluid Flow*, 2021; *ahead-of-print*.

26. Mebarek-Oudina, F. Convective heat transfer of Titania nanofluids of different base fluids in cylindrical annulus with discrete heat source. *Heat Transf. Asian Res.* **2019**, *48*, 135–147. [CrossRef]

27. Dhif, K.; Mebarek-Oudina, F.; Chouf, S.; Vaidya, H.; Chamkha, A.J. Thermal Analysis of the Solar Collector Cum Storage System Using a Hybrid-Nanofluids. *J. Nanofluids* **2021**, *10*, 616–626. [CrossRef]

28. Zamzari, F.; Mehrez, Z.; Cafsi, A.E.; Belghith, A.; Quéré, P.L. Entropy generation and mixed convection in a horizontal channel with an open cavity. *Int. J. Exergy* **2015**, *17*, 219–239. [CrossRef]

29. Li, Y.X.; Al-Khaled, K.; Khan, S.U.; Sun, T.C.; Khan, M.I.; Malik, M.Y. Bio-convective Darcy-Forchheimer periodically accelerated flow of non-Newtonian nanofluid with Cattaneo–Christov and Prandtl effective approach. *Case Stud. Therm. Eng.* **2021**, *26*, 101102. [CrossRef]

30. Mehrez, Z.; El Cafsi, A. Heat exchange enhancement of ferrofluid flow into rectangular channel in the presence of a magnetic field. *Appl. Math. Comput.* **2021**, *391*, 125634. [CrossRef]

31. Khashi'ie, N.S.; Arifin, N.M.; Sheremet, M.; Pop, I. Shape factor effect of radiative Cu–Al$_2$O$_3$/H$_2$O hybrid nanofluid flow towards an EMHD plate. *Case Stud. Therm. Eng.* **2021**, *26*, 101199. [CrossRef]

32. Esfe, M.H.; Esfandeh, S.; Kamyab, M.H.; Toghraie, D. Analysis of rheological behavior of MWCNT-Al$_2$O$_3$ (10: 90)/5W50 hybrid non-Newtonian nanofluid with considering viscosity as a three-variable function. *J. Mol. Liq.* **2021**, *341*, 117375. [CrossRef]

33. Mehrez, Z.; El Cafsi, A. Thermodynamic Analysis of Al$_2$O$_3$–Water Nanofluid Flow in an Open Cavity Under Pulsating Inlet Condition. *Int. J. Appl. Comput. Math.* **2017**, *3*, 489–510. [CrossRef]

34. Algehyne, E.A.; El-Zahar, E.R.; Sohail, M.; Nazir, U.; AL-bonsrulah, H.A.Z.; Veeman, D.; Felemban, B.F.; Alharbi, F.M. Thermal Improvement in Pseudo-Plastic Material Using Ternary Hybrid Nanoparticles via Non-Fourier's Law over Porous Heated Surface. *Energies* **2021**, *14*, 8115. [CrossRef]

35. Sakiadis, B.C. Boundary-layer behavior on continuous solid surfaces: II. The boundary layer on a continuous flat surface. *AiChE J.* **1961**, *7*, 221–225. [CrossRef]

36. Fox, V.G.; Erickson, L.E.; Fan, L.T. Methods for solving the boundary layer equations for moving continuous flat surfaces with suction and injection. *AIChE J.* **1968**, *14*, 726–736. [CrossRef]

37. Chen, C.H. Forced convection over a continuous sheet with suction or injection moving in a flowing fluid. *Acta Mech.* **1999**, *138*, 1–11. [CrossRef]

 *micromachines*

Article

Computational Assessment of Thermal and Solute Mechanisms in Carreau–Yasuda Hybrid Nanoparticles Involving Soret and Dufour Effects over Porous Surface

Enran Hou [1,*]**, Fuzhang Wang** [2,3]**, Essam Roshdy El-Zahar** [4,5]**, Umar Nazir** [6] **and Muhammad Sohail** [6,*]

[1] College of Mathematics, Huaibei Normal University, Huaibei 235000, China
[2] Nanchang Institute of Technology, Nanchang 330044, China; wangfuzhang1984@163.com
[3] School of Mathematical and Statistics, Xuzhou University of Technology, Xuzhou 221018, China
[4] Department of Mathematics, College of Science and Humanities in Al-Kharj, Prince Sattam Bin Abdulaziz University, P.O. Box 83, Al-Kharj 11942, Saudi Arabia; er.elzahar@psau.edu.sa
[5] Department of Basic Engineering Science, Faculty of Engineering, Menoufia University, Shebin El-Kom 32511, Egypt
[6] Department of Applied Mathematics and Statistics, Institute of Space Technology, P.O. Box 2750, Islamabad 44000, Pakistan; nazir_u2563@yahoo.com
*   Correspondence: houenran@163.com (E.H.); muhammad_sohail111@yahoo.com (M.S.)

Abstract: Engineers, scientists and mathematicians are greatly concerned about the thermal stability/instability of any physical system. Current contemplation discusses the role of the Soret and Dufour effects in hydro-magnetized Carreau–Yasuda liquid passed over a permeable stretched surface. Several important effects were considered while modelling the thermal transport, including Joule heating, viscous dissipation, and heat generation/absorption. Mass transportation is presented in the presence of a chemical reaction. Different nanoparticle types were mixed in the Carreau–Yasuda liquid in order to study thermal performance. Initially, governing laws were modelled in the form of PDEs. Suitable transformation was engaged for conversion into ODEs and then the resulting ODEs were handled via FEM (Finite Element Method). Grid independent analysis was performed to determine the effectiveness of the chosen methodology. Several important physical effects were explored by augmenting the values of the influential parameters. Heat and mass transfer rates were computed against different parameters and discussed in detail.

Keywords: viscous dissipation; chemical reaction; finite element procedure; hybrid nanoparticles; heat and mass transfer rates; joule heating

Citation: Hou, E.; Wang, F.; El-Zahar, E.R.; Nazir, U.; Sohail, M. Computational Assessment of Thermal and Solute Mechanisms in Carreau–Yasuda Hybrid Nanoparticles Involving Soret and Dufour Effects over Porous Surface. *Micromachines* **2021**, *12*, 1302. https://doi.org/10.3390/mi12111302

Academic Editor: Jinyuan Qian

Received: 29 July 2021
Accepted: 21 October 2021
Published: 23 October 2021

Publisher's Note: MDPI stays neutral with regard to jurisdictional claims in published maps and institutional affiliations.

1. Introduction

The mechanism of transport phenomenon in different materials has received reasonable attention recently due to its wider applications in industry and different medical processes. Several important materials exist for the support of these mechanisms. Due to their different characteristics, these materials cannot be explained through one constitutive relation. Carreau–Yasuda is one such important material which has the following constitute relation.

For $Y = 0$ or $n = 1$, the Newtonian model is recovered. This model predicts the relation of shear stress with frequency. Several important contributions have been made by

$$\eta_{CY}(\dot{\gamma}) = \mu_\infty + (\mu_0 - \mu_\infty)\left[1 + \left(Y\dot{\gamma}\right)^d\right]^{\frac{n-1}{d}}. \tag{1}$$

considering this material. For example, Zare et al. [1] discussed this model by experimentally considering the complex viscosity relationship. In their investigation, they found an excellent settlement of frequency data. They considered the involvement of carbon nanotubes in the mixture of Carreau–Yasuda material. Kayani et el. [2] reported on the behavior of wall properties on the peristaltic flow of the Carreau–Yasuda model in a sinusoidal channel by considering the Hall effect. Governing laws for the transport of species, heat and momentum were modeled under the low-Reynolds-number assumption along with the long wavelength approach. After implementing the scaling group transformation, the transformed problem was approximated numerically via the ND-Solve tool in the MATHEMATICA 15.0 symbolic package. The authors conducted a parametric analysis and their findings were shown in several graphs. They noticed a decline in the thermal field for the Biot number and recorded an enhancement in the mounting values of the Brinkman number. The unsteady rheology of the Carreau–Yasuda material in a circular tube was examined by Rana and Murthy [3]. In their investigation, they reported the wall absorption effect. They retrieved different flow behavior cases by considering different values of material parameters. Sochi [4] presented the modelling of Cross and Carreau liquid through a circular pipe. Analytical and numerical schemes were jointly implemented for the solution of flow equations and an excellent settlement was monitored. The rheology of the Carreau–Yasuda model in a cavity at high-Reynolds-number was examined by Shamekhi and Aliabadi [5] via the mesh-free algorithm. The phenomenon of blood flow via the Carreau–Yasuda model was reported by Jahangiri [6] through the FEM package.

The involvement of nanoparticles enhanced the thermal performance and heat transportation rate. Several models of the nanomaterials are available and frequently used to study the thermal performance of different materials. Several researchers have paid attention to these materials due to their wider applications and usage. For instance, Gorla and Gireesha [7] developed the modeling of steady viscoelastic material with convective heat transport. The Buongiorno model is utilized to capture the characteristics of Brownian diffusion and thermophoresis. Modelling of heat transport is carried out by considering thermal radiation and heat generation. They solved boundary layer equations via a shooting procedure in the MATLAB symbolic package. The impact of several pertinent parameters were displayed through graphs, and tabular results were prepared to demonstrate the effectiveness and applicability of the shooting scheme for a large set of nonlinear data arising in the mechanical engineering problem. Muhammad et al. [8] modelled the squeezed flow of a viscous nanofluid with an updated mass and heat fluxes between parallel plates and handled the resulting expressions analytically via the OHAM in MATHEMATICA 15.0 computational tool. They noticed the enhancement in fluid velocity for the larger squeezing parameter. Rashid et al. [9] presented the exact solution of a water-based mixture of aligned nanoparticle materials under the radiation effect. They recorded the depreciation in heat transfer rate against the radiation and slip parameter. The double stratification phenomenon in the buoyancy-driven flow of a micro-polar viscous nanofluid was examined by Ramzan et al. [10]. They analyzed the depreciation in velocity for the buoyancy parameter and an enhancement was recorded against the micro-polar parameter. Hezma et al. [11] studied the behavior of SWCNTs in order to investigate the mechanical properties of polyvinyl chloride. Upadhay and Raju [12] examined the inclusion of dust particles in Eyring–Powell material over a stretched sheet. They studied the thermal and mass transport in dusty Eyring–Powell nanofluid by engaging the revised definitions of mass and heat fluxes. They found the numerical solutions for nonlinear modeled equations via the shooting scheme. Several important pieces of research on the transport phenomenon are reported in [13–17] and the references therein. Nazir et al. [18] studied comparison analysis among hybrid nanoparticles and nanoparticles in hyperbolic tangent liquid past a stretching surface. They adopted a finite element approach to conduct numerical results. Chu et al. [19] modeled correlations between nanoparticles and hybrid nanoparticles, considering activation energy and chemical reaction. They noticed thermal aspects past a parabolic surface using a finite element scheme. Cui et al. [20] simulated effects related to radius and roughness of inserting nanoparticles. Awais et al. [21] applied the KKL model in the transfer of energy using nanoparticles. Nazir et al. [22] discussed the numerical results of the Carreau–Yasuda liquid in heat/energy considering the hybrid nanoparticles and nanomaterials, numerically solved by the finite element approach.

In the above cited literature, no study deals with the combined behavior of the following: mass, heat transport in hydro-magnetized Carreau–Yasuda material using Joule heating, viscous dissipation, heat generation, chemical reaction and the Soret and Dufour influences in the Darcy–Forchheimer porous stretching sheet. This report fills the gap in this discussion and should be used as a foundation for researchers to work further on this model. The inclusion of nanoparticles in the Carreau–Yasuda material is attractive to researchers. Organization of this research is divided in the following way: the literature survey is reported in Section 1, modelling with important physical assumptions are covered in Section 2, Section 3 covers a detailed description of the finite element procedure with a grid independent survey, a detailed description of the solution and the influence of several emerging parameters are explained in Section 4 and important findings of the reported study are listed in Section 5. Figure 1 reveals the division of the base fluids, hybrid nanoparticles and nanoparticles. In this Figure, Ag, Cu, Al_2O_3, Ni and MoS_2 are known as nanoparticles whereas H_2O, ethylene glycol and oils are called base fluids. In this current analysis, ethylene glycol is considered as a base fluid. Mixtures of MoS_2 and SiO_2 are hybrid nanoparticles.

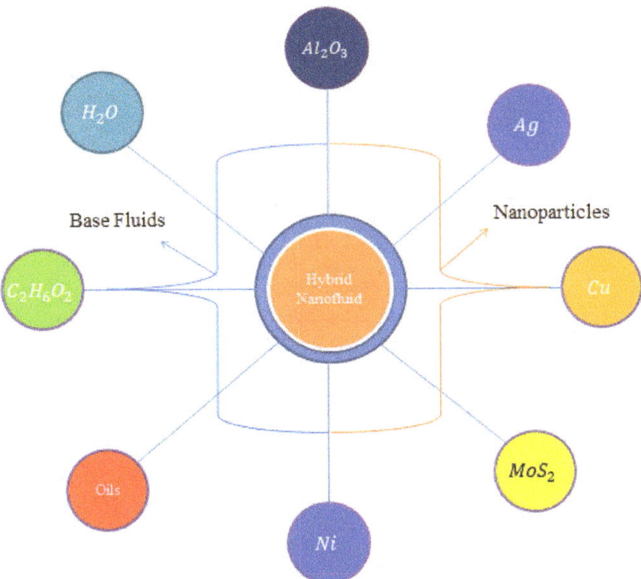

Figure 1. The sketching behavior of hybrid nanoparticles.

2. Development of the Flow Model

An enhancement in the thermal and solute performance of Carreau–Yasuda rheology, inserting the impact of nanoparticles and hybrid nanoparticles, is considered as shown in Figure 1. The flow runs towards the stretching surface under the action of a constant magnetic field. Heat takes place due to Joule heating and viscous dissipation. The Soret and Dufour influences are captured with heat generation and chemical reaction. Forchheimer's porous theory is imposed in the transport phenomenon. The geometrical flow diagram is considered in Figure 2 and the thermal properties of the nanoparticles are shown in Figure 3.

The non-linear PDEs are developed according to the physical happenings and the boundary layer approximations.

$$\frac{\partial \tilde{u}}{\partial x} + \frac{\partial \tilde{v}}{\partial y} = 0, \tag{2}$$

$$\tilde{u}\frac{\partial \tilde{u}}{\partial x} + \tilde{v}\frac{\partial \tilde{u}}{\partial y} + \frac{v_{hnf}}{k^a}F_s\tilde{u} + \frac{F_s}{(k^a)^{1/2}}(\tilde{u})^2 = v_{hnf}\left[\frac{\partial^2 \tilde{u}}{\partial y^2} + (\Lambda)^d\left(\frac{m-1}{d}\right)(d+1)\frac{\partial^2 \tilde{u}}{\partial y^2}\left(\frac{\partial \tilde{u}}{\partial y}\right)^d\right] \tag{3}$$

$$-\frac{B_0^2\sigma_{hnf}}{\rho_{hnf}}\tilde{u}\sin^2\alpha, \tag{4}$$

$$\tilde{u}\frac{\partial \tilde{T}}{\partial x} + \tilde{v}\frac{\partial \tilde{T}}{\partial y} - \frac{Q}{(\rho C_p)_{hnf}}\left(\tilde{T} - T_\infty\right) - \frac{D_{hnf}k_T}{(C_p)_f C_s}\frac{\partial^2 \tilde{C}}{\partial y^2} = \frac{k_{hnf}}{(\rho C_p)_{hnf}}\frac{\partial^2 \tilde{T}}{\partial y^2} + \frac{B_0^2\sigma_{hnf}}{(\rho C_p)_{hnf}}\sin^2\alpha(\tilde{u})^2 \tag{5}$$

$$+\frac{\mu_{hnf}}{(\rho C_p)_{hnf}}\left[(\Lambda)^d\left(\frac{m-1}{d}\right)\left(\frac{\partial \tilde{u}}{\partial y}\right)^d\right]\left(\frac{\partial \tilde{u}}{\partial y}\right)^2, \tag{6}$$

$$\tilde{u}\frac{\partial \tilde{C}}{\partial x} + \tilde{v}\frac{\partial \tilde{C}}{\partial y} = D_{hnf}\frac{\partial^2 \tilde{C}}{\partial y^2} - k_0\left(\tilde{C} - C_\infty\right) + \frac{D_{hnf}k_T}{T_m}\frac{\partial^2 \tilde{T}}{\partial y^2}, \tag{7}$$

where $(\tilde{u}, \tilde{v})$ is velocity componets, space coordinates are (x, y), kinamtic viscosity is v, F_s is inertia cofficient (porous medium), permeability (porou medium) is k^a, power law index number is m, time constant is Λ, magnetic induction is B_0, electrical conductivity is σ, temprature is $\tilde{T}$, $\tilde{C}$ is the concentration, heat source is Q, T_∞ is ambient temprature, fluid density is ρ, $\tilde{C}$ is concentarion, mass diffusion is D, T_m is fluid mean temprature, k_T is thermal diffusion, C_s is concentration susceptibility and k_0 is chemical reaction number.

The no-slip theory provides the required boundary conditions of the current model:

$$\tilde{u} = ax, \; \tilde{v} = 0, \; \tilde{C} = C_w, \; \tilde{T} = T_w : y = 0,$$

$$\tilde{u} \to 0, \; \tilde{C} \to C_\infty, \; \tilde{T} \to T_\infty : y \to \infty. \tag{8}$$

Change in variables is constructed as:

$$\tilde{u} = axF', \; \tilde{v} = -\left(av_f\right)^{\frac{1}{2}}F, \; \zeta = \left(\frac{a}{v_f}\right)^{\frac{1}{2}}y, \tag{9}$$

$$\theta(T_w - T_\infty) = \tilde{T} - T_\infty, \; (C_w - C_\infty)\phi = \tilde{C} - C_\infty. \tag{10}$$

Transformations are used in Equations (1)–(5) and the system of non-linear PDEs are converted to following ODEs

$$\left.\begin{array}{c} F_{\zeta\zeta\zeta} + (We)^d\frac{(m-1)(d+1)}{d}F_{\zeta\zeta\zeta}\left(F_{\zeta\zeta}\right)^d + A_1\left(FF_{\zeta\zeta} - F_\zeta^2\right) - \epsilon F_\zeta - A_1 F_r\left(F_\zeta\right)^2 \\ -A_2 M^2 \sin^2\alpha F_\zeta = 0, \\ F(0) = 0, \; F_\zeta(0) = 1, F_\zeta(\infty) = 1, \end{array}\right\}, \tag{11}$$

$$\left.\begin{array}{c} \theta_{\zeta\zeta} + A_3 Pr F\theta_\zeta + A_4 H_s\theta + A_4 M^2 Ec\sin^2\alpha\left(F_\zeta\right)^2 + A_5 PrEC\left[1 + (We)^d\frac{(m-1)}{d}\left(F_{\zeta\zeta}\right)^d\right]\left(F_{\zeta\zeta}\right)^2 \\ +A_4 A_6 PrD_f\phi_{\zeta\zeta} = 0, \\ \theta(0) = 1, \; \theta(\infty) = 0, \end{array}\right\}, \tag{12}$$

$$\left.\begin{array}{c} \varphi_{\zeta\zeta} + \frac{Sc}{(1-\phi_2)^{2.5}(1-\phi_1)^{2.5}}F\varphi_\zeta - \frac{K_c Sc}{(1-\phi_2)^{2.5}(1-\phi_1)^{2.5}}\varphi + ScSr\theta_{\zeta\zeta} = 0, \\ \varphi(0) = 1, \; \varphi(\infty) = 0, \end{array}\right\}. \tag{13}$$

Here, A_1, A_2, A_3, A_4, A_5 and H_1 are involved parameters (representing the correlation of nanoparticles and hybrid nanostructures) in the above equations which are defined as

$$A_1 = (1-\phi_1)^{2.5}(1-\phi_2)^{2.5}\left[(1-\phi_2)\left\{(1-\phi_1) + \phi_1\frac{\rho_{s_1}}{\rho_f}\right\}\right] + \phi_2\frac{\rho_{s_2}}{\rho_f}, \tag{14}$$

$$A_2 = (1 - \phi_1)^{2.5}(1 - \phi_2)^{2.5}, \; A_4 = \frac{k_f}{k_{hnf}}, \; A_5 = \frac{k_f}{(1 - \phi_1)^{2.5}(1 - \phi_2)^{2.5}k_{hnf}}, \tag{15}$$

$$A_3 = A_4 \left[(1 - \phi_2) \left\{ (1 - \phi_1) + \phi_1 \frac{(\rho C_p)_{s_1}}{(\rho C_p)_f} \right\} \right] + \phi_1 \frac{(\rho C_p)_{s_2}}{(\rho C_p)_f}, \tag{16}$$

$$A_6 = (1 - \phi_1)^{2.5}(1 - \phi_2)^{2.5} \left[(1 - \phi_2) \left\{ (1 - \phi_1) + \phi_1 \frac{(\rho C_p)_{s_1}}{(\rho C_p)_f} \right\} \right] + \phi_1 \frac{(\rho C_p)_{s_2}}{(\rho C_p)_f}, \tag{17}$$

$$\frac{k_{hnf}}{k_{bf}} = \left\{ \frac{k_{s_2} + (n-1)k_{bf} - (n-1)\phi_2\left(k_{bf} - k_{s_2}\right)}{k_{s_2} + (n-1)k_{bf} - \phi_2\left(k_{bf} - k_{s_2}\right)} \right\}. \tag{18}$$

Figure 3 demonstrates the thermal properties of density, electrical conductivity, thermal conductivity and specific heat capacitance for ethylene glycol, MoS_2/SiO_2 and MoS_2. Density of $C_2H_6O_2$ is 1113.5, density of MoS_2 is 2650, density of MoS_2/SiO_2 is 5060, thermal conductivity of $C_2H_6O_2$ is 0.253, thermal conductivity of MoS_2 is 1.5, thermal conductivity of MoS_2/SiO_2 is 34.5, electrical conductivity of MoS_2 is 0.0005, electrical conductivity of $C_2H_6O_2$ is 4.3×10^{-5}, electrical conductivity of MoS_2/SiO_2 is 1×10^{-18}, C_p of $C_2H_6O_2$ is 2430, C_p of MoS_2 is 730 and C_p of MoS_2/SiO_2 is 397.746, respectively. Here, the Weissenberg number is $We = \left(\frac{\Lambda x a^{3/2}}{(\nu_f)^{1/2}} \right)$, the magnetic number is $M^2\left(= \frac{B_0^2 \sigma_f}{a \rho_f} \right)$, the porosity number is $\epsilon\left(= \frac{F_s \nu_f}{a} \right)$, the Forchheimer number is $F_r = \left(\frac{x F_s}{\sqrt{k^*}} \right)$, the Prandtl number is $Pr\left(= \frac{\mu_f (c_p)_f}{k_f} \right)$, the Eckert number is $Ec\left(= \frac{(U_w)^2}{(T_w - T_\infty)(c_p)_f} \right)$, the heat generation number is $H_s\left(= \frac{Q}{a \rho_f (c_p)_f} \right)$, the Schmidt number is $Sc\left(= \frac{\nu_f}{D_f} \right)$, the chemical reaction number is $K_c\left(\frac{k_c}{a} \right)$, the Dufour number is $D_f\left(= \frac{(C_w - C_\infty)D_f k_t}{C_a \nu_f (c_p)_f} \right)$ and the Soret number is $Sr\left(= \frac{D_f (T_w - T_\infty)k_t}{\nu_f (C_w - C_\infty)T_m} \right)$. The surface force in attendance of the Carreau–Yasuda liquid at the wall of the melting surface is

$$(Re)^{\frac{1}{2}}C_f = \frac{-1}{(1 - \phi_1)^{2.5}(1 - \phi_2)^{2.5}} \left[1 + \frac{m-1}{d}\left(We F_{\xi\xi}(0) \right)^2 \right] F_{\xi\xi}(0), \tag{19}$$

The temperature gradient because of the nano and hybrid nanoparticles is

$$Nu = \frac{x Q_w}{k_f(T - T_\infty)}, \; Q_w = -k_{hnf}\frac{\partial T}{\partial y}, \tag{20}$$

$$(Re)^{-1/2}Nu = \frac{-k_{hnf}}{k_f}\theta_\xi(0). \tag{21}$$

Concentration gradient at the surface of the melting sheet is

$$Sh = \frac{x l_m}{\left(\widetilde{C_w} - \widetilde{C_\infty} \right)D_{hnf}}, \; l_m = -D_{hnf}\frac{\partial \widetilde{C}}{\partial y}\Big|_{y=0}, \; (Re)^{-\frac{1}{2}}Sh = -\frac{\varphi'(0)}{(1 - \phi_1)^{2.5}(1 - \phi_2)^{2.5}}, \tag{22}$$

The local Reynolds number is $Re = \frac{a x^2}{\nu_f}$.

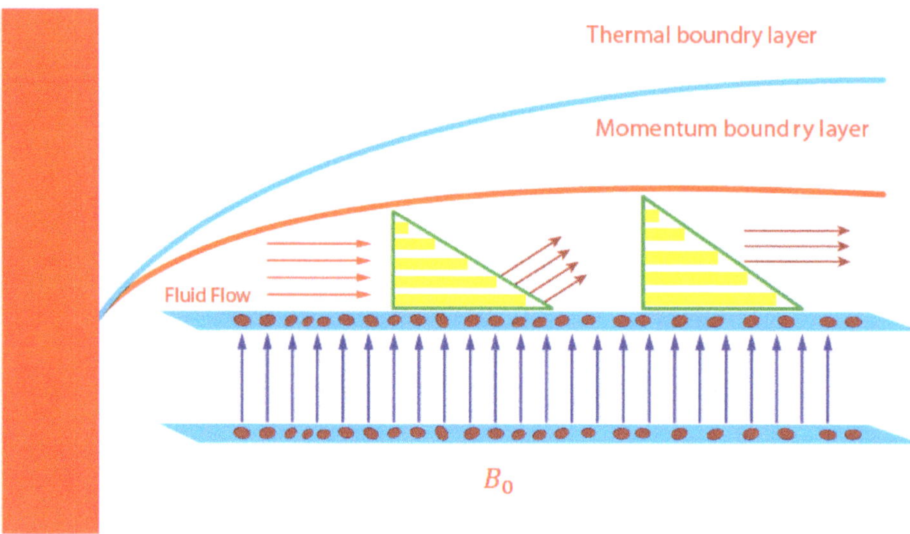

Figure 2. Geometry of transport phenomenon in Carreau–Yasuda fluid.

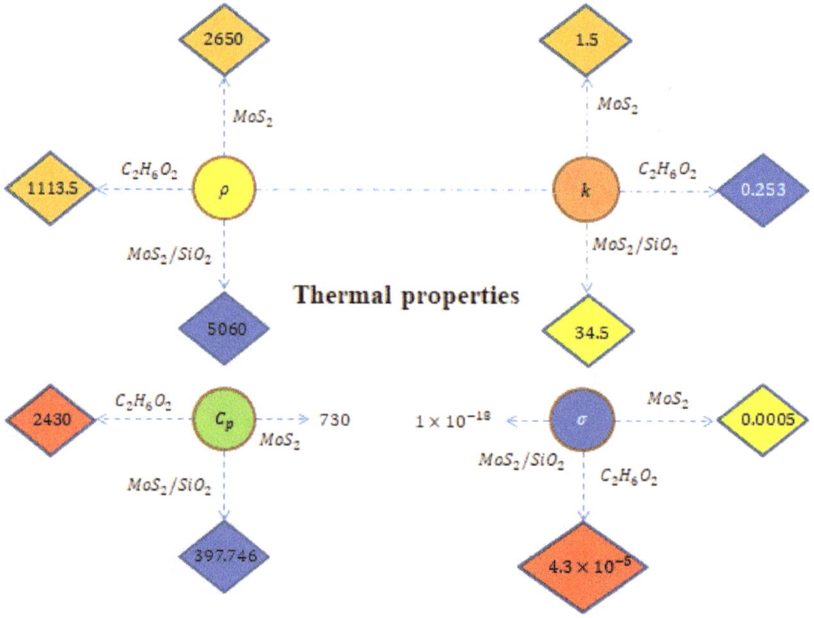

Figure 3. Thermal properties of nanoparticles and hybrid nanoparticles in base liquid.

3. Numerical Approach and Convergence Analysis

The finite element method is an effective method in the view of accuracy and the convergence of a problem compared with other numerical approaches. There are many advantages to FEM but some are discussed here:

➢ FEM has the ability to handle various complex geometries;

> This numerical method is thought to be most significant in solving physical problems with wide ranges;
> FEM requires less investment in the view of time and resources;
> A main advantage of FEM is its handling of various types of boundary conditions and
> It has a good ability with regards to the discretization (of derivative) problems into small elements;
> The Working scheme of finite element method has been shown with the help of Figure 4.

The numerical approach called finite element scheme is used to simulate the numerical results of highly non-linear PDEs and numerous applications of FEM are found in CFD (computational fluid dynamics) problems. The FEM approach is explained in the following steps:

Step I: The division of a problem domain into a finite number of elements and residuals. The weak form is captured from the strong form due to residuals. The approximation result is simulated using shape functions and the approximation simulations of the variables are:

$$H = \sum_{l=1}^{2}(H_1\omega_j), \ F = \sum_{l=1}^{2}(F_1\omega_j), \ \theta = \sum_{l=1}^{2}(\theta_1\omega_j), \ \phi = \sum_{l=1}^{2}(\phi_1\omega_j). \tag{23}$$

Here F_ζ is H and the shape function is defined as:

$$\omega_j = (-1)^{l-1}\frac{1 - \frac{\zeta}{\zeta_{l-1}}}{1 - \frac{\zeta_l}{\zeta_{l-1}}}, \ l = 1, 2. \tag{24}$$

Step II: In this step, the matrices are stiffness, vector and boundary (integral vector). The global stiffness (matrix) is obtained whereas the Picard (linearization approach) is utilized to obtain a linear system of equations that are defined as:

$$\overline{H} = \sum_{l=1}^{2}\omega\overline{H}_l, \ \overline{F} = \sum_{l=1}^{2}\omega\overline{F}_l. \tag{25}$$

Here $\overline{F}_l$ and $\overline{H}_l$ are variables (nodal values).

Step III: The algebraic equations (non-linear) resulting from the assembly process are:

$$Mat(F, \ H, \ \theta, \phi)\begin{pmatrix} F \\ \theta \\ \phi \end{pmatrix} = (F), \tag{26}$$

where (Mat) is the global stiffness matrix, (F) is the force vector and the nodal values (variables) are $\begin{pmatrix} F \\ \theta \\ \phi \end{pmatrix}$. The Equation (18) related to the residual form is

$$(R) = \left[M\left(F^{(r-1)}, \ H^{(r-1)}, \theta^{(r-1)}, \varphi^{(r-1)}\right)\right]\begin{bmatrix} F^r \\ \theta^r \\ \varphi^r \end{bmatrix} = [F], \tag{27}$$

$$\frac{\left(\sum_{l=1}^{N}\left(T^r - T^{r-1}\right)\right)^{1/2}}{\left(\sum_{l=1}^{N}|T^r|^2\right)^{1/2}} < \frac{1}{10^8}. \tag{28}$$

Step IV: The computational domain is considered as $[0, \ 8]$ while mesh-free analysis is computed along with 270 elements. The problem is converged at mid of each of the

270 elements. Hence, simulations of the problem are performed along with the 270 elements. Table 1 reveals the study of convergence analysis. The solution to the problem is converged after simulations of 210 to 270 elements. It is observed that 270 elements are ensured for the convergence of the problem. Outcomes are provided for velocity, concentration and temperature at mid of each of the 270 elements. All numerical simulations related to tables and graphs are captured for the 270 elements.

Comparative analysis: The numerical result of the current problem is verified with published results [23] by the disappearing effects of $We = \epsilon = F_r = H_s = Ec = M = D_f = \varphi_1$ and $\varphi_2 = 0$. Numerical values of the Nusselt number are computed against the distribution in Prandtl number. Good agreements among the results of the present problem and published work [23] are presented in Table 2.

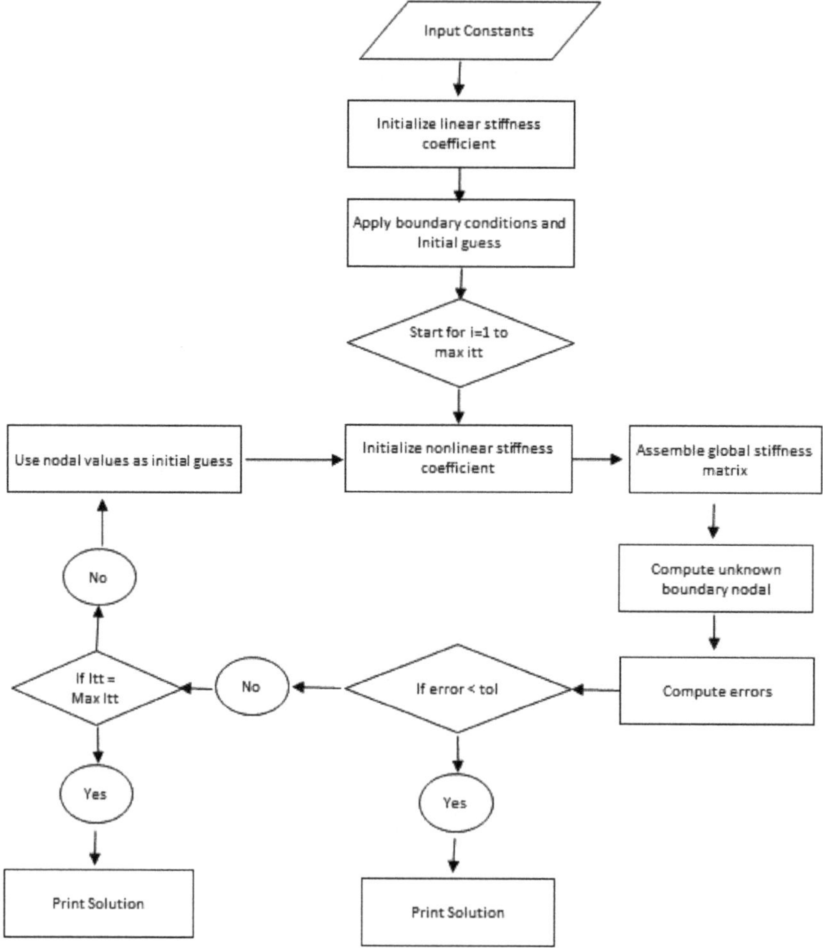

Figure 4. Working scheme of finite element method.

Table 1. Mesh-free analysis of velocity, mass diffusion and thermal energy considering 270 elements.

Number of Elements	$F_\zeta(\frac{\zeta_\infty}{2})$	$\theta(\frac{\zeta_\infty}{2})$	$\varphi(\frac{\zeta_\infty}{2})$
30	0.001723308133	0.3009272684	0.5332200981
60	0.001476057680	0.2862125651	0.5165526300
90	0.001389387659	0.2814243572	0.5109967127
120	0.001345741043	0.2790514389	0.5082194520
150	0.001319507594	0.2776354377	0.5065521700
180	0.001302011520	0.2766938280	0.5054410754
210	0.001289514516	0.2760222981	0.5046476756
240	0.001280142826	0.2755193140	0.5040526977
270	0.001272856820	0.2751285026	0.5035909496

Table 2. Comparative simulations of Nusselt number considering by $We = \epsilon = F_r = H_s = Ec = M = D_f = \varphi_1 = \varphi_2 = 0..$

	Bilal et al. [23]	Present Results
Pr	$(Re)^{-1/2}Nu$	$(Re)^{-1/2}Nu$
0.07	0.0663	0.0662110383
0.20	0.1619	0.1619120330
0.70	0.4539	0.4529370132
2.00	0.9113	0.9112098201

4. Results and Discussion

Mechanisms of velocity, thermal energy and diffusion of mass influenced by chemical reaction are addressed over a stretched melting surface. Correlations between silicon dioxide and Molybdenum dioxide in EG (ethylene glycol) are used in the presence of the Carreau–Yasuda liquid. Various kinds of influences (Soret, Dufour, viscous dissipation, Joule heating and magnetic field) are also addressed. As such, the complex transport phenomenon is simulated with the help of a numerical approach (FEM). The graphical computational investigations are captured in graphs and tables. The detailed outcomes are discussed below:

Graphical investigations of velocity against distribution in various parameters: The change in Weissenberg, power law index, Forchheimer numbers and Carreau–Yasuda variables are addressed in the motion of fluid particles considered in Figures 5–8. Figure 5 is plotted to measure the role of We in the motion of hybrid nanoparticles. It is estimated that the motion of hybrid and fluid - nanoparticles is slowed down by applying higher We Values. The Weissenberg number is constructed in the current model due to the consideration of the rheology of the Carreau–Yasuda fluid while We is defined as a ration of elastic and viscous forces. An increase in We brings the declination in motion of fluid particles in the presence of nanoparticles and hybrid nanoparticles. Hence, a reduction is noticed versus the change in We. Moreover, the thickness of the momentum boundary layers decline when We is increased. The flow for Newtonian fluid is the dominated flow for a case of non-Newtonian fluid. The relationship between the velocity and power law index number is shown in Figure 6. The decreasing phenomenon of motion in fluid particles is captured and m is created due to tensor of the Carreau–Yasuda liquid. The numerical values of m are decided by the category of fluids (shear thinning or shear thickening). The fluid becomes thick in the case of large m values. Hence, the power law index number is not a significant parameter in the case of an enhancement in flow involvement of nanoparticles and hybrid nanoparticles. Parameter related to the power law number has a significant impact on adjusting the momentum boundary layer thickness. The role of F_r is noticeable in the flow of nanoparticles and hybrid nanoparticles (see Figure 7). It is demonstrated that the parameter related to F_r occurs in the momentum equation because of the Forchheimer porous. This kind of parameter behaves like a non-linear-type function in the flow of nanoparticles. In this case, the retardation force is created in fluid motion and brings resistance of the fluid particles into motion. Moreover, the thickness related to the boundary layer is reduced when large values of F_r are applied. Further, the motion created by the Forchheimer porous is less than the motion created in the particles, excepting the involvement of Forchheimer porous media. The parameter associated with d is called the fluid variable and the change in d versus the velocity is captured in Figure 8. The large values of d create the resistance force during the flow of hybrid nanoparticles and nanoparticles. Meanwhile, the motion of fluid particles declines against the higher values of d. Momentum boundary layers have a decreasing function versus the impact of d.

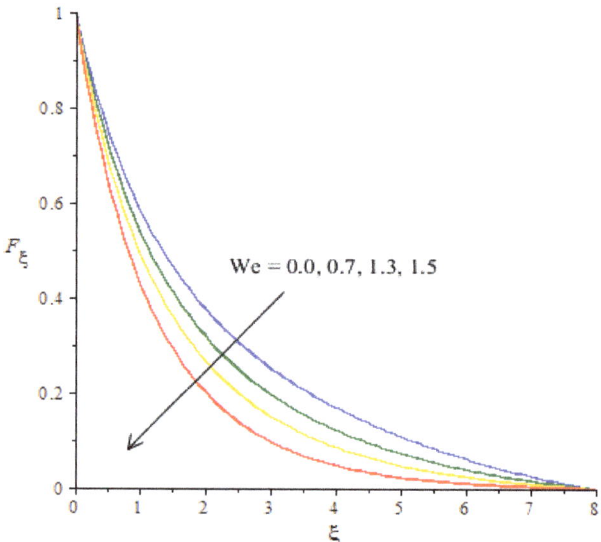

Figure 5. Change in velocity against *We*.

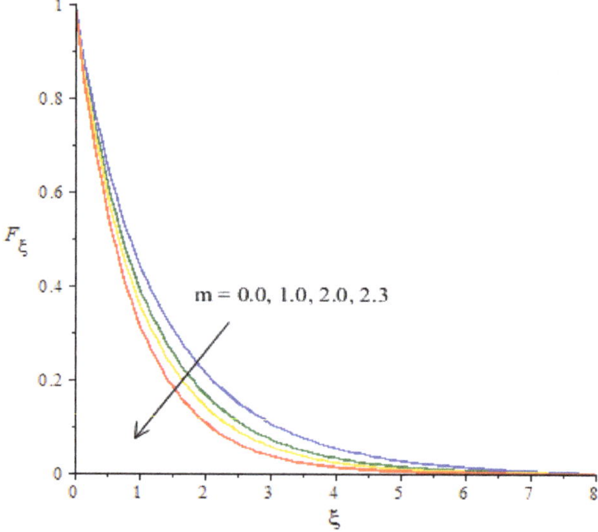

Figure 6. Change in velocity against *m*.

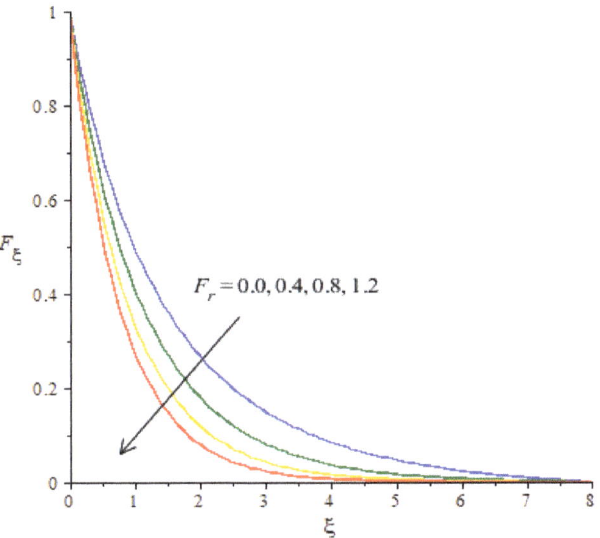

Figure 7. Change in velocity against F_r.

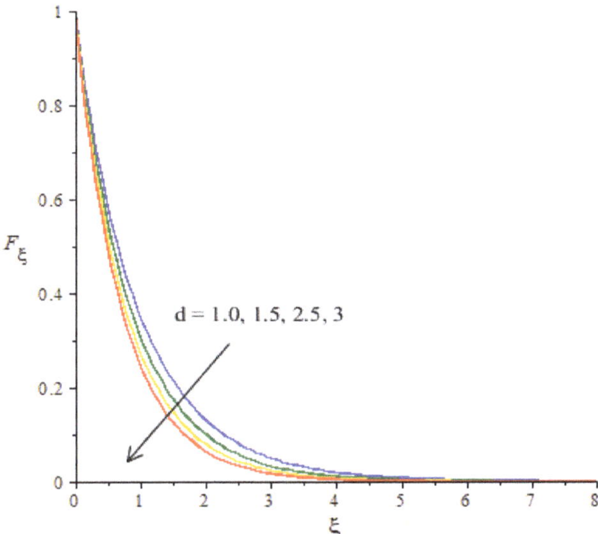

Figure 8. Change in velocity against d.

Graphical investigations of heat energy against distribution in various parameters: Figures 9–12 reveal the characterization of thermal energy phenomenon versus the variation of H_s, F_r, Ec and D_f. The thermal energy performance is measured with respect to the variation in heat generation number while this phenomena is shown in Figure 9. The production of heat energy is at its maximum when using higher values of H_s. The external heat source at the sheet surface results in the maximum production into heat energy. to Thickness related to thermal layers is also enhanced due to the large values of heat generation number. Thus, the heat generation number is a significant parameter for the maximum production of heat energy. Moreover, an inclination in thermal energy is created due to the direct relation to thermal energy. It is demonstrated that the positive values for H_s are present due to the phenomena of heat generation. As such, the impact of heat generation is visualized in our analysis. Figure 10 plots the enhancement in heat energy against the change in Eckert number. Physically, an enhancement in heat energy due to viscous dissipation is simulated. A direct relationship between viscous dissipation and the kinetic energy phenomenon was found. The temperature of the fluid particles is enhanced due to the Eckert number. The work-done rate is enhanced by the particle heat energy when viscous dissipation occurs. The relationship between F_r and temperature profile is visualized in Figure 11. This figure captures the better performance in heat energy, including the appearance of F_r. In fact, F_r is generated due to the Forchheimer porous while F_r generates more heat energy in the presence of nanoparticles and hybrid nanoparticles. Moreover, the concept of Dufour's number is characterized in the dimensionless heat equation due to the first law of thermodynamics while terms related to thermodynamics refer to the concept of Dufour (heat energy) because of the concentration gradient. The concentration gradient is enhanced using large values of D_f. The fluid particles absorb more heat energy when D_f is increased. Hence, the temperature profile is increased when there are higher values of D_f (see Figure 12). Further, thickness of the thermal boundary layers are controlled by the impacts of F_r and D_f.

Graphical investigations of mass diffusion against distribution in various parameters: Figures 13–15 have been plotted to visualize the transport of diffusion against the change in Sc, K_c and Sr. The measurement of mass diffusion is captured in Figure 13, considering the influence of Sc. The diffusion of mass decreases when Sc is enhanced. We can observe that this reduction into mass diffusion happens due to the definition of Sc. Physically, Sc is rationed among mass and momentum diffusivities. According to the concept of Sc, the concentration curves are reduced when Sc is inclined. A similar situation occurs in terms of the boundary layer thickness in relation to concentration. Figure 14 visualizes the effect of the chemical reaction number on the transport of mass diffusion We found that the parameter related to K_c revealed the coefficient of thermal energy along with the chemical reaction. The positive values of K_c correspond to the destructive chemical reaction. In this case, a reduction is captured in the diffusion of the mass species. In the current flow model, the case related to a destructive chemical reaction is used. Thickness associated with the concentration layers is reduced when K_c is increased. The decreasing graph is plotted between the Soret number and diffusion of mass as shown in Figure 15. The concept of Sr (fractioned between the difference in temperature and concentration) appears due to the temperature gradient in the concentration equation. Using Soret's theory, the solute diffusion is enhanced due to thermal energy.

Mechanisms of gradient temperature, surface force and mass diffusion rate versus the distribution of various parameters: The computational analysis of surface force (skin friction coefficient), gradient temperature (Nusselt number) and rate of mass diffusion versus the variation of D_f, Sc, H_s, F_r and We is simulated in Table 3. Table 3 reveals that the drag force (skin friction coefficient) declines when We and F_r are increased. However, we see a reduction in the skin friction coefficient when the heat generation number is increased. When D_f is enhanced, the constant variation is simulated in surface force at the wall. The temperature gradient decreases using the large values of D_f, H_s, F_r and We. The parameter related to Sc plays a vital impact in maximizing the rate of thermal energy. As for the concentration gradient, it shows the same behavior as the temperature gradient.

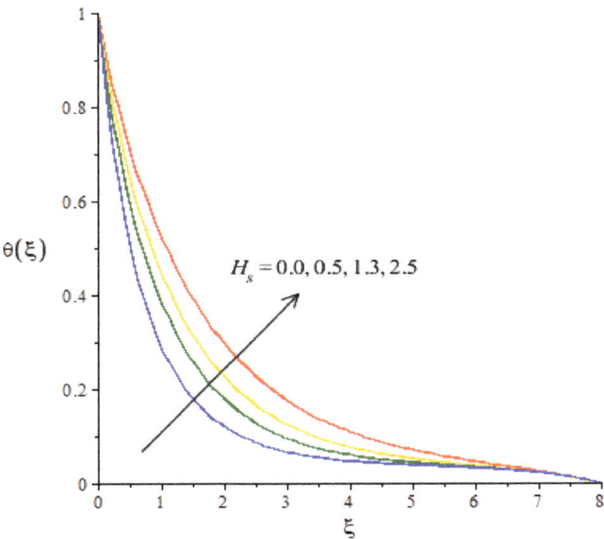

Figure 9. Change in thermal energy against H_s.

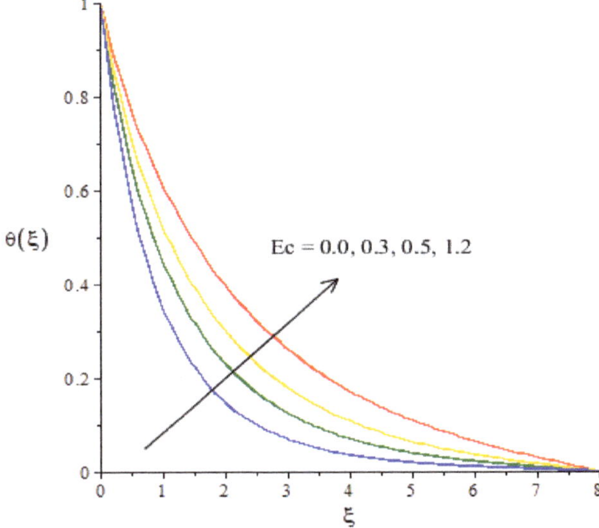

Figure 10. Change in thermal energy against Ec.

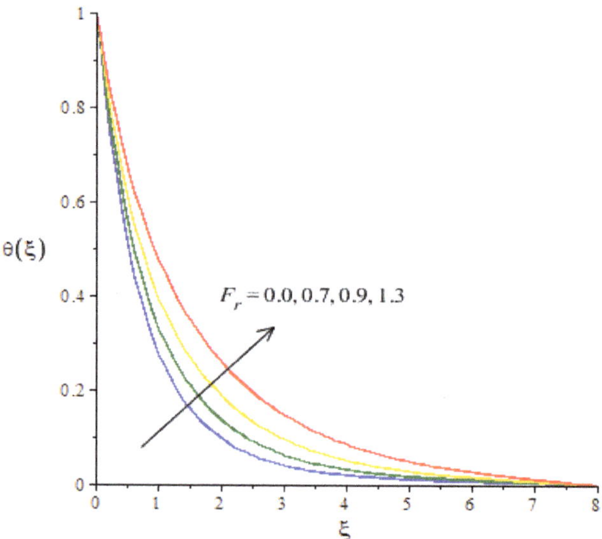

Figure 11. Change in thermal energy against F_r.

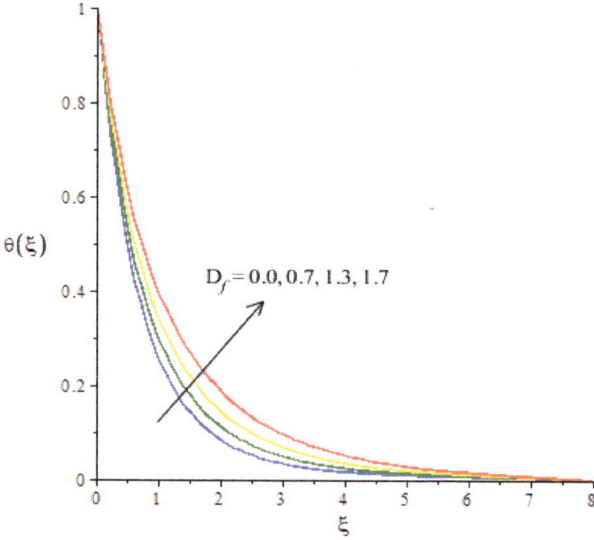

Figure 12. Change in thermal energy against D_f.

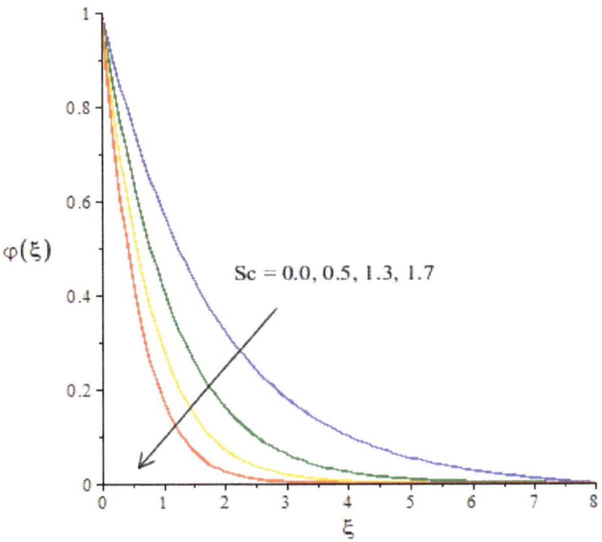

Figure 13. Change in concentration against *Sc*.

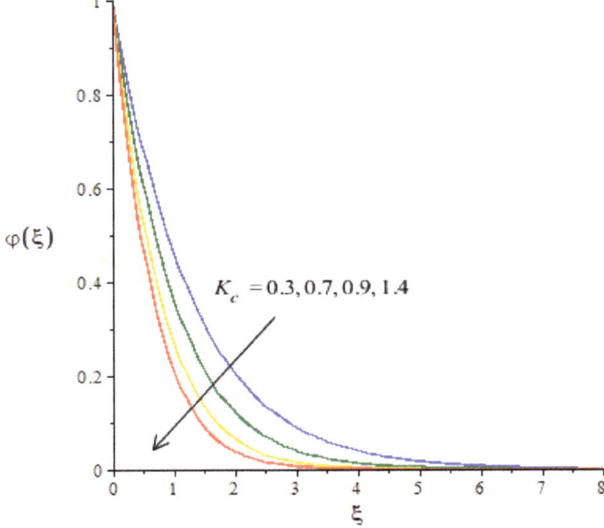

Figure 14. Change in concentration against K_c.

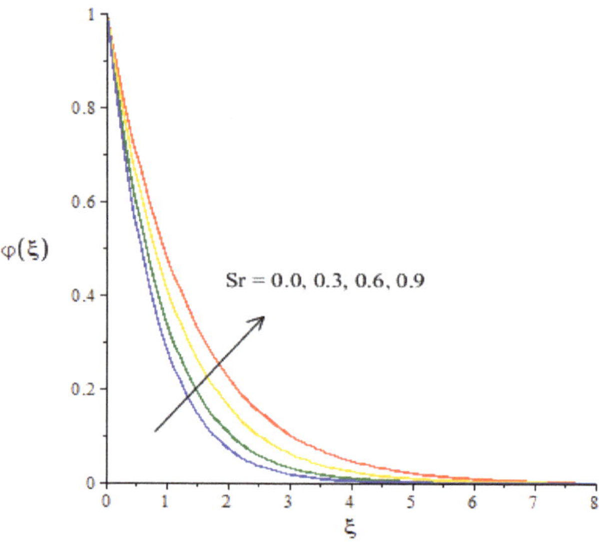

Figure 15. Change in concentration against *Sr*.

Table 3. Computational analysis of surface force, gradient temperature and rate of mass diffusion versus the variation of D_f, Sc, H_s, F_r and We in view of hybrid nanoparticles and nanoparticles.

Parameters		$(Re)^{\frac{1}{2}}C_f$	$(Re)^{-1/2}Nu$	$(Re)^{-\frac{1}{2}}Sh$
	0.0	0.2565506571	0.9069968916	0.4191080595
We	0.4	0.3021743934	0.5709115739	0.3046713057
	0.8	0.6496811069	0.4087628063	0.1059290192
	0.3	0.3500005761	2.800587924	0.3350535080
F_r	0.5	0.5935103805	2.331989057	0.2339179505
	1.2	0.7690074135	1.272381650	0.11317865862
	0.0	0.5541216982	0.4657146871	0.99748124590
H_s	0.7	0.3541216982	0.3493805142	0.89748124590
	1.3	0.1541216982	0.1173513733	0.69748124590
	0.2	0.5541216982	0.1186981159	0.04850074987
Sc	0.4	0.5541216982	0.1186575085	0.36213534820
	0.8	0.5541216982	0.1181890841	0.57775563972
	0.3	0.5541216982	0.7182669750	0.37626972734
D_f	0.6	0.5541216982	0.5183838165	0.27404085874
	1.3	0.5541216982	0.3177686869	0.18920433118

5. Prime Consequences of the Problem

The transport features in the rheology of Carreau–Yasuda liquid and involvement of nanostructures and hybrid nanoparticles over a heated surface have been visualized. The Dufour and Soret effects under the action of a magnetic field have been addressed. Forchheimer porous media was also considered. The simulations of the current model were computed by finite element approach. The prime findings are captured below:

- Convergence of the problem is ensured at 270 elements;
- The motion of nanoparticles and hybrid nanoparticles in ethylene glycol is boosted versus the enhancement in fluid variable, power law index number, Weissenberg number and Forchheimer porous number;
- Significant production of heat energy versus higher values of heat generation, Eckert, Dufour and Forchheimer porous numbers;

- The transportation of solute particles declines versus the large values of Schmidt and chemical reaction numbers, but solute particles accelerate against higher values of Soret number;
- Surface force is increased via large values of Weissenberg and Forchheimer porous numbers but surface force is decreased versus the large values of heat generation number and

The role of the Schmidt number is significant in the development of temperature and concentration gradient.

Author Contributions: Conceptualization, U.N. and M.S.; methodology, U.N.; software, E.H.; validation, F.W., E.R.E.-Z. and M.S.; formal analysis, E.R.E.-Z.; investigation, E.H.; resources, F.W.; data curation, M.S.; writing—original draft preparation, M.S.; writing—review and editing, U.N.; visualization, E.R.E.-Z.; supervision, M.S.; project administration, M.S.; funding acquisition, E.H.; F.W. and E.R.E.-Z. All authors have read and agreed to the published version of the manuscript.

Funding: This work was supported by the University Natural Science Research Project of Anhui Province (Project nos. KJ2020B06 and KJ2020ZD008).

Data Availability Statement: The data used to support this study are included in the Manuscript.

Conflicts of Interest: Authors declare that they have no conflict of interest.

References

1. Zare, Y.; Park, S.P.; Rhee, K.Y. Analysis of complex viscosity and shear thinning behavior in poly (lactic acid)/poly (ethylene oxide)/carbon nanotubes biosensor based on Carreau–Yasuda model. *Results Phys.* **2019**, *13*, 102245. [CrossRef]
2. Kayani, S.M.; Hina, S.; Mustafa, M. A new model and analysis for peristalsis of Carreau–Yasuda (CY) nanofluid subject to wall properties. *Arab. J. Sci. Eng.* **2020**, *45*, 5179–5190. [CrossRef]
3. Rana, J.; Murthy, P.V.S.N. Unsteady solute dispersion in non-Newtonian fluid flow in a tube with wall absorption. Proceedings of the Royal Society A: Mathematical. *Phys. Eng. Sci.* **2016**, *472*, 20160294.
4. Sochi, T. Analytical solutions for the flow of Carreau and Cross fluids in circular pipes and thin slits. *Rheol. Acta* **2015**, *54*, 745–756. [CrossRef]
5. Shamekhi, A.; Aliabadi, A. Non-Newtonian lid-driven cavity flow simulation by mesh free method. *ICCES* **2009**, *11*, 67.
6. Jahangiri, M.; Haghani, A.; Ghaderi, R.; Harat, S.M.H. Effect of non-Newtonian models on blood flow in artery with different consecutive stenosis. *ADMT J.* **2018**, *11*, 79–86.
7. Gorla, R.S.R.; Gireesha, B.J. Convective heat transfer in three-dimensional boundary-layer flow of viscoelastic nanofluid. *J. Thermophys. Heat Transf.* **2016**, *30*, 334–341. [CrossRef]
8. Muhammad, N.; Nadeem, S.; Mustafa, T. Squeezed flow of a nanofluid with Cattaneo–Christov heat and mass fluxes. *Results Phys.* **2017**, *7*, 862–869. [CrossRef]
9. Rashid, I.; Haq, R.U.; Al-Mdallal, Q.M. Aligned magnetic field effects on water based metallic nanoparticles over a stretching sheet with PST and thermal radiation effects. *Phys. E Low-Dimens. Syst. Nanostruct.* **2017**, *89*, 33–42. [CrossRef]
10. Ramzan, M.; Ullah, N.; Chung, J.D.; Lu, D.; Farooq, U. Buoyancy effects on the radiative magneto Micropolar nanofluid flow with double stratification, activation energy and binary chemical reaction. *Sci. Rep.* **2017**, *7*, 12901. [CrossRef]
11. Hezma, A.M.; Elashmawi, I.S.; Abdelrazek, E.M.; Rajeh, A.; Kamal, M. Enhancement of the thermal and mechanical properties of polyurethane/polyvinyl chloride blend by loading single walled carbon nanotubes. *Prog. Nat. Sci. Mater. Int.* **2017**, *27*, 338–343. [CrossRef]
12. Upadhay, M.S.; Raju, C.S.K. Cattaneo-Christov on heat and mass transfer of unsteady Eyring Powell dusty nanofluid over sheet with heat and mass flux conditions. *Inform. Med. Unlocked* **2017**, *9*, 76–85. [CrossRef]
13. Sohail, M.; Ali, U.; Al-Mdallal, Q.; Thounthong, P.; Sherif, E.S.M.; Alrabaiah, H.; Abdelmalek, Z. Theoretical and numerical investigation of entropy for the variable thermophysical characteristics of couple stress material: Applications to optimization. *Alex. Eng. J.* **2020**, *59*, 4365–4375. [CrossRef]
14. Nawaz, M.; Nazir, U. An enhancement in thermal performance of partially ionized fluid due to hybrid nano-structures exposed to magnetic field. *AIP Adv.* **2019**, *9*, 085024. [CrossRef]
15. Sohail, M.; Shah, Z.; Tassaddiq, A.; Kumam, P.; Roy, P. Entropy generation in MHD Casson fluid flow with variable heat conductance and thermal conductivity over non-linear bi-directional stretching surface. *Sci. Rep.* **2020**, *10*, 12530. [CrossRef] [PubMed]
16. Khan, R.M.; Ashraf, W.; Sohail, M.; Yao, S.W.; Al-Kouz, W. On Behavioral Response of Microstructural Slip on the Development of Magnetohydrodynamic Micropolar Boundary Layer Flow. *Complexity* **2020**, *2020*, 8885749. [CrossRef]

17. Abdelmalek, Z.; Nazir, U.; Nawaz, M.; Alebraheem, J.; Elmoasry, A. Double diffusion in Carreau liquid suspended with hybrid nanoparticles in the presence of heat generation and chemical reaction. *Int. Commun. Heat Mass Transf.* **2020**, *119*, 104932. [CrossRef]
18. Nazir, U.; Sohail, M.; Alrabaiah, H.; Selim, M.M.; Thounthong, P.; Park, C. Inclusion of hybrid nanoparticles in hyperbolic tangent material to explore thermal transportation via finite element approach engaging Cattaneo-Christov heat flux. *PLoS ONE* **2021**, *16*, e0256302. [CrossRef] [PubMed]
19. Chu, Y.M.; Nazir, U.; Sohail, M.; Selim, M.M.; Lee, J.R. Enhancement in Thermal Energy and Solute Particles Using Hybrid Nanoparticles by Engaging Activation Energy and Chemical Reaction over a Parabolic Surface via Finite Element Approach. *Fractal Fract.* **2021**, *5*, 119. [CrossRef]
20. Cui, H.; Saleem, S.; Jam, J.E.; Beni, M.H.; Hekmatifar, M.; Toghraie, D.; Sabetvand, R. Effects of roughness and radius of nanoparticles on the condensation of nanofluid structures with molecular dynamics simulation: Statistical approach. *J. Taiwan Inst. Chem. Eng.* **2021**, *128*, 346–353. [CrossRef]
21. Awais, M.; Ehsan Awan, S.; Raja, M.A.; Nawaz, M.; Ullah Khan, W.; Yousaf Malik, M.; He, Y. Heat transfer in nanomaterial suspension (CuO and Al2O3) using KKL model. *Coatings* **2021**, *11*, 417. [CrossRef]
22. Nazir, U.; Abu-Hamdeh, N.H.; Nawaz, M.; Alharbi, S.O.; Khan, W. Numerical study of thermal and mass enhancement in the flow of Carreau-Yasuda fluid with hybrid nanoparticles. *Case Stud. Therm. Eng.* **2021**, *27*, 101256. [CrossRef]
23. Bilal, S.; Rehman, K.U.; Malik, M.Y.; Hussain, A.; Awais, M. Effect logs of double diffusion on MHD Prandtl nano fluid adjacent to stretching surface by way of numerical approach. *Results Phys.* **2017**, *7*, 470–479. [CrossRef]

Article

Significant Involvement of Double Diffusion Theories on Viscoelastic Fluid Comprising Variable Thermophysical Properties

Muhammad Sohail [1,*], Umar Nazir [1], Omar Bazighifan [2], Rami Ahmad El-Nabulsi [3,4,5,*], Mahmoud M. Selim [6,7], Hussam Alrabaiah [8,9] and Phatiphat Thounthong [10]

1   Department of Applied Mathematics and Statistics, Institute of Space Technology, P.O. Box 2750, Islamabad 44000, Pakistan; nazir_u2563@yahoo.com
2   Section of Mathematics, International Telematic University Uninettuno, CorsoVittorio Emanuele II, 39, 00186 Roma, Italy; o.bazighifan@gmail.com
3   Research Center for Quantum Technology, Faculty of Science, Chiang Mai University, Chiang Mai 50200, Thailand
4   Department of Physics and Materials Science, Faculty of Science, Chiang Mai University, Chiang Mai 50200, Thailand
5   Athens Institute for Education and Research, Mathematics and Physics Divisions, 8 Valaoritou Street, Kolonaki, 10671 Athens, Greece
6   Department of Mathematics, Al-Aflaj College of Science and Humanities Studies, Prince Sattam Bin Abdulaziz University, Al-Aflaj 710-11912, Saudi Arabia; m.selim@psau.edu.sa
7   Department of Mathematics, Suez Faculty of Science, Suez University, Suez 34891, Egypt
8   College of Engineering, Al Ain University, Al Ain P.O. Box 15551, United Arab Emirates; hussam.alrabaiah@aau.ac.ae
9   Department of Mathematics, Tafila Technical University, Tafila 66110, Jordan
10 Renewable Energy Research Centre, Department of Teacher Training in Electrical Engineering, Faculty of Technical Education, King Mongkut's University of Technology North Bangkok, 1518 Pracharat 1 Road, Bang Sue, Bangkok 10800, Thailand; phtt@kmutnb.ac.th
*   Correspondence: muhammad_sohail111@yahoo.com (M.S.); el-nabulsi@atiner.gr or nabulsiahmadrami@yahoo.fr (R.A.E.-N.)

Citation: Sohail, M.; Nazir, U.; Bazighifan, O.; El-Nabulsi, R.A.; Selim, M.M.; Alrabaiah, H.; Thounthong, P. Significant Involvement of Double Diffusion Theories on Viscoelastic Fluid Comprising Variable Thermophysical Properties. *Micromachines* **2021**, *12*, 951. https://doi.org/10.3390/mi12080951

Academic Editors: Lanju Mei and Shizhi Qian

Received: 21 June 2021
Accepted: 27 July 2021
Published: 12 August 2021

Abstract: This report examines the heat and mass transfer in three-dimensional second grade non-Newtonian fluid in the presence of a variable magnetic field. Heat transfer is presented with the involvement of thermal relaxation time and variable thermal conductivity. The generalized theory for mass flux with variable mass diffusion coefficient is considered in the transport of species. The conservation laws are modeled in simplified form via boundary layer theory which results as a system of coupled non-linear partial differential equations. Group similarity analysis is engaged for the conversion of derived conservation laws in the form of highly non-linear ordinary differential equations. The solution is obtained vial optimal homotopy procedure (OHP). The convergence of the scheme is shown through error analysis. The obtained solution is displayed through graphs and tables for different influential parameters.

Keywords: viscoelastic material; group similarity analysis; thermal relaxation time; parametric investigation; variable magnetic field; error analysis

1. Introduction

Fluid flows over stretched surfaces have applications in several fields and has significant involvement in the practical usage of several items. Scientists and engineers have made efforts to explore their features and usage in different processes. The mathematical relations of non-Newtonian materials are different as compared with Newtonian material. These materials are divided into different categories according to their properties. An important non-Newtonian fluid is a second grade fluid [1–7]. It has the following constitutive relation:

$$\tau^* = -PI + \mu F_1 + \beta_1 F_2 + \beta_2 F_1 * F_1, \ \beta_1 \geq 0, \ \mu \geq 0, \ \beta_1 + \beta_2 = 0.$$

Hayat et al. [1] studied the mixed convection in the second grade model over a stretching cylinder. They modeled the problem in two dimensions with thermal transport by taking the variable thermal conductivity. They used a homotopy method for the solution. They studied the contribution of several emerging parameters on the solution through graphs. They noticed the decrease in velocity field for a mixed convection parameter. Massoudi et al. [2] reported the study on the second grade model with temperature dependent viscosity between parallel plates. They presented a comparative study for the validity of obtained solution through tabular data. An exact solution through an oscillating sphere for a second grade model was computed by Fetecau et al. [3]. They found that the solution is periodic and it is independent of initial data. Hankel and Laplace transforms were engaged by Kamran et al. [4] to handle the modeled equations for fractional second grade model in cylindrical coordinates. They presented that the fractional model present the fluid flow phenomenon more accurately as compared with the ordinary derivatives. Chauhan and Kumar [5] studied the unsteady mechanics for second grade model in partially filled porous channel. They used a Laplace transform technique to analyze the solution. They observed the increase in velocity field against time parameter. A rotating viscoelastic model with ramped wall temperature condition for exact solution was reported by Mohamad et al. [6]. They plotted the solution against numerous emerging parameters. They noticed the dual behavior of velocity against the rotation parameter. Hayat et al. [7] examined the comportment of chemical reaction with solutal and thermal transport in a second grade model passed over a bi-directional stretched surface. They found the increase in dimensionless stress against ratio parameter and viscoelastic parameter. Moon et al. [8] discussed the phenomenon of heat transfer and Weber number including droplets of xanthan gum solution (non-Newtonian) and DI-water (Newtonian) over a heated surface. They noted that the DI-water droplet has higher spreading diameter as compared with the non-Newtonian (droplet) because of variation in fluid difference. German and Bertola [9] studied free-fall related to the liquid drops due gravity of force. They imagined high speed of drops based on viscoelastic fluids. They found that shape of the drop is changed under the action of yield stress. An and Lee [10] experimentally discussed the oscillations (free falling) of drops (shear-thinning) due to the force of gravity based on viscoelastic fluids passed over a solid surface. Moon et al. [11] suggested a mixed regime (coalescence occurs) using sequential images and mixed regime is exaggerated due to volumes of static droplet volumes, static droplets and Weber numbers. They estimated the film thickness (between two drops) via lubrication theory. Zhao and Khayat [12] discussed the flow behavior in view of shear-thickening and shear-thinning of a jet (non-Newtonian) over a flat plate via a hydraulic jump. Moon et al. [13] considered the features of droplets (non-Newtonian) on solid surfaces considering various Weber numbers. They considered xanthan gum solution to produce the droplets (non-Newtonian) measured via spreading diameters and camera (high speed).

Transportation of heat has applications in different engineering aspects and it is now a hot topic for researchers working in the field of engineering and applied mathematics. Several researchers are working on transport phenomena actively. For instance, Naseem et al. [14] worked on third grade nanofluid passed over a Riga plate. They considered several physical effects while modelling the transport equations and the resulting equations were simplified via boundary layer theory with the solution approximated analytically. They noticed the rise in velocity profile for modified magnetic parameter and Reynolds number. Sahoo and Poncet [15] studied the Blasius flow of a fourth grade model with heat transfer in a porous permeable stretching surface. Heat transfer in MHD viscous stagnation point dusty fluid with a non-uniform heat source in a porous stretching sheet was studied by Ramesh et al. [16]. They presented a comparative analysis and solved the resulting equations with the help of a shooting method. They discussed the contribution of several parameters on velocity and temperature fields. They noticed the decline in thermal field against higher Prandtl number. Qiu et al. [17] studied the thermal transport in channel via finite volume technique. They also reported the analysis of entropy in their findings. They discussed the impact of nanoparticles volume fraction on flow and entropy. Khan et al. [18] studied the heat transfer in a stagnation point Powell–Eyring model in a stretching cylinder with variable properties analytically. They observed the enhancement in temperature and velocity fields against the growing values or curvature parameter. Also they listed the numerical values for heat transfer coefficient against different parameters. Few recent studies covering non-linear transport problems with different effects have been reported in [19–25].

The objective of current inspection is to analyze the comportment of variable properties in heat and mass transportation in a second grade steady incompressible model past over a bi-directional elongating surface. This article is organized as follows: Section 1 contains the literature survey; modeling of considered problem is included in Section 2 with important physical quantities; Section 3 comprises methodology and results with key findings covered in Sections 4 and 5 respectively.

2. Mathematical Drafting of Viscoelastic Fluid with Thermal and Mass Transport

An analysis of the transport phenomenon on a second grade fluid [7,14] over a bi-directional elastic surface is presented in Figure 1. It is assumed that the sheet is stretched along x- and y- directions, respectively, and flow occupies the region normal to x- and y-axis. The sheet is kept at temperature "T_w" and concentration "C_w". Along x-axis, the velocity is "$U_W = ax$" and "$V_W = by$" is along the y-axis. The following important considerations have been adopted to derive the conservation laws

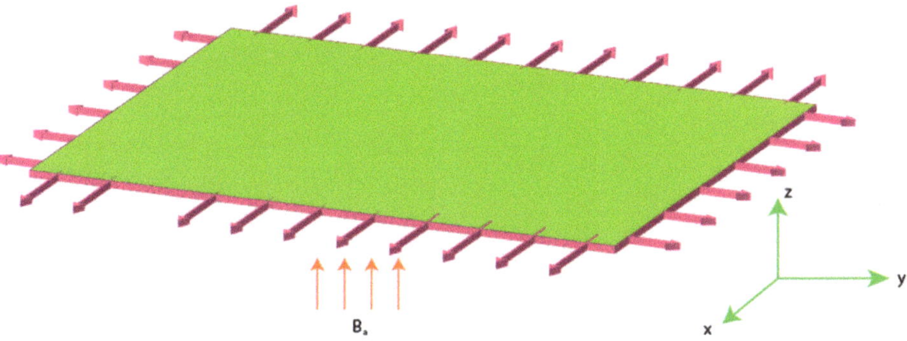

Figure 1. Geometry of second grade fluid model.

- ❖ Three-dimensional flow;
- ❖ Bi-directional elastic surface;
- ❖ Incompressible fluid;
- ❖ Steady flow;
- ❖ Viscoelastic second grade fluid;
- ❖ Heat flux via generalized theory of Cattaneo–Christov;
- ❖ Temperature-dependent thermal conductivity model;
- ❖ Space-dependent magnetic field;
- ❖ Updated mass flux model with temperature dependent diffusion coefficient;

The flowing resulting equations [7] appears by using the above stated assumptions

$$u_x + v_y + w_z = 0, \tag{1}$$

$$
\begin{aligned}
uu_x + v\frac{\partial u}{\partial y} + w\frac{\partial u}{\partial z} - \vartheta\frac{\partial^2 u}{\partial z^2} \\
+ \alpha_0\left[u\frac{\partial^3 u}{\partial x\partial z^2} + w\frac{\partial^3 u}{\partial z^3} - \frac{\partial u}{\partial x}\frac{\partial^2 u}{\partial z^2} - \frac{\partial u}{\partial z}\frac{\partial^2 w}{\partial z^2} - 2\frac{\partial u}{\partial z}\frac{\partial^2 u}{\partial x\partial z} \right. \\
\left. - 2\frac{\partial w}{\partial z}\frac{\partial^2 u}{\partial z^2} \right] + \frac{\sigma}{\rho}B_a^2(x,y)u = 0,
\end{aligned}
\tag{2}
$$

$$uv_x + v\frac{\partial v}{\partial y} + w\frac{\partial v}{\partial z} - \vartheta\frac{\partial^2 v}{\partial z^2}$$
$$+\alpha_0\left[v\frac{\partial^3 v}{\partial x\partial z^2} + w\frac{\partial^3 v}{\partial z^3} - \frac{\partial v}{\partial x}\frac{\partial^2 v}{\partial z^2} - \frac{\partial v}{\partial z}\frac{\partial^2 w}{\partial z^2} - 2\frac{\partial v}{\partial z}\frac{\partial^2 v}{\partial x\partial z}\right. \tag{3}$$
$$\left. -2\frac{\partial w}{\partial z}\frac{\partial^2 v}{\partial z^2}\right] + \frac{\sigma}{\rho}B_a^2(x,y)v = 0,$$

$$uT_x + vT_y + wT_z + \alpha_a\left[\begin{array}{c}(uu_x + vu_y + wu_z)T_x + \left(uv_x + v\frac{\partial v}{\partial y} + w\frac{\partial v}{\partial z}\right)T_y \\ +\left(uw_x + v\frac{\partial w}{\partial y} + w\frac{\partial w}{\partial z}\right)T_z + 2uvT_{xy} + 2vwT_{yz} \\ +2uwT_{xz} + u^2 T_{xx} + v^2 T_{yy} + w^2 T_{zz}\end{array}\right]$$
$$-\nabla[K_A(T)\nabla T] = 0 \tag{4}$$

$$uC_x + vC_y + wC_z + \alpha_b\left[\begin{array}{c}\left(uu_x + v\frac{\partial u}{\partial y} + w\frac{\partial u}{\partial z}\right)C_x + \left(uv_x + v\frac{\partial v}{\partial y} + w\frac{\partial v}{\partial z}\right)C_y \\ +\left(uw_x + v\frac{\partial w}{\partial y} + w\frac{\partial w}{\partial z}\right)C_z + 2uvC_{xy} + 2vwC_{yz} \\ +2uwC_{xz} + u^2 C_{xx} + v^2 C_{yy} + w^2 C_{zz}\end{array}\right]$$
$$-\nabla[D_A(T)\nabla C] = 0. \tag{5}$$

Boundary conditions for the dimensional problem are

$$\begin{cases} u = U_W = ax, \ v = V_W = by, w = 0, \ T = T_w, \ C = C_w \text{ at } z = 0. \\ u \to 0, \ v \to 0, \ T \to T_\infty, \ C \to C_\infty \text{ for } z \to \infty. \end{cases} \tag{6}$$

With the use of the following similarity variables, the governing law reduce to

$$\begin{cases} u = axf'[\eta], \ v = ayg'[\eta], w = -(a\vartheta)^{\frac{1}{2}}[f[\eta] + g[\eta]], \\ \theta[\eta] = \frac{T-T_\infty}{T_w-T_\infty}, \ \phi[\eta] = \frac{C-C_\infty}{C_w-C_\infty}, \ [\eta] = \left(\frac{a}{\vartheta}\right)^{\frac{1}{2}}z, \end{cases} \tag{7}$$

$$-Mf'[\eta] - f'[\eta]^2 + (f[\eta] + g[\eta])f''[\eta] + f^{(3)}[\eta] + R(f''[\eta](f''[\eta] - g''[\eta])$$
$$-2(f'[\eta] + g'[\eta])f^{(3)}[\eta] + (f[\eta] + g[\eta])f^{(4)}[\eta]) = 0, \tag{8}$$

$$-Mg'[\eta] - g'[\eta]^2 + (f[\eta] + g[\eta])g''[\eta] + g^{(3)}[\eta] + R((f''[\eta] - g''[\eta])g''[\eta]$$
$$-2(f'[\eta] + g'[\eta])g^{(3)}[\eta] + (f[\eta] + g[\eta])g^{(4)}[\eta]) = 0, \tag{9}$$

$$(f[\eta] + g[\eta])\theta'[\eta] + \frac{1}{Pr}(1 + \gamma_1\theta[\eta])\theta''[\eta] - \alpha_1(f[\eta] + g[\eta])(f'[\eta]$$
$$+g'[\eta]\theta'[\eta] + (f[\eta] + g[\eta])\theta''[\eta]) = 0, \tag{10}$$

$$(f[\eta] + g[\eta])\phi'[\eta] + \frac{1}{Sc}(1 + \gamma_2\theta[\eta])\phi''[\eta] - \alpha_2(f[\eta] + g[\eta])(f'[\eta]$$
$$+g'[\eta]\phi'[\eta] + (f[\eta] + g[\eta])\phi''[\eta]) = 0, \tag{11}$$

$$\begin{cases} f(0) = 0, \ g(0) = 0, \ f'(0) = 1, \ g'(0) = \delta, \ \theta(0) = 1, \ \phi(0) = 1, \\ f'(\infty) = 0, \ g'(\infty) = 0, \ \theta(\infty) = 0, \ \phi(\infty) = 0. \end{cases} \tag{12}$$

Physical Quantities

The study of heat, mass transfer rates and dimensionless stress at the boundary has significant applications and usage in industry. Therefore, scientists and engineers are keenly observing their features against different physical parameters which influence them directly. These quantities are defined as:

$$C_{XF} = \frac{\tau_{xz}|_z = 0}{\rho(u_w)^2}, C_{YF} = \frac{\tau_{yz}|_z = 0}{\rho(v_w)^2}, \tag{13}$$

$$\tau_{xz} = \left[\mu\frac{\partial u}{\partial z} + \alpha_0\left[u\frac{\partial^2 u}{\partial x\partial z} + v\frac{\partial^2 u}{\partial y\partial z} + w\frac{\partial^2 u}{\partial z^2} + \frac{\partial u}{\partial x}\frac{\partial u}{\partial z} + \frac{\partial v}{\partial z}\frac{\partial v}{\partial x} - \frac{\partial w}{\partial z}\frac{\partial u}{\partial y}\right]\right]_{z=0} \tag{14}$$

$$\tau_{yz} = \left[\mu \frac{\partial v}{\partial z} + \alpha_0 \left[u \frac{\partial^2 v}{\partial x \partial z} + v \frac{\partial^2 v}{\partial y \partial z} + w \frac{\partial^2 v}{\partial z^2} + \frac{\partial u}{\partial y}\frac{\partial u}{\partial z} + \frac{\partial v}{\partial z}\frac{\partial v}{\partial y} - \frac{\partial w}{\partial z}\frac{\partial v}{\partial y} \right]\right]_{z=0}, \tag{15}$$

$$Nu_{xy} = \frac{(x+y)Q_w^*}{K(T)[T_w - T_\infty]}, \quad Q_w^* = -K(T)\frac{\partial T}{\partial z}|_{z=0}, \tag{16}$$

$$Su_{xy} = \frac{(x+y)M_w^*}{D_B(T)[C_w - C_\infty]}, \quad M_w^* = -D_B(T)\frac{\partial C}{\partial z}|_{z=0} \tag{17}$$

After boundary layer theory, the dimensionless form is:

$$C_{Fx}^* = f''(0) - R[f(0) + g(0)]f'''(0) + R[f'(0) + g'(0)]f'(0) + 2Rf'(0)f''(0), \tag{18}$$

$$C_{Fy}^* = g''(0) - R[f(0) + g(0)]g'''(0) + R[f'(0) + g'(0)]g''(0) + 2Rf'(0)g'(0) \tag{19}$$

$$H_{xy}^* = -\left(R_{exy}\right)^{\frac{1}{2}}\theta'(0), \quad M_{xy}^* = -\left(R_{exy}\right)^{\frac{1}{2}}\phi'(0). \tag{20}$$

3. Numerical Method for Solution

Modelling of the fluid flow problems results in the form of a set of coupled non-linear differential equations. The derived problem is highly non-linear and coupled. Due to high non-linearity, an exact solution is not possible. Researchers proposed several schemes to handle the non-linear complex differential equations. Here, optimal homotopy analysis procedure (OHAP) [7,14,19–23] is engaged due to its several advantages.

This section covers the necessary steps for the adopted procedure. It has the following steps:

❖ Linear operator selection;
❖ Using the boundary data;
❖ Determination of unknown constants;
❖ Adopting of initial guesses;

The linear operators with initial guesses are:

$$\begin{cases} L_f^* = \frac{D^3}{D\eta^3} - \frac{D}{D\eta}, \ L_g^* = \frac{D^3}{D\eta^3} - \frac{D}{D\eta}, \\ L_t^* = \frac{D^2}{D\eta^2} - 1, \ L_c^* = \frac{D^2}{D\eta^2} - 1, \end{cases} \tag{21}$$

$$\begin{cases} f_q(\eta) = 1 - e^{-\eta}, \ g_q = \gamma[1 - e^{-\eta}], \\ \theta_q(\eta) = e^{-\eta}, \ \phi_q(\eta) = e^{-\eta}, \end{cases} \tag{22}$$

The operators in Equation (21) obeys:

$$L_f^*[Q_1 + Q_2 e^{-\eta} + Q_3 e^{\eta}] = 0,$$
$$L_g^*[Q_4 + Q_5 e^{-\eta} + Q_6 e^{\eta}] = 0,$$
$$L_t^*[Q_7 e^{-\eta} + Q_8 e^{\eta}] = 0, \tag{23}$$
$$L_c^*[Q_9 e^{\eta} + Q_{10} e^{-\eta}] = 0.$$

where $Q_b (b = 1, 2, \ldots, 10)$ are unknowns.

Using the concepts of minimization of average squared residual error [7–14,19–23]:

$$\delta_m^f = \frac{1}{B+1} \sum_{r=0}^{B} \left[S_f\left(\sum_{L=0}^{a} \hat{f}(\eta), \sum_{L=0}^{a} \hat{g}(\eta) \right) \right]^2, \tag{24}$$

$$\delta_m^g = \frac{1}{B+1} \sum_{r=0}^{B} \left[S_g\left(\sum_{L=0}^{a} \hat{f}(\eta), \sum_{L=0}^{a} \hat{g}(\eta) \right) \right]^2, \tag{25}$$

$$\delta_m^\theta = \frac{1}{B+1} \sum_{r=0}^{B} \left[S_\theta \left(\sum_{L=0}^{a} \hat{f}(\eta), \ \sum_{L=0}^{a} \hat{g}(\eta), \ \sum_{L=0}^{a} \hat{\theta}(\eta) \right) \right]^2, \tag{26}$$

$$\delta_m^\phi = \frac{1}{B+1} \sum_{r=0}^{B} \left[S_\phi \left(\sum_{L=0}^{a} \hat{f}(\eta), \ \sum_{L=0}^{a} \hat{g}(\eta), \ \sum_{L=0}^{a} \hat{\theta}(\eta), \ \sum_{L=0}^{a} \hat{\phi}(\eta) \right) \right]^2, \tag{27}$$

where

$$\delta_i^t = \delta_i^f + \delta_i^g + \delta_i^\theta + \delta_i^\phi. \tag{28}$$

The minimum error at second order is 0.00031589832079710834 and optimal values at third order are $B_f = -1.2160$, $B_g = -1.1638$, $B_\theta = -0.9509$, $B_\phi = -0.5871$, by fixing the involved parameters as $R = 0.1$, $Sc = 0.6$, $Pr = 1.1$, $\alpha_1 = 0.2 = \alpha_2$, $\gamma_1 = 0.3 = \gamma_2$, $M = 0.1$, $\delta = 0.8$.

$$\begin{aligned}
f = {} & 1.0 - 1.0e^{-z} - 0.6080M^2 + 0.6080e^{-z}M^2 - 0.6080R + 0.6080e^{-z}R \\
& + 0.6080e^{-z}M^2 z + 0.6080e^{-z}Rz - 0.4053\delta \\
& + 0.2026e^{-2z}\delta + 0.2026e^{-z}\delta - 1.0133R\delta \\
& - 0.4053e^{-2z}R\delta + 1.4187e^{-z}R\delta + 0.6080e^{-z}z\delta \\
& + 0.6080e^{-z}Rz\delta,
\end{aligned} \tag{29}$$

$$\begin{aligned}
g = {} & 1.1939\delta + 0.1939e^{-2z}\delta - 1.3879e^{-z}\delta - 0.5819M^2\delta + 0.5819e^{-z}M^2\delta \\
& - 0.5819R\delta + 0.5819e^{-z}R\delta + 0.5819e^{-z}M^2 z\delta \\
& + 0.5819e^{-z}Rz\delta - 0.5819\delta^2 + 0.5819e^{-z}\delta^2 - 0.9698R\delta^2 \\
& - 0.3879e^{-2z}R\delta^2 + 1.3578e^{-z}R\delta^2 + 0.5819e^{-z}z\delta^2 \\
& + 0.5819e^{-z}Rz\delta^2,
\end{aligned} \tag{30}$$

$$\begin{aligned}
\theta = {} & -0.3169e^{-2z} + 1.3169e^{-z} - 0.4754e^{-z}z + \frac{0.47543e^{-z}z}{Pr} - 0.3169e^{-2z}\delta \\
& + 0.3169e^{-z}\delta - 0.4754e^{-z}z\delta + 0.11887e^{-3z}\alpha_1 \\
& - 0.95097e^{-2z}\alpha_1 + 0.8321e^{-z}\alpha_1 - 0.9509e^{-z}z\alpha_1 \\
& + 0.3566e^{-3z}\delta\alpha_1 - 1.9019e^{-2z}\delta\alpha_1 + 1.5453e^{-z}\delta\alpha_1 \\
& - 1.4264e^{-z}z\delta\alpha_1 + 0.2377e^{-3z}\delta^2\alpha_1 - 0.9509e^{-2z}\delta^2\alpha_1 \\
& + 0.7132e^{-z}\delta^2\alpha_1 - 0.475497e^{-z}z\delta^2\alpha_1 - \frac{0.3169e^{-2z}\gamma_1}{Pr} \\
& + \frac{0.3169e^{-z}\gamma_1}{Pr},
\end{aligned} \tag{31}$$

$$\begin{aligned}
= {} & -0.1957e^{-2z} + 1.1957e^{-z} - 0.2935e^{-z}z + \frac{0.2935e^{-z}z}{Sc} - 0.1957e^{-2z}\delta \\
& + 0.1957e^{-z}\delta - 0.2935e^{-z}z\delta + 0.0733e^{-3z}\alpha_2 \\
& - 0.5871e^{-2z}\alpha_2 + 0.5137e^{-z}\alpha_2 - 0.5871e^{-z}z\alpha_2 \\
& + 0.2201e^{-3z}\delta\alpha_2 - 1.1742e^{-2z}\delta\alpha_2 + 0.9540e^{-z}\delta\alpha_2 \\
& - 0.8806e^{-z}z\delta\alpha_2 + 0.1467e^{-3z}\delta^2\alpha_2 - 0.5871e^{-2z}\delta^2\alpha_2 \\
& + 0.4403e^{-z}\delta^2\alpha_2 - 0.2935e^{-z}z\delta^2\alpha_2 - \frac{0.1957e^{-2z}\gamma_2}{Sc} \\
& + \frac{0.1957e^{-z}\gamma_2}{Sc}.
\end{aligned} \tag{32}$$

4. Analysis and Discussion

The assessments of vital applications of thermal energy and mass transport (using second grade liquid) in industrial and engineering areas are addressed, including various features. In this current problem, non-Fourier's theory is investigated in energy and mass transport equations along with the concept of variable properties (mass diffusion and thermal conductivity). The motion in nanoparticles is induced because of movement of a melting 3D-surface. The simulations of temperature, diffusion of mass and flow behavior are captured in the form of graphs and tables via an analytical approach. The error analysis is presented with the help of Table 1. The graphical discussions are addressed below:

Table 1. Computation of averaged squared residuals errors of velocity, temperature, and concentration solution.

(B)	δ_B^f	δ_B^g	δ_B^θ	δ_B^ϕ
1	0.0003	0.0001	0.0008	0.0130
4	6.612×10^{-6}	3.231×10^{-6}	5.461×10^{-6}	0.00017
8	3.653×10^{-7}	1.934×10^{-7}	1.787×10^{-7}	8.412×10^{-6}
12	3.068×10^{-8}	1.661×10^{-8}	8.074×10^{-9}	7.117×10^{-7}
16	4.169×10^{-9}	2.265×10^{-9}	1.063×10^{-9}	1.076×10^{-7}
20	5.760×10^{-10}	3.108×10^{-10}	1.522×10^{-10}	1.850×10^{-8}

Assessments of flow phenomena via physical parameters: the distribution of flow phenomena is analyzed with respect to magnetic number (M), second grade fluid number (R) and velocity stretching ratio number (δ). In this current problem, M is considered as variable magnetic number for measurement of flow behavior in both directions (vertical and horizontal). The variation strength of a magnetic field is considered during the flow of particles at the surface of the melting sheet. This effect is analyzed by Figures 2 and 3. The decreasing character in the motion of particles is noticed via enhancement strength regarding magnetic field. The declination in flow is due to a negative force called Lorentz force appeared in momentum equations. The retardation forces play a reducing role in motion of fluid particles. Therefore, a magnetic number is used to adjust (MBL) momentum boundary layer thickness. It is noticed that the last terms of dimensionless momentum equations represent negative Lorentz forces. These negative Lorentz forces generate hindrance during the flow of fluid particles. The reduction is investigated in velocity profiles for $M = 0.0, 0.2, 0.4, 0.6, 0.8$. Therefore, a local minimum trend is noticed for $M = 0.0, 0.2, 0.4, 0.6, 0.8$. The parameter related to R is known as a second grade number whereas the character of R is simulated on the flow behavior. In this case, the reduction in flow mechanism is conducted through the numerical values of R and this graphical impact is considered in Figures 4 and 5. It is noticed that the appearance of R is modeled due to the appearance of second grade fluid in the current model. Meanwhile, flow in the horizontal and vertical directions is reduced, taking higher values of R. Moreover, flow in the absence of R is higher as compared with flow situation in the presence of R. Simply, it is included that flow for the case Newtonian liquid becomes higher than as compared flow for the case of non-Newtonian liquid. Figure 6 provides the role of δ on distribution of velocities. The enhancement in velocity profiles is conducted using higher values of δ while the parameter related to δ is addressed as velocity ration parameter. From these figures, fluid speed is enhanced via more stretching of melting surface. The large stretching of the surface is the reason for the more enhancements in fluid motion. Hence, δ is favorable to achieve the enhancement in motion of the fluid particles.

Assessments of heat energy via physical parameters: the characterization of the thermal energy mechanism against the variation in Pr, α_1, Y_1 and R is conducted in Figures 7–10. In Figure 7, the role of second grade fluid number is considered. The mechanism of thermal energy is reduced using higher values of R. The thickness of thermal boundary is also decreased. The characterization of heat energy against the distribution of very small number (Y_1) based on thermal conductivity is measured by Figure 8. The production of heat energy becomes higher via higher values of Y_1. Hence, Y_1 plays an essential role in the enhancement of maximum production of energy. The performance of Prandtl number versus thermal energy is visualized by Figure 9. The temperature profile is noticed as reducing the role versus the variation of Pr. This reducing role happens because of the definition of the Prandtl number. According to the definition of Pr, MBL is enhanced while the thickness of TBL is increased. Meanwhile, the production of heat energy is reduced with respect to large values of Pr. Figure 10 illustrates the trend of α_1 versus the temperature profile and α_1 is denoted as a relation time parameter. It is investigated how α_1

has appeared due to Cattaneo heat flux theory in the energy equation. The thermal energy is reduced due to attendance of non-Fourier's concept whereas thermal energy for the case of Fourier's law is higher than thermal energy for the case of the non-Fourier's concept.

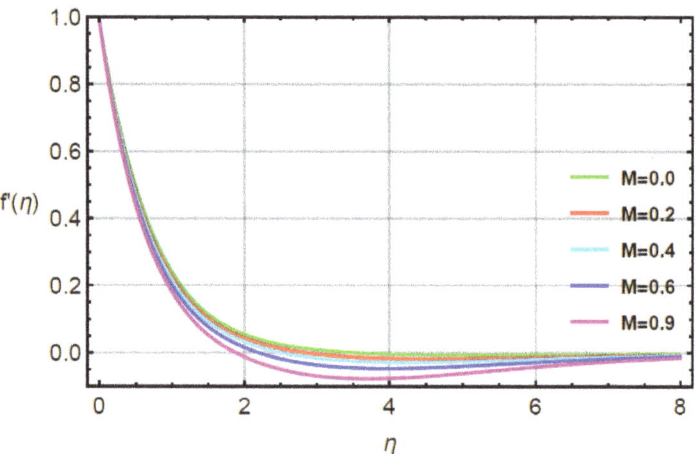

Figure 2. Behavior of $f'(\eta)$ for M when $R = 0.1, Sc = 0.6, Pr = 1.1, \; \alpha_1 = 0.2 = \alpha_2, \gamma_1 = 0.3 = \gamma_2, \delta = 0.8$.

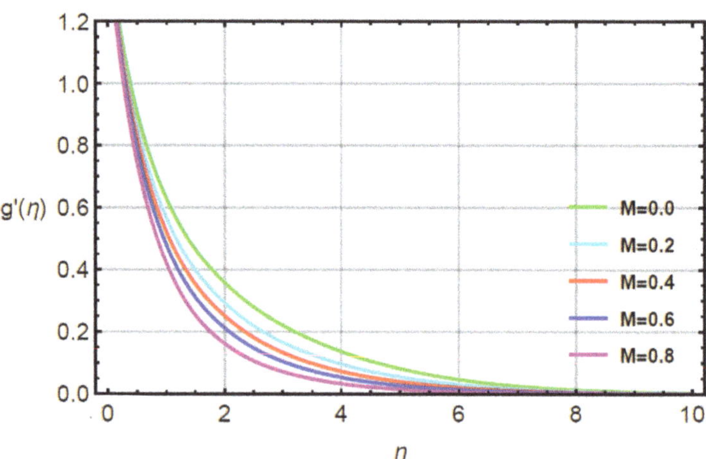

Figure 3. Behavior of $g'(\eta)$ for M when $R = 0.1, Sc = 0.6, Pr = 1.1, \; \alpha_1 = 0.2 = \alpha_2, \; \gamma_1 = 0.3 = \gamma_2, \delta = 0.8$.

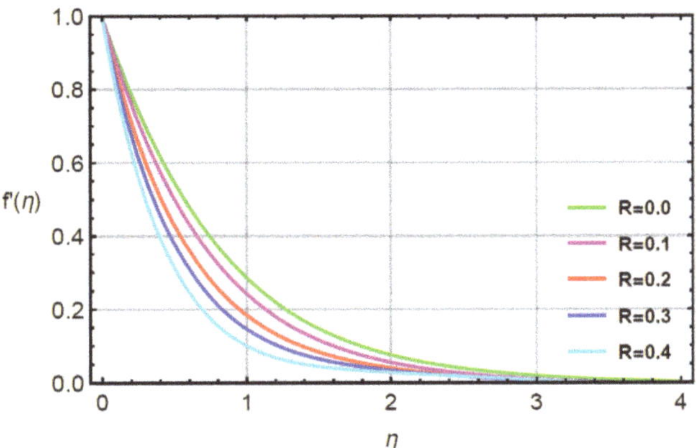

Figure 4. Behavior of $f'(\eta)$ for R when $Sc = 0.6, Pr = 1.1, \ \alpha_1 = 0.2 = \alpha_2, \gamma_1 = 0.3 = \gamma_2, M = 0.1, \delta = 0.8$.

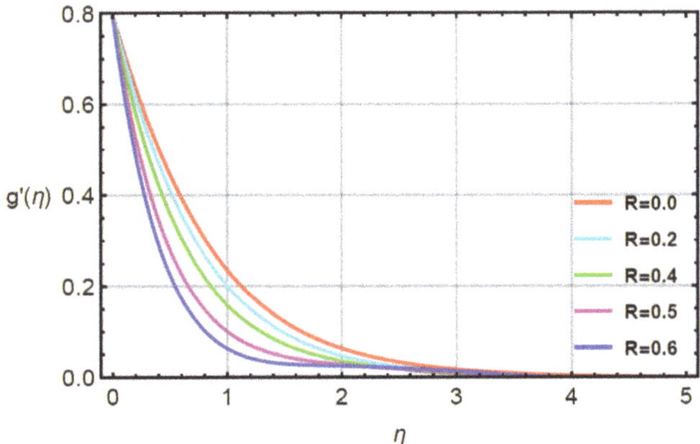

Figure 5. Behavior of $g'(\eta)$ for R when $Sc = 0.6, Pr = 1.1, \ \alpha_1 = 0.2 = \alpha_2, \gamma_1 = 0.3 = \gamma_2, M = 0.1, \delta = 0.8$.

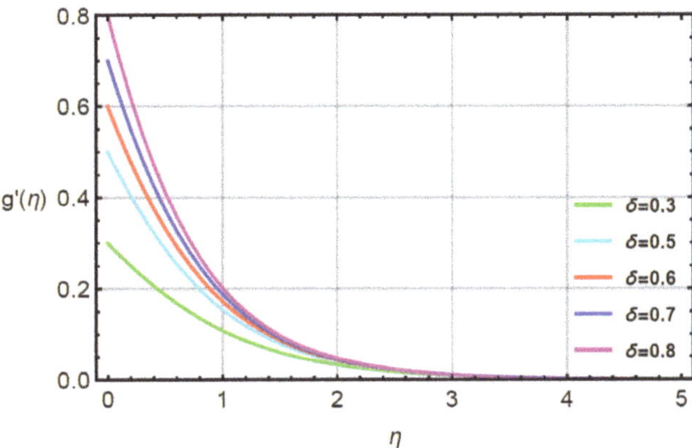

Figure 6. Behavior of $g'(\eta)$ for δ when $R = 0.1, Sc = 0.6, Pr = 1.1, \ \alpha_1 = 0.2 = \alpha_2, \gamma_1 = 0.3 = \gamma_2, M = 0.1$.

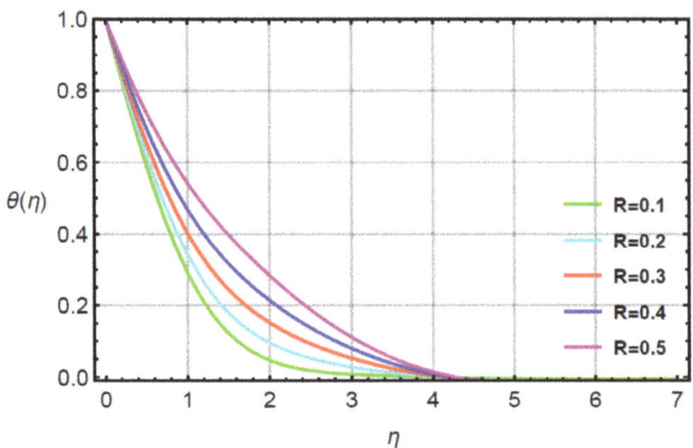

Figure 7. Behavior of $\theta(\eta)$ for R when $Sc = 0.6, Pr = 1.1, \alpha_1 = 0.2 = \alpha_2, \; \gamma_1 = 0.3 = \gamma_2, M = 0.1, \delta = 0.8$.

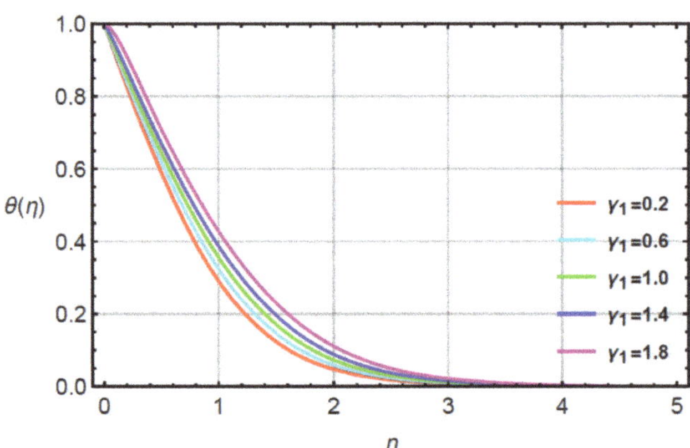

Figure 8. Behavior of $\theta(\eta)$ for γ_1 when $R = 0.1, Sc = 0.6, Pr = 1.1, \alpha_1 = 0.2 = \alpha_2, \gamma_2 = 0.3, M = 0.1, \delta = 0.8$.

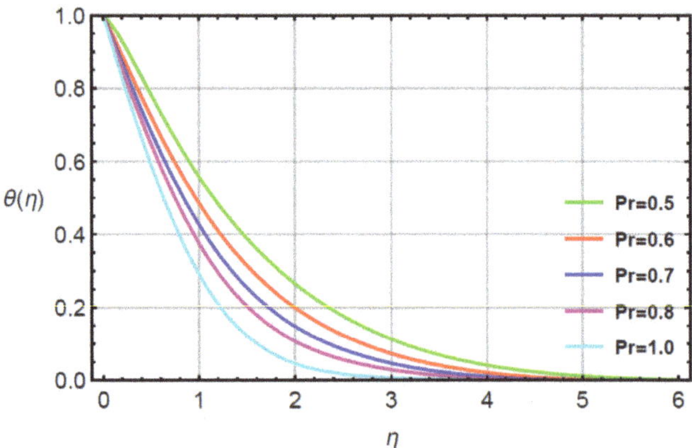

Figure 9. Behavior of $\theta(\eta)$ for Pr when $R = 0.1, Sc = 0.6, \; \alpha_1 = 0.2 = \alpha_2, \; \gamma_2 = 0.3, M = 0.1, \delta = 0.8$.

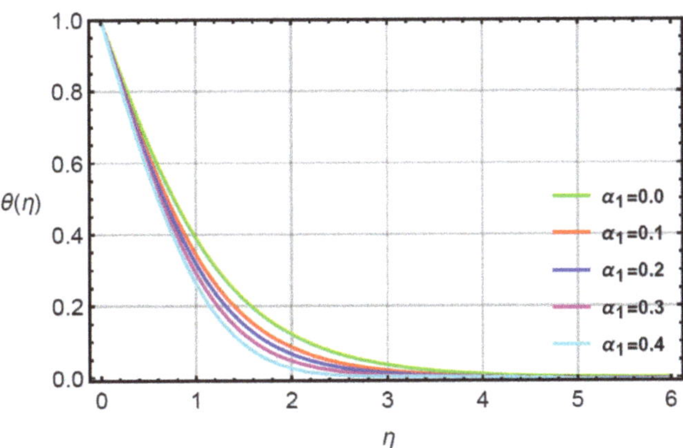

Figure 10. Behavior of $\theta(\eta)$ for α_1 when $R = 0.1, Sc = 0.6, \alpha_2 = 0.2, \gamma_2 = 0.3, M = 0.1, \delta = 0.8$.

Assessments of mass diffusion via physical parameters: the distribution in transport of mass diffusion is measured against the variation in Sc, α_2 and γ_2 via Figures 11–14. The mechanism related to mass diffusion versus Sc called the Schmidt number is conducted by Figure 11. From this figure, declination is measured in view of the transport of mass using large values of the Schmidt number. The transport of mass becomes slow due to the concept of Sc. According to this definition, the diffusion of mass (ratio between momentum and mass diffusivities) has an inverse relation with respect to Sc. Hence, the solute slows down considering enlargement in Sc. Due to this depreciation occurs in the concentration field. Moreover, the combined enlargement in the Schmidt number and solutal relaxation time lessen the concentration profile. The decreasing graph of the concentration profile is observed versus the values of Sc. This decreasing trend and local minimum in solute particles is occurred due to reduction of mass diffusivity. Physical, large values of Sc reduce mass diffusivity and this reduction in mass diffusivity is a reason for the local minimum in solute particles. The better performance of solute is the investigated against variation in R considering by Figure 12. The range of R is $0.1 \leq R \geq 0.8$ is used to obtain maximum production of solute. Figure 13 simulates the variation in mass transport with respect to values of α_2. The decrement in solute is verified considering large values of α_2. This physical parameter has the ability to restore a state of equilibrium resulting in the solute becoming slow. Figure 14 captures the role of γ_2 versus the diffusion of mass. In this figure, diffusion of mass becomes fast using large values of γ_2.

Assessments of Nusselt number, divergent velocity and Sherwood number via physical parameters: the characterization of surface force, temperature gradient and rate of solute is analyzed considering large values of second grade fluid, stretching ration and time relaxation numbers. These simulations are captured by Tables 2 and 3. From Table 2, it can be noted that an enhancement is addressed in the case of positive values of second grade fluid and time relaxation numbers. Table 2 presents the comparative analysis of current model. The remarkable simulations are verified with published work [7]. Table 4 illustrates the impact of time relaxation numbers against the diffusion of mass. The reparable increasing role of the Sherwood number is measured against positive values of the time relaxation number.

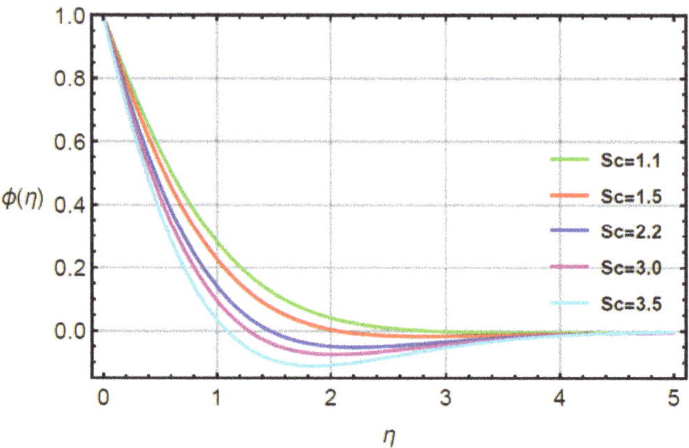

Figure 11. Behavior of $\phi(\eta)$ for Sc when $R = 0.1$, $\alpha_1 = 0.2 = \alpha_2$, $\gamma_2 = 0.3$, $M = 0.1$, $\delta = 0.8$.

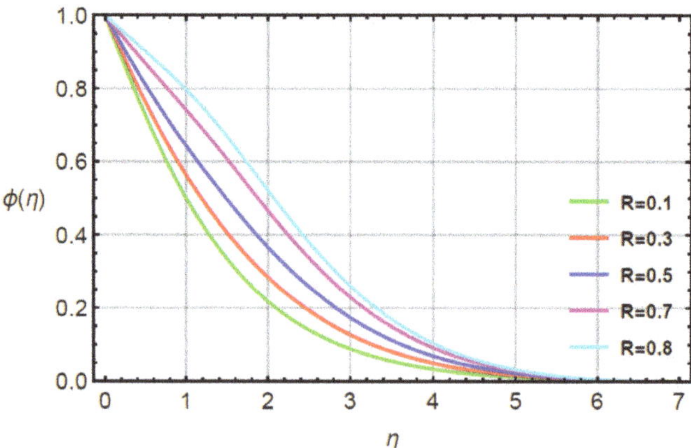

Figure 12. Behavior of $\phi(\eta)$ for R when $Sc = 0.6$, $\alpha_1 = 0.2 = \alpha_2$, $\gamma_2 = 0.3$, $M = 0.1$, $\delta = 0.8$.

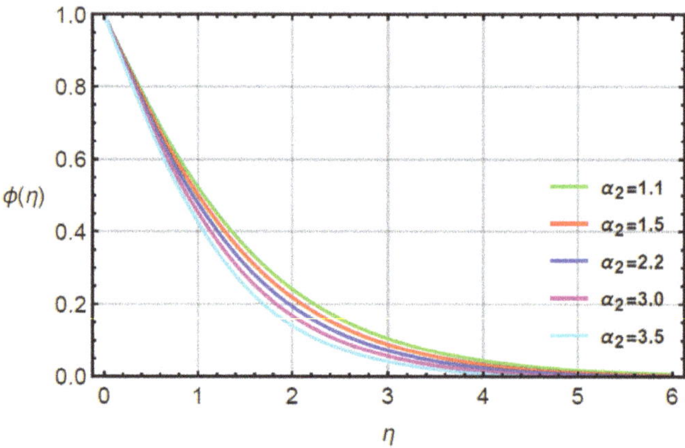

Figure 13. Behavior of $\phi(\eta)$ for α_2 when $R = 0.1$, $Sc = 0.6$, $\alpha_1 = 0.2$, $\gamma_2 = 0.3$, $M = 0.1$, $\delta = 0.8$.

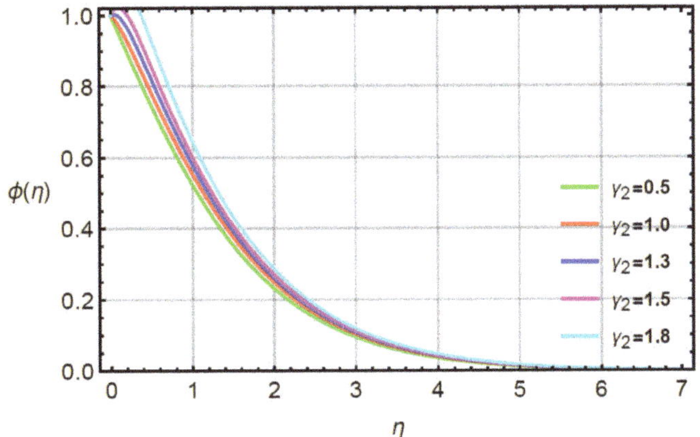

Figure 14. Behavior of $\phi(\eta)$ for γ_2 when $R = 0.1, Sc = 0.6, \alpha_1 = 0.2 = \alpha_2, M = 0.1, \delta = 0.8$.

Table 2. Comparative analysis for dimensionless stress against R and γ by fixing the other parameters.

R	γ	$-\left(R_{e_{xy}}\right)^{\frac{1}{2}} C_{Fx}^{*}$ [7]	Present	$-\left(R_{e_{xy}}\right)^{\frac{1}{2}} C_{Fx}^{*}$ [7]	Present
0.0	0.1	1.0203	1.0209	0.0669	0.0668
0.15	-	1.6510	1.6519	0.0785	0.0784
0.2	-	1.8970	1.8975	0.0806	0.0803
0.1	0.0	1.3703	1.3707	0.0000	0.0000
-	0.2	1.4798	1.4798	0.1762	0.1769
-	0.5	1.6510	1.6516	0.6317	0.6312

Table 3. Comparative investigation for heat transfer rate against α_1.

α_1	$-\left(R_{e_{xy}}\right)^{\frac{1}{2}} \theta'(0)$ [7]	Present Results
0.0	0.6051	0.6059
0.2	0.6258	0.6256
0.4	0.6483	0.6489
0.6	0.6727	0.6729

Table 4. Comparative investigation for mass transfer rate against α_2.

α_2	$-\left(R_{e_{xy}}\right)^{\frac{1}{2}} \phi'(0)$ [7]	Present Results
0.0	0.3668	0.3669
0.2	0.3764	0.3760
0.4	0.3864	0.3862
0.6	0.3973	0.3978

5. Conclusions and Key Findings of the Investigation Performed

The physical occurrence of solute, heat energy and flow phenomena in a second grade liquid were visualized passing a 3D melting moving surface. The theory of non-

Fourier's law was imposed in the current flow model inserting variable properties. The current complex model was simulated with the help of an analytical scheme. The main consequences of the current problem are addressed below:

- The improvement in motion of fluid particles was captured via large values of second grade fluid and stretching ratio numbers while a decrement in flow behavior was conducted via enlargement in magnetic number;
- The mechanism of heat energy became maximum using higher values of second grade fluid number but an opposite trend was captured via large values of time relaxation, Prandtl and very small numbers;
- The solute became fast considering large values of second grade fluid, time relaxation and very small numbers. The reduction in solute became slow against variation in Schmidt number;
- An incline in rate of solute and gradient temperature was addressed against higher values of time relaxation numbers;
- The surface force was enhanced near the wall of the hot surface via large values of second grade liquid and flow stretching parameters.

Author Contributions: Conceptualization, M.S. and U.N.; methodology, M.S.; software, U.N.; validation, R.A.E.-N., M.M.S. and O.B.; formal analysis, O.B.; investigation, P.T.; resources, H.A.; data curation, U.N.; writing—original draft preparation, M.S.; writing—review and editing, M.S.; visualization, U.N.; supervision, M.S.; project administration, M.S.; funding acquisition, R.A.E.-N. All authors have read and agreed to the published version of the manuscript.

Funding: This research received no external funding.

Data Availability Statement: The data used to support this study are included in the manuscript.

Conflicts of Interest: The authors declare that they have no conflicts of interest.

Nomenclature

Symbols	Used for	Symbols	Used for
"$U_W = ax$" and "$V_W = by$"	Velocity at wall	x, y, z	Space coordinates
u, v, w	Dimensional velocity	ϑ	Kinematic viscosity
$B_a^2(x,y)$	Non-uniform magnetic field	σ	Electrical conductivity
ρ	Fluid density	δ	stretching ratio number
$-\left(R_{e_{xy}}\right)^{\frac{1}{2}}\theta'(0)$	Heat transfer rate	$-\left(R_{e_{xy}}\right)^{\frac{1}{2}}\phi'(0)$	Mass transportation rate
TBL	Thermal boundary layer	MBL	Momentum boundary layer
$L_f^* = \frac{D^3}{D\eta^3} - \frac{D}{D\eta}$, $L_g^* = \frac{D^3}{D\eta^3} - \frac{D}{D\eta}$, $L_t^* = \frac{D^2}{D\eta^2} - 1$, $L_c^* = \frac{D^2}{D\eta^2} - 1$,	Linear operators	$f_q(\eta) = 1 - e^{-\eta}$, $g_q = \gamma[1 - e^{-\eta}]$, $\theta_q(\eta) = e^{-\eta}$, $\phi_q(\eta) = e^{-\eta}$,	Initial guesses
$\phi(\eta)$	Dimensionless concentration	$\theta(\eta)$	Dimensionless temperature
η	Dimensionless independent variable	Pr	Prandtl number
α_b	Concentration relaxation time	α_a	Temperature relaxation time
Sc	Schmidt number	$f'[\eta], g'[\eta], f[\eta], g[\eta]$	Dimensionless velocity
R	Fluid parameter	M	Magnetic parameter

References

1. Hayat, T.; Anwar, M.S.; Farooq, M.; Alsaedi, A. Mixed convection flow of viscoelastic fluid by a stretching cylinder with heat transfer. *PLoS ONE* **2015**, *10*, e0118815. [CrossRef]
2. Massoudi, M.; Vaidya, A.; Wulandana, R. Natural convection flow of a generalized second grade fluid between two vertical walls. *Nonlinear Anal. Real World Appl.* **2008**, *9*, 80–93. [CrossRef]
3. Fetecau, C.; Nazar, A.; Khan, I.; Shah, N.A. First exact solution for a second grade fluid subject to an oscillating shear stress induced by a sphere. *J. Nonlinear Sci.* **2018**, *19*, 186–195.
4. Kamran, M.; Imran, M.; Athar, M. Analytical Solutions for the Flow of a Fractional Second Grade Fluid due to a Rotational Constantly Accelerating Shear. *Int. Sch. Res. Not.* **2012**, *2012*. [CrossRef]
5. Mohamad, A.Q.; Khan, I.; Zin, N.A.M.; Ismail, Z.; Shafie, S. The unsteady mixed convection flow of rotating second grade fluid in porous medium with ramped wall temperature. *Malays. J. Fundam. Appl. Sci.* **2017**, *13*, 798–802. [CrossRef]
6. Chauhan, D.S.; Kumar, V. Unsteady flow of a non-Newtonian second grade fluid in a channel partially filled by a porous medium. *Adv. Appl. Sci. Res.* **2012**, *2*, 1–2.
7. Hayat, T.; Ayub, T.; Muhammad, T.; Alsaedi, A. Three-dimensional flow with Cattaneo–Christov double diffusion and homogeneous-heterogeneous reactions. *Results Phys.* **2017**, *7*, 2812–2820. [CrossRef]
8. Moon, J.H.; Kim, D.Y.; Lee, S.H. Spreading and receding characteristics of a non-Newtonian droplet impinging on a heated surface. *Exp. Therm. Fluid Sci.* **2014**, *57*, 94–101. [CrossRef]
9. German, G.; Bertola, V. The free-fall of viscoplastic drops. *J. Non Newton. Fluid Mech.* **2010**, *165*, 825–828. [CrossRef]
10. An, S.M.; Lee, S.Y. Observation of the spreading and receding behavior of a shear-thinning liquid drop impacting on dry solid surfaces. *Exp. Therm. Fluid Sci.* **2012**, *37*, 37–45. [CrossRef]
11. Moon, J.H.; Choi, C.K.; Allen, J.S.; Lee, S.H. Observation of a mixed regime for an impinging droplet on a sessile droplet. *Int. J. Heat Mass Transf.* **2018**, *127*, 130–135. [CrossRef]
12. Zhao, J.; Khayat, R.E. Spread of a non-Newtonian liquid jet over a horizontal plate. *J. Fluid Mech.* **2008**, *613*, 411–443. [CrossRef]
13. Moon, J.H.; Lee, J.B.; Lee, S.H. Dynamic behavior of non-Newtonian droplets impinging on solid surfaces. *Mater. Trans.* **2013**, *54*, 260–265. [CrossRef]
14. Naseem, A.; Shafiq, A.; Zhao, L.; Farooq, M.U. Analytical investigation of third grade nanofluidic flow over a riga plate using Cattaneo-Christov model. *Results Phys.* **2018**, *9*, 961–969. [CrossRef]
15. Sahoo, B.; Poncet, S. Blasius flow and heat transfer of fourth-grade fluid with slip. *Appl. Math. Mech.* **2013**, *34*, 1465–1480. [CrossRef]
16. Ramesh, G.K.; Gireesha, B.J.; Bagewadi, C.S. MHD flow of a dusty fluid near the stagnation point over a permeable stretching sheet with non-uniform source/sink. *Int. J. Heat Mass Transf.* **2012**, *55*, 4900–4907. [CrossRef]
17. Qiu, S.; Xie, Z.; Chen, L.; Yang, A.; Zhou, J. Entropy generation analysis for convective heat transfer of nanofluids in tree-shaped network flowing channels. *Therm. Sci. Eng. Prog.* **2018**, *5*, 546–554. [CrossRef]
18. Khan, M.I.; Kiyani, M.Z.; Malik, M.Y.; Yasmeen, T.; Khan, M.W.A.; Abbas, T. Numerical investigation of magnetohydrodynamic stagnation point flow with variable properties. *Alex. Eng. J.* **2016**, *55*, 2367–2373. [CrossRef]
19. Sohail, M.; Naz, R. Modified heat and mass transmission models in the magnetohydrodynamic flow of Sutterby nanofluid in stretching cylinder. *Phys. A Stat. Mech. Appl.* **2020**, *549*, 124088. [CrossRef]
20. Sohail, M.; Raza, R. Analysis of radiative magneto nano pseudo-plastic material over three dimensional nonlinear stretched surface with passive control of mass flux and chemically responsive species. *Multidiscip. Modeling Mater. Struct.* **2020**, *16*, 1061–1083. [CrossRef]
21. Sohail, M.; Tariq, S. Dynamical and optimal procedure to analyze the attributes of yield exhibiting material with double diffusion theories. *Multidiscip. Modeling Mater. Struct.* **2019**, *16*, 557–580. [CrossRef]
22. Sohail, M.; Naz, R.; Bilal, S. Thermal performance of an MHD radiative Oldroyd-B nanofluid by utilizing generalized models for heat and mass fluxes in the presence of bioconvective gyrotactic microorganisms and variable thermal conductivity. *Heat Transf. Asian Res.* **2019**, *48*, 2659–2675. [CrossRef]
23. El-Nabulsi, R.A. Non-standard Lagrangians in rotational dynamics and the modified Navier–Stokes equation. *Nonlinear Dyn.* **2015**, *79*, 2055–2068. [CrossRef]
24. Sohail, M.; Nazir, U.; Chu, Y.M.; Alrabaiah, H.; Al-Kouz, W.; Thounthong, P. Computational exploration for radiative flow of Sutterby nanofluid with variable temperature-dependent thermal conductivity and diffusion coefficient. *Open Phys.* **2020**, *18*, 1073–1083. [CrossRef]
25. El-Nabulsi, R.A. Fractional Navier-Stokes equation from fractional velocity arguments and its implications in fluid flows and microfilaments. *Int. J. Nonlinear Sci. Numer. Simul.* **2019**, *20*, 449–459. [CrossRef]

micromachines

Article

Entropy Generation Analysis and Radiated Heat Transfer in MHD (Al$_2$O$_3$-Cu/Water) Hybrid Nanofluid Flow

Nabeela Parveen [1], Muhammad Awais [1], Saeed Ehsan Awan [2], Wasim Ullah Khan [3,*], Yigang He [3,*] and Muhammad Yousaf Malik [4]

[1] Department of Mathematics, Attock Campus, COMSATS University Islamabad, Attock 43600, Pakistan; nabeela.mpa2016@gmail.com (N.P.); awais@ciit-attock.edu.pk (M.A.)

[2] Department of Electrical and Computer Engineering, Attock Campus, COMSATS University Islamabad, Attock 43600, Pakistan; saeed.ehsan@cuiatk.edu.pk

[3] School of Electrical Engineering and Automation, Wuhan University, Wuhan 430072, China

[4] Department of Mathematics, College of Sciences, King Khalid University, Abha 61413, Saudi Arabia; drmymalik@hotmail.com

* Correspondence: kwasim814@whu.edu.cn (W.U.K.); yghe1221@whu.edu.cn (Y.H.)

Abstract: This research concerns the heat transfer and entropy generation analysis in the MHD axisymmetric flow of Al$_2$O$_3$-Cu/H$_2$O hybrid nanofluid. The magnetic induction effect is considered for large magnetic Reynolds number. The influences of thermal radiations, viscous dissipation and convective temperature conditions over flow are studied. The problem is modeled using boundary layer theory, Maxwell's equations and Fourier's conduction law along with defined physical factors. Similarity transformations are utilized for model simplification which is analytically solved with the homotopy analysis method. The h-curves up to 20th order for solutions establishes the stability and convergence of the adopted computational method. Rheological impacts of involved parameters on flow variables and entropy generation number are demonstrated via graphs and tables. The study reveals that entropy in system of hybrid nanofluid affected by magnetic induction declines for β while it enhances for Bi, R and λ. Moreover, heat transfer rate elevates for large Bi with convective conditions at surface.

Keywords: hybrid nanofluid; entropy generation; induced magnetic field; convective boundary conditions; thermal radiations; stretching disk

Citation: Parveen, N.; Awais, M.; Awan, S.E.; Khan, W.U.; He, Y.; Malik, M.Y. Entropy Generation Analysis and Radiated Heat Transfer in MHD (Al$_2$O$_3$-Cu/Water) Hybrid Nanofluid Flow. *Micromachines* **2021**, *12*, 887. https://doi.org/10.3390/mi12080887

Academic Editors: Jin-yuan Qian and Lanju Mei

Received: 25 May 2021
Accepted: 19 July 2021
Published: 27 July 2021

Publisher's Note: MDPI stays neutral with regard to jurisdictional claims in published maps and institutional affiliations.

1. Introduction

MHD boundary layer flows of electrically conducting fluids and heat transfer caused by a stretching sheet have gained immense importance due to their ample applications and significant bearings on several engineering and technological processes. Major applications include heat exchangers, metals' spinning, and power generators, spinning of fiber, aerodynamic extrusion of plastic sheets, polymer industries, and condensation process of metallic sheets inside cooling glass. The quality of the resulting product greatly depends on the stretching process as well as its rate of cooling. Boundary-layer flow combined with heat transfer and followed by a stretching sheet was primarily proposed by Sakiadis [1]. Crane [2] also investigated flow caused by stretching sheets of plastic material. Boundary layer equations describing air motion due to a plate were solved analytically. Rates of flow and heat transfer were analyzed in terms of coefficients of friction and thermal conductivity, respectively. Since then, many attempts have been made in regard to their application in different areas. The analysis of viscous dissipation, thermal radiations and convective wall conditions in fluid flow has its importance in view of its broad coverage including chemical, mechanical, and aerospace engineering, paper production, continuous casting, stretching of plastic films, cooling of electronic chips and crystal growing etc. Khan et al. [3] explored analysis of heat and mass transfer in three-dimensional nanofluids flowing on a linear stretching sheet under convective wall conditions and thermal radiations. It was deduced that heat and mass transfer rates enhance with the stretching parameter. Gireesha et al. [4] investigated the influence of non-linear radiation on MHD boundary layer dynamics of a Jeffrey nanofluid past a non-linear porous stretching plate. Three-dimensional flow of fluid was considered. It was found that magnetic field diminishes the fluid velocity while it enhances temperature. Hayat et al. [5] analyzed the influence of magnetic induction on dynamics of second-grade nanofluid due to stretching sheet with convective wall conditions. The viscoelastic parameter was observed to upsurge the fluid parameter. Rafiq et al. [6] exposed the effects of non-linear thermal radiation towards boundary layer dusty fluid dynamics close to a rotating isothermally heated blunt-nosed object similar to a hemisphere. It was determined that skin friction coefficient shows an asymptotic trend while heat transfer coefficient increases significantly corresponding to large radiation parameter. Khan et al. [7] analytically studied a mixed convection hybrid nanofluid consisting of Al_2O_3 and Ag nanoparticles affected by induced magnetic field in order to analyze entropy generation under viscous dissipation and heat generation effects. They observed that viscous dissipation dominantly increases flow and heat transfer rate due to the no-slip surface condition. Few other attempts in this regime are cited here [8–14].

Technological developments and increasing demand of optical and electronic equipment required an improved cooling performance of the products which is acquired by utilizing heat transfer fluids with the improvements in their thermo physical characteristics. In order to obtain modified results, there have been plenty of endeavors by researchers to synthesize these fluids for an efficient heat transfer rate using the composition of several fluids as well as the dispersion of metallic particles of different sizes and shapes etc. Recently, an upsurge of embedding thermal resistive and conductive nano-particles, initially introduced by Choi [15], has been implemented to enhance thermal characteristics of working fluids. Liquids containing suspended nanoparticles, named as nanofluids, were experimentally guaranteed to possess their higher thermal conductivities than that of base fluids and may enable the operation of cooling systems for practical use in many fields such as in micro-electromechanical systems, pharmaceutical procedures, heat transfer, hybrid-powered engines and in field of nanotechnology. Dynamics of a magneto convective Casson nanofluid caused by Stefen blowing with bio active mixers was theoretically inspected by Puneeth et al. [16]. Awais et al. [17] theoretically exposed the rheological behavior of copper-water nanofluid peristaltic flowthrough generalized compliant walls in order to inspect influence of Hall and slip with temperature dependent viscosity. It was deduced that first and second order velocity slip parameters significantly increase flow velocity whereas rate of heat transfer is maximum in the vicinity of channel boundaries. An experimental study on the rheological characteristics of nanofluids manufactured by dispersing multi-walled carbon nanotubes in liquid paraffin was carried out by Liu et al. [18]. It was revealed that velocity components enlarge for velocity slip parameters while temperature-dependent viscosity has shown an impact of increasing temperature. In order to characterize the solar energy storage, improvement in thermal capacity of binary nitrate eutectic salt-silica nanofluid was studied by Hu et al. [19]. Relevant literature in the regime of nanofluids under several aspects can be found in [20–22]. Regardless of the noteworthy consequences of researchers' endeavors, authentic applications require transaction in dissimilar characteristics of nanofluids and therefore hybrid nanofluids were synthesized by embedding special nanoparticles in base fluid. Such fluids possess improved physical and chemical properties in a homogeneous phase. Waini et al. [23] inspected MHD flow dynamics and heat transport of a hybrid nanofluid over a porous stretching/shrinking wedge. A drop of the heat transport rate was determined with the rise in radiation parameter. The temporal stability analysis was presented to evaluate the dual solutions' stability, and it was revealed that one of the dual stables is reliable physically. Parveen et al. [24] utilized computational intelligence techniques in order to analyze heat transfer rate and pressure rise behavior in hybrid nanofluid dynamics influenced by magnetic induction effects past an endoscope. It was shown that coefficient of heat transfer accelerates toward Br and χ. Accuracy and stability of experimental data were established by employing neural network algorithm and very reliable results were obtained. Radhika et al. [25] explained the effects of magnetic field and melting heat transfer in the dynamics of dusty fluid suspended with hybrid nanoparticles. It was revealed that thermal gradient enhances for high values of magnetic parameter and Prandtl number. Reddy et al. [26] carried out theoretical analysis for heat transfer in dusty fluid dynamics suspended with hybrid nanoparticles by taking the Cattaneo-Christov heat flux model. Khan et al. [27] studied sustainability based performance evaluation of hybrid nanofluid assisted machining. Their theoretical study showed that a very small portion of nanoparticles affect the cost of each industrial product and the study was in complete accordance with the industrial applications of nanofluids. Literature in this area is shown in the references [28–31]. Moreover, the shape factor can approximately portray the difference of shape among non-spherical and spherical nanoparticles. In general, the nanoparticles possess polyhedral shapes, and their surface is made up of various planes. Thermophysical properties of nanoparticles directly depend on shape of nanoparticles. Therefore, flow and heat transport rates are examined in terms of coefficient of skin friction and Nusselt number for the nanoparticles of platelets shape with $s = 5.7$ in this analysis.

Entropy generation analysis is one of the vital factors in fluid mechanics. Performance of thermal devices directly depends upon the available amount of work which degrades by flow irreversibility and causes more disorder. Therefore, the study of dynamics behind entropy production becomes necessary in order to optimize thermal efficiencies of devices. Dogonchi et al. [32] inspected entropy generation behavior in natural convective hybrid nanofluid rheology by considering the effects of applied magnetic field past a porous cavity with wavy walls embedded in three circular cylinders. Sahoo et al. [33] carried out the analysis of entropy optimization, with dissipative heat transfer, in mixed convective MHD Casson nanofluid dynamics under the influences of Hall current and thermal radiation. Results showed that entropy generation amplifies significantly for diffusive variable, Brinkman number, and concentration ratio parameter whereas Bejan number decreases for all these parameters. In this regard, some investigations on entropy generation analysis for different flows and geometries under various physical aspects are reviewed (see articles [34–36]). Moreover, use of an analytical technique for the solution of the mathematical model is aimed by using homotopy analysis method (HAM). HAM, intended by Shi Jun Liao in 1992, depends on the fundamental concept of topology and differential geometry, homotopy. Being an analytical technique, HAMs are able to solve algebraic, ordinary/partial differential and differential-integral, and linear/non-linear equations in terms of series sum. It provides a broader way for selection of its arguments like initial guess, linear operator and convergence control parameter, which can be highly effective to control convergence rate of the solutions. This characteristic of HAM makes it preferable toother analytical techniques.

The objectives of this study are to theoretically analyze entropy generation and rate of heat transfer in steady flow of (Al_2O_3-Cu/H_2O) hybrid nanofluid induced due to radially stretching disk by imposing convective-type thermal conditions. Flow is axisymmetric in which all the flow variables are independent of angular dimension. The influence of induced magnetic field is taken into account. Flow is considered in the presence of viscous dissipation and thermal radiation effects. The complete mechanism of the present study is explored in terms of a workflow diagram in Figure 1.

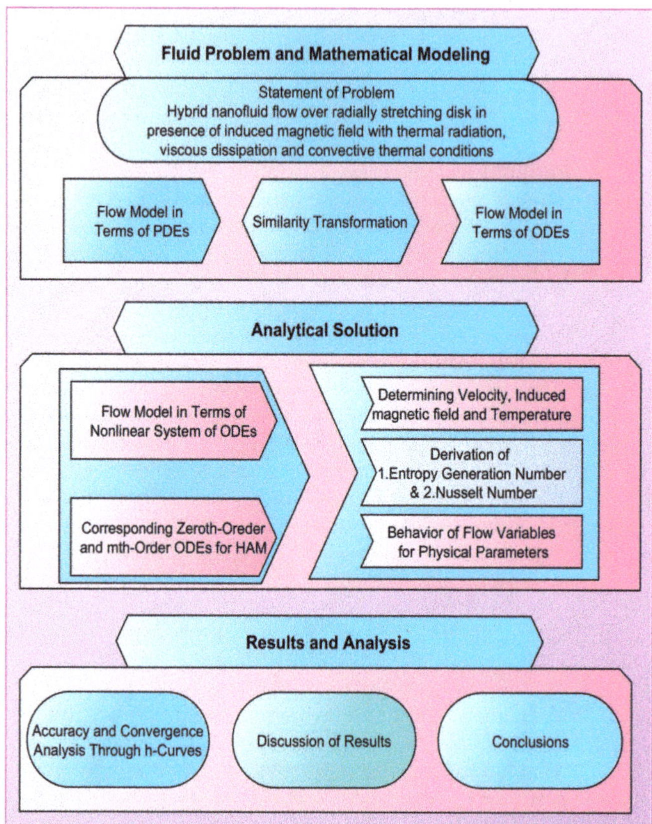

Figure 1. Workflow chart.

2. Modeling and Problem

Steady, boundary layer flow of viscous (Al_2O_3-Cu/H_2O) hybrid nanofluid induced due to stretching disk in radial directions is assumed. The volume concentration of Al_2O_3 and Cu nanomaterials is taken to be 0.05%. The stretching velocity of the disk surface is $U_w(r) = ar$ where a represents positive constant. The surface of disk is convectively heated by fluid having temperature T and held in plane $z = 0$ while hybrid nanofluid is flowing in the region $z > 0$ as shown in Figure 2. Moreover, a scheme for manufacturing of nanofluid and hybrid nanofluid for the nanoparticles in the present study is presented in Figure 3.

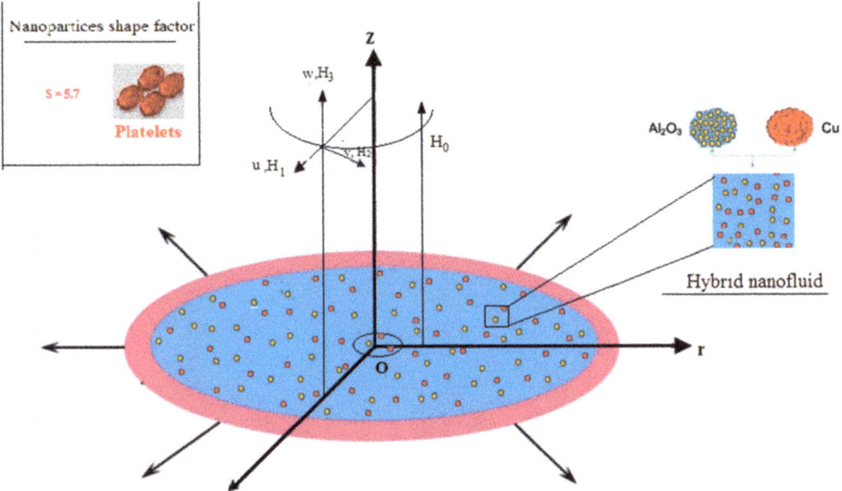

Figure 2. Schematic representation.

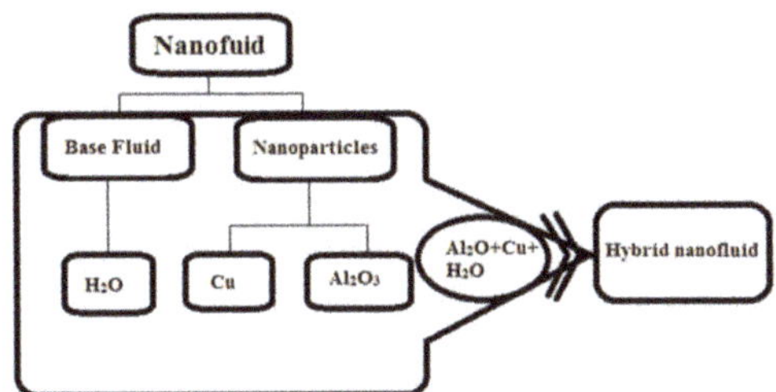

Figure 3. Preparation scheme of nanofluid and hybrid nanofluids.

The constitutive governing model along with induced magnetic field, thermal radiations and viscous dissipation effects under boundary layer approximation is:

$$\frac{\partial u}{\partial r} + \frac{u}{r} + \frac{\partial w}{\partial z} = 0, \tag{1}$$

$$\frac{\partial H_1}{\partial r} + \frac{H_1}{r} + \frac{\partial H_3}{\partial z} = 0, \tag{2}$$

$$u\frac{\partial u}{\partial r} + w\frac{\partial u}{\partial z} - \frac{\hat{\mu}}{4\pi\rho_f}\left(H_1\frac{\partial H_1}{\partial r} + H_3\frac{\partial H_1}{\partial z}\right) = -\frac{\hat{\mu}}{4\pi\rho_f}H_e\frac{dH_e}{dr} + \left(\frac{\mu_{hnf}}{\rho_{hnf}}\right)\frac{\partial^2 u}{\partial z^2}, \tag{3}$$

$$u\frac{\partial H_1}{\partial r} + w\frac{\partial H_1}{\partial z} - H_1\frac{\partial u}{\partial r} - H_3\frac{\partial u}{\partial z} = \mu_e\frac{\partial^2 H_1}{\partial z^2}, \tag{4}$$

$$u\frac{\partial T}{\partial r} + w\frac{\partial T}{\partial z} = \alpha_{hnf}\frac{\partial^2 T}{\partial z^2} + \left(\frac{\mu_{hnf}}{(\rho c_p)_{hnf}}\right)\left(\frac{\partial u}{\partial z}\right)^2 - \left(\frac{1}{(\rho c_p)_{hnf}}\right)\frac{\partial q_r}{\partial z}, \tag{5}$$

Radial, axial and azimuthal components of the induced magnetic field vector are, H_1, H_2 and H_3, respectively. Corresponding boundary conditions are:

$$u = U_w(r) = ar, \quad -\kappa_{hnf}\frac{\partial T}{\partial z} = h(T_w - T), \quad w = 0, \quad \frac{\partial H_1}{\partial z} = 0,$$
$$H_2 = H_3 = 0, \; at \; z = 0, \tag{6}$$
$$u = w = 0, \quad T \to T_\infty, \quad H_1 = H_e(r) = H_0 r, \; as \; z \to \infty.$$

In Equation (5), expression for q_r by using Roseland approximation and Taylor series expansion of T^4 about T_∞ can be expressed as:

$$q_r = -\frac{4\sigma}{3k}\frac{\partial T^4}{\partial z} = -\frac{16\sigma T_\infty^3}{3k}\frac{\partial T}{\partial z}, \tag{7}$$

Now, using similarity transformation:

$$u(r,z) = arf'(\eta), \quad w(r,z) = -2\sqrt{a\nu_f}f(\eta), \quad \eta = \sqrt{\frac{a}{\nu_f}}z,$$
$$H_1 = H_0 rg'(\eta), \quad H_3 = -2\sqrt{a\nu_f}g(\eta), \quad \theta(\eta) = \frac{T - T_\infty}{T_w - T_\infty}. \tag{8}$$

For the above transformations, Equations (1) and (2) are identically satisfied while Equations (3)–(5) within boundary conditions (6) and Equation (7) gives:

$$f''' - \Phi_1\Phi_2\left\{f'^2 - 2ff'' - \beta(g'^2 - 2gg'' - 1)\right\} = 0, \tag{9}$$

$$\lambda g''' + 2fg'' - 2f''g = 0, \tag{10}$$

$$\left(\Phi_4 + \frac{4}{3}R\right)\theta'' + 2\Phi_3 Pr.f\theta' + \frac{Pr \cdot Ec}{\Phi_1}f''^2 = 0 \tag{11}$$

The transformed boundary conditions are:

$$f(\eta) = g(\eta) = 0, \quad f'(\eta) = 1, \quad g''(\eta) = 0, \; \theta'(\eta) = -\frac{Bi}{\Phi_4}(1 - \theta(0)), \; at \; \eta = 0,$$
$$f'(\eta) \to 0, \quad g'(\eta) \to 1, \quad \theta(\eta) \to 0, \quad as \; \eta \to \infty. \tag{12}$$

where prime represents differentiation with η; respect to
Moreover, expressions for thermophysical properties are:

$$\Phi_1 = (1 - \varphi_1)^{2.5}(1 - \varphi_2)^{2.5}, \quad \Phi_2 = (1 - \varphi_2)\left[(1 - \varphi_1) + \varphi_1\left(\frac{\rho_{s_1}}{\rho_f}\right)\right] + \varphi_2\left(\frac{\rho_{s_2}}{\rho_f}\right),$$
$$\Phi_3 = (1 - \varphi_2)\left[(1 - \varphi_1) + \varphi_1\left(\frac{(\rho c_p)_{s_1}}{(\rho c_p)_f}\right)\right] + \varphi_2\frac{(\rho c_p)_{s_2}}{(\rho c_p)_f}, \tag{13}$$
$$\Phi_4 = \frac{\kappa_{s_2} + (s-1)\kappa_{bf} - (s-1)\varphi_2(\kappa_{bf} - \kappa_{s_2})}{\kappa_{s_2} + (s-1)\kappa_{bf} + \varphi_2(\kappa_{bf} - \kappa_{s_2})}\frac{\kappa_{s_1} + (s-1)\kappa_f - (s-1)\varphi_1(\kappa_f - \kappa_{s_1})}{\kappa_{s_1} + (s-1)\kappa_f + \varphi_1(\kappa_f - \kappa_{s_1})},$$

Moreover, all the thermophysical characteristics of nanoparticles and base fluid are mentioned in Table 1 while expressions for dimensionless parameters are:

$$Pr = \frac{(\mu c_p)_f}{\kappa_f}, \quad \lambda = \frac{\mu_e}{\nu_f}, \quad \beta = \frac{\mu}{4\pi\rho_f}\left(\frac{H_0}{a}\right)^2,$$
$$Ec = \frac{U_w^2}{(c_p)_f(T_w - T_\infty)}, \quad R = \frac{4\sigma T_\infty^3}{k\kappa_f}, \quad Bi = \frac{h}{\kappa_f}\sqrt{\frac{\nu_f}{a}}, \tag{14}$$

Table 1. Experimental values of various thermophysical properties for base fluid and nanoparticles [27].

Properties\Constituents	H_2O	Al_2O_3	Cu
Density, ρ (Kg/m^3)	997	3970	8933
Specific heat, C_p (J/kg K)	4179	765	385
Thermal conductivity, κ (W/m K)	0.613	40	401

Expressions for coefficient of skin friction C_f and local Nusselt number Nu are:

$$C_f = \frac{\tau_w}{\rho_f (U_w)^2} \quad, Nu = -\frac{r q_w}{\kappa_f (T_w - T_\infty)}. \tag{15}$$

In the above expressions, τ_w and q_w symbolize shear stress and heat flux for wall, respectively. The dimensionless form by substituting Equation (8) is:

$$Re_r^{\frac{1}{2}} C_f = \frac{-1}{\Phi_1} f''(0), \quad Re_r^{\frac{-1}{2}} Nu = -\left(\Phi_4 + \frac{4}{3}R\right)\theta'(0), \tag{16}$$

where, $Re_r = \frac{U_w r}{v_f}$ indicates local Reynolds number.

By adopting the second law of thermodynamics, volumetric entropy generation rate in existence of radiative and dissipative factors can be expressed as:

$$\dot{S}'''_{Gen} = \left(\Phi_4 + \frac{4}{3}R\right)\frac{\kappa_f}{T_\infty^2}\left(\frac{\partial T}{\partial y}\right)^2 + \frac{\mu_f}{\Phi_1 T_\infty}\left(\frac{\partial u}{\partial y}\right)^2 + \frac{\mu_f}{\Phi_1 T_\infty}\left(\frac{\hat{\mu}}{4\pi\rho_f}\right)\left(\frac{\partial H_1}{\partial y}\right)^2. \tag{17}$$

The expression for characteristic entropy generation rate is:

$$\dot{S}'''_0 = \frac{\kappa_f}{T_\infty^2 L^2}(T_w - T_\infty)^2. \tag{18}$$

Utilizing similarity transformation with Equation (18) in Equation (17), we have:

$$N_S = \frac{\dot{S}'''_{Gen}}{\dot{S}'''_0} = \left(\Phi_4 + \frac{4}{3}R\right)Re_L.\theta'^2 + \frac{\varepsilon.Pr.Ec.Re_L}{\Phi_1}f''^2 + \frac{\varepsilon.Pr.Ec.Re_L.\beta}{\Phi_1}g''^2. \tag{19}$$

where, Ns is non-dimensional form of entropy generation number, $Re_L = \frac{aL^2}{v_f}$ and $\varepsilon = \frac{T_\infty}{T_w - T_\infty}$ demonstrate local Reynolds number and temperature ratio parameter, respectively.

3. Homotopy Analysis Method and Convergence of Solutions

The dimensionless governing model mentioned in Equations (9)–(11) and subjected boundary conditions of Equation (12) for boundary layer flow of hybrid nanofluid over the stretching disk is analytically solved by employing the homotopy analysis method [7,37–39]. For flow variables, initial guesses are:

$$f_0(\eta) = 1 - \exp(-\eta), g_0(\eta) = \eta, \ \theta_0(\eta) = \frac{Bi}{1 + Bi}\exp(-\eta). \tag{20}$$

Expressions for linear operators are:

$$T_1(\eta) = f'''(\eta) - f'(\eta), \ T_2(\eta) = g'''(\eta) + g''(\eta), \ T_3(\eta) = \theta''(\eta). \tag{21}$$

and

$$T_1\left(A_1 + A_2 e^{-\eta} - A_3 e^{-\eta}\right) = T_2\left(A_4 + A_5\eta + A_6 e^{-\eta}\right) = T_3(A_7 + A_8\eta) = 0. \tag{22}$$

where, A_1–A_8 represents constants in general solutions.

3.1. Convergence-Control Parameters

Suppose an h is auxiliary parameter in the frame of HAM which directly affects the convergence of solutions. Let $\xi \in [0,1]$ be an embedding parameter, then the problem for zeroth order deformation is constructed as:

$$(1 - \xi)T_1[f(\eta,\xi) - f_0(\eta)] = \xi h R_1[f(\eta,\xi), g(\eta,\xi), \theta(\eta,\xi)], \tag{23}$$

$$(1 - \xi)T_2[g(\eta,\xi) - g_0(\eta)] = \xi h R_2[f(\eta,\xi), g(\eta,\xi), \theta(\eta,\xi)], \tag{24}$$

$$(1 - \xi)T_3[\theta(\eta,\xi) - \theta_0(\eta)] = \xi h R_3[f(\eta,\xi), g(\eta,\xi), \theta(\eta,\xi)]. \tag{25}$$

Furthermore,

$$\text{at } \eta = \eta_1 = 0, \ f(\eta_1,\xi) = g(\eta_1,\xi) = 0, \ f'(\eta_1,\xi) = 1, \ g''(\eta_1,\xi) = 0, \ \theta'(\eta_1,\xi) = -\frac{Bi}{A_4}(1 - \theta(\eta_1,\xi)).$$
$$\text{At } \eta = \eta_2 \to \infty, f'(\eta_2,\xi) \to 0, g'(\eta_2,\xi) \to 1, \ \theta(\eta_2,\xi) \to 0. \tag{26}$$

Using the aforementioned quantities, the mth order solution series is constructed as:

$$T_1[f_m(\eta,\xi) - \varsigma_m f_{m-1}(\eta,\xi)] = h S_m^1[(\eta,\xi)], \tag{27}$$

$$T_2[g_m(\eta,\xi) - \varsigma_m g_{m-1}(\eta,\xi)] = h S_m^2[(\eta,\xi)], \tag{28}$$

$$T_3[\theta_m(\eta,\xi) - \varsigma_m \theta_{m-1}(\eta,\xi)] = h S_m^3[(\eta,\xi)]. \tag{29}$$

In the above equations,

$$\varsigma_m = \left. \begin{array}{l} 0 m \leq 1 \\ 1 m > 1 \end{array} \right]$$

Subjected boundary conditions are:

$$\text{At } \eta = \eta_1 = 0, \ f_m(\eta_1,\xi) = g_m(\eta_1,\xi) = 0, \ f'{}_m(\eta_1,\xi) = 0, \ g''{}_m(\eta_1,\xi) = 0,$$
$$\theta_m(\eta_1,\xi) = 0,$$
$$\text{At } \eta = \eta_2 \to \infty, f'{}_m(\eta_2) \to 0, g'{}_m(\eta_2) \to 0, \ \theta_m(\eta_2) \to 0. \tag{30}$$

Then, we can write:

$$\begin{aligned} f_m(\eta,\xi) &= F_m(\eta,\xi) + A_1 + A_2 e^{-\eta} - A_3 e^{-\eta}, \\ g_m(\eta,\xi) &= G_m(\eta,\xi) + A_4 + A_5 \eta + A_6 e^{-\eta}, \\ \theta_m(\eta,\xi) &= \Theta_m(\eta,\xi) + A_7 + A_8 \eta. \end{aligned} \tag{31}$$

3.2. Convergence of Solutions

For stability and convergence of analytical solutions, h-curves for, $f'(0)$ $g'(0)$ and $\theta(0)$ are prepared upto 20th order solutions by plotting the interval of convergence and are shown in Figures 4–6. Convergence control parameter h is able to control and adjust convergence region of HAM solutions. It is observed from convergence analysis that valid convergence region is $R_h = [-0.4, 0.2]$. For $f'(0)$, $R_h = [-0.35, 0.2]$ for $g'(0)$, and $R_h = [-0.35, 0.15]$ for $\theta(0)$. Moreover, testing various values of h from corresponding regions, the error is minimum at $h = -0.1$ which guarantees that calculated solutions are very accurate with a negligible error and become more accurate for higher-order solutions.

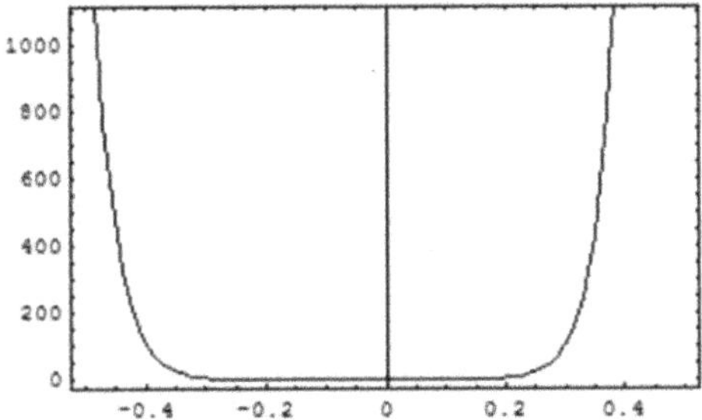

Figure 4. h-curves for solution of $f'(0)$.

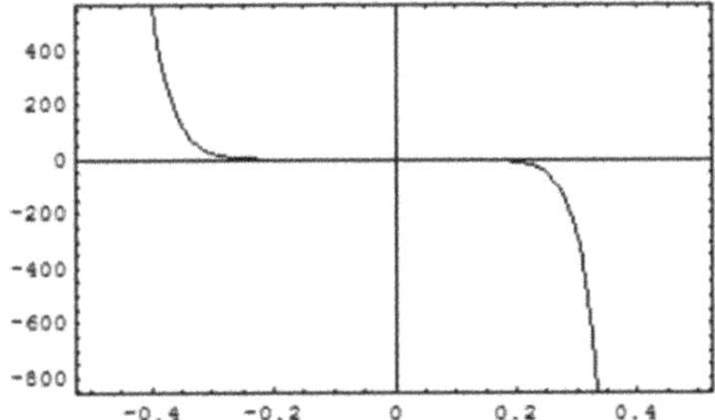

Figure 5. h-curves for solution of $g'(0)$.

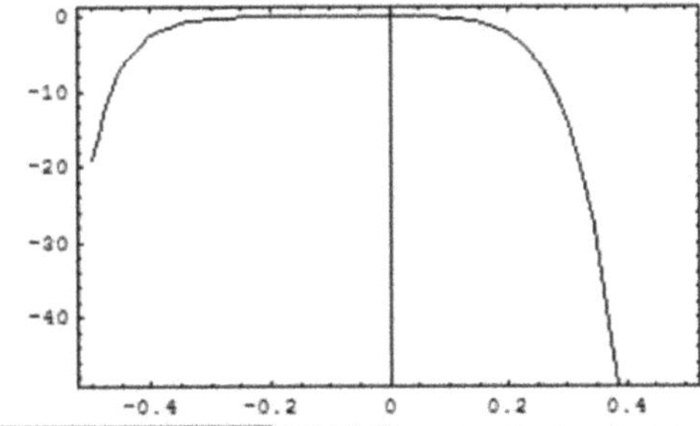

Figure 6. h-curves for the solution of $\theta(0)$.

4. Discussion of Results

Analytical solutions for the Al_2O_3-Cu/water hybrid nanofluid flow obtained by HAM are discussed in this section. Figures 7 and 8 plot the variation in magnetic parameter β and the reciprocal of magnetic Prandtl number λ for the velocity profile. The graph indicates that rise in values of β enhances $f'(\eta)$ due to more dominant induction effects than magnetic diffusion and the flow rate increases. A drop in $f'(\eta)$ is noticed for λ because magnetic diffusivity rises with rise in λ. This effect causes enhancement of frictional force and the velocity boundary layer thickness reduces.

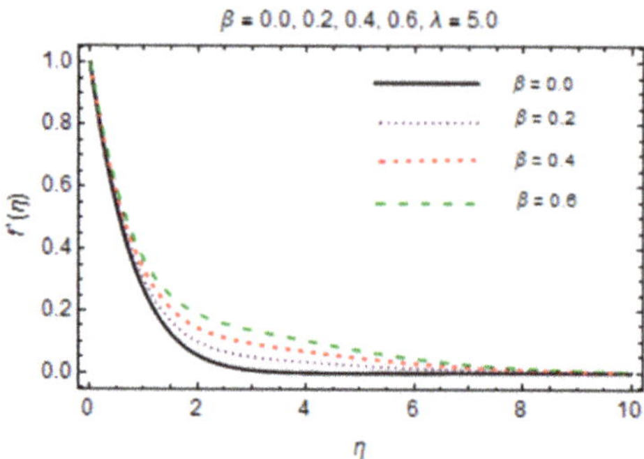

Figure 7. β verses velocity profile.

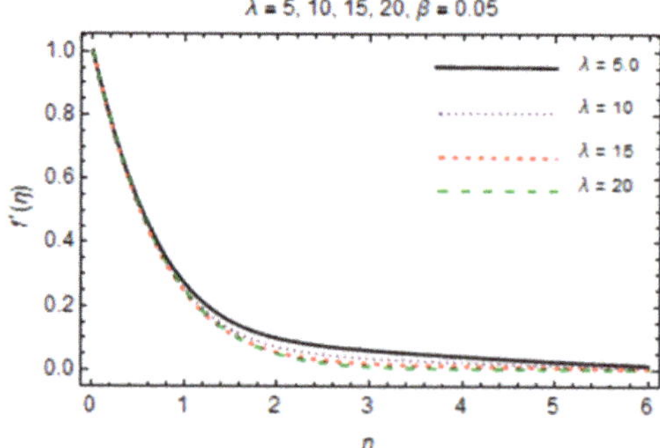

Figure 8. λ verses velocity profile.

The behavior of induced magnetic field profile against parameters β and λ is explored in Figures 9 and 10, respectively. Graphs demonstrate that with augmentation in β, magnetic effects become strong as due to fast advection process, therefore increasing flow rate amplifies magnetic induction profile. Consecutively, $g'(\eta)$ is a decreasing function of λ which is mainly due to enhancing magnetic diffusivity with high values of λ. It is noteworthy that the influences of parameters on $f'(\eta)$ and $g'(\eta)$ are more prominent at the interface and very little variation is shown near stretching disk due to the no slip boundary condition.

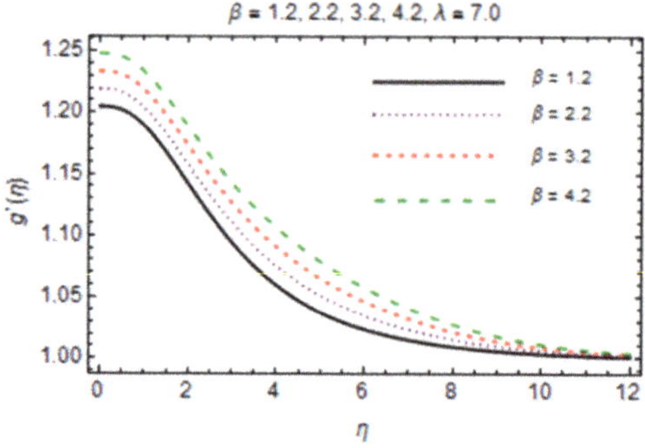

Figure 9. β verses induced magnetic field profile.

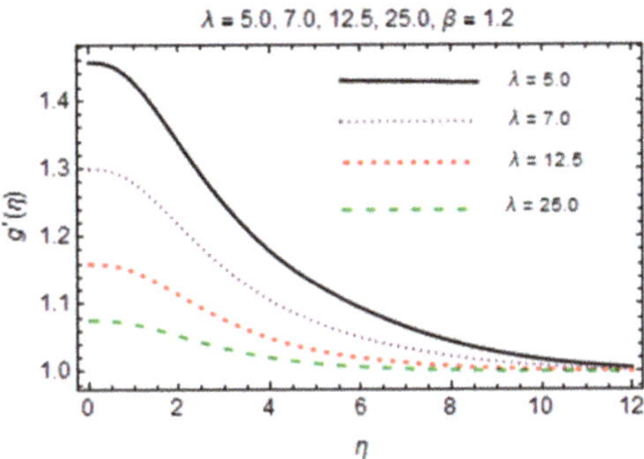

Figure 10. λ verses induced magnetic field profile.

Figures 11–14 are plotted in order to express the variation in temperature against magnetic parameter, reciprocal of magnetic Prandtl number, radiation parameter and Biot number. Results reveal that $\theta(\eta)$ decreases with increment in β which is caused by increasing heat transfer rate near the disk in existence of induced magnetic field due to adding the flow mechanism as explored in Figure 11. The impact of radiation parameter in Figure 12 depicts the dual behavior of the $\theta(\eta)$ profile which is depressed close to the disk and elevated away from it showing dominant effects at free stream. Since the disk is convectively heated and these effects trim down away from its surface, temperature enhances at the disk surface due to rising intensity of convective heating in Figure 13.

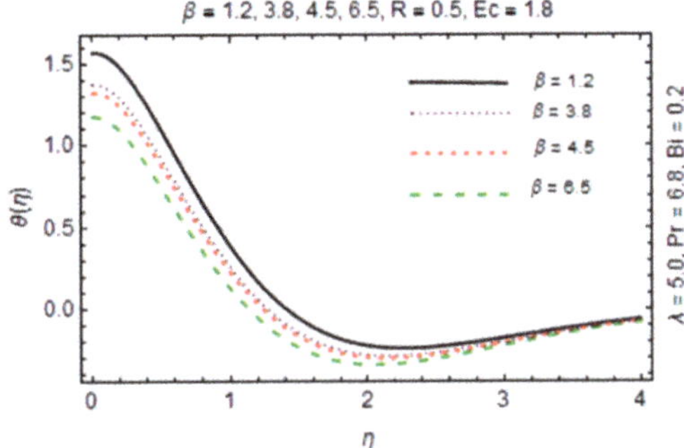

Figure 11. β verses temperature profile.

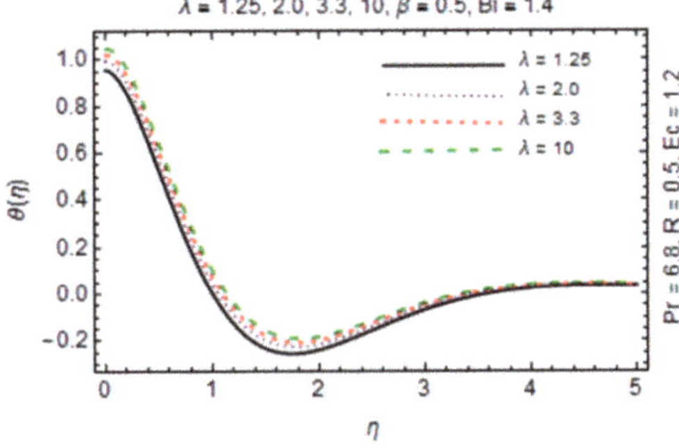

Figure 12. λ verses temperature profile.

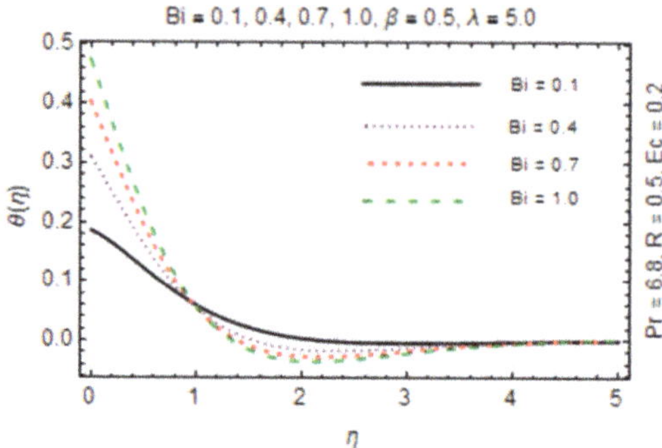

Figure 13. *Bi* verses temperature profile.

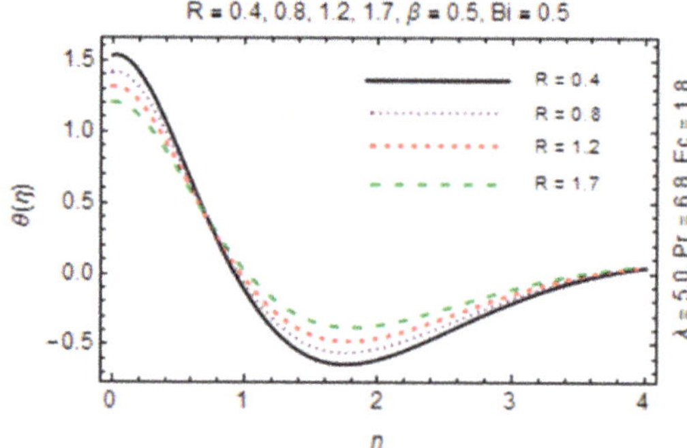

Figure 14. *R* verses temperature profile.

An opposite trend is observed in the free stream region due to the fact that additional heat released to the coolant at the surface. The impact of λ on temperature of hybrid nanofluid is exposed in Figure 14 which signifies that temperature rises against λ. The reason is high diffusivity and a small flow rate due to low magnetic induction for large magnitude of λ. Moreover, variation in $\theta(\eta)$ directly associated with values of parameters Bi, R and Ec which are taken to be small with $Pr = 6.8$ in this study.

The most persuasive part of this section is entropy generation analysis. The variation in entropy generation number Ns against emerging parameters is pointed out graphically in Figures 15–18. Impact of magnetic parameter in Figure 15 demonstrates that rise in β decreases entropy production as strong effects of magnetic field close to the disk reduce the frictional effects and rise in heat transfer rate. It is of true significance physically since the rate of fluid flow enhances instantly and heat transfer rate increases. The number slightly increases in the free stream region due to decreasing fluid flow. Figure 16 expresses an enhancement in Ns as values of reciprocal of magnetic Prandtl number rises. This is because of the fact that as λ enhances, reduction in viscosity occurs while magnetic diffusivity accelerates, which produce disorder in the system. In Figure 17, entropy generation number is plotted against Bi which illustrates that for somewhat large values of parameter Bi, Ns have high magnitude due to strong influences of thermal convection and maximum radial gradient. Additionally, thermal radiation effects on Ns in Figure 18 displays it as an increasing function of R due to increasing emitting radiations which boosts frictional irreversibility that encourage entropy generation. Furthermore, variations in entropy generation number are drawn for small values of radiation parameter R, Eckert number Ec and Biot number Bi which are directly related to entropy production. Also, extensive behavior for Ns against parameters is observed at the interface due to large velocity and temperature gradients caused by no slip at the surface and convective wall conditions but it is rarely affected by these parameters away from it.

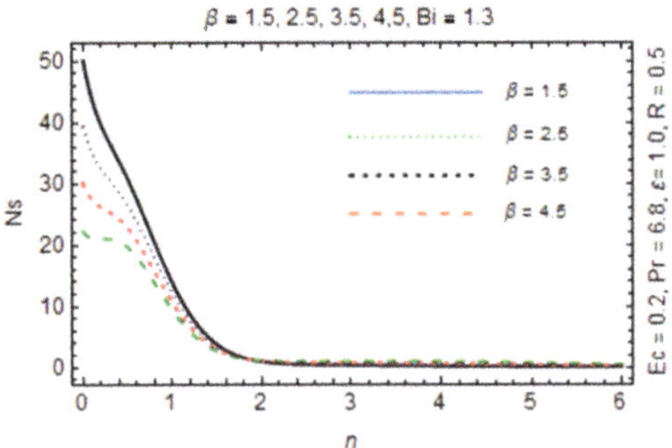

Figure 15. β verses entropy generation number.

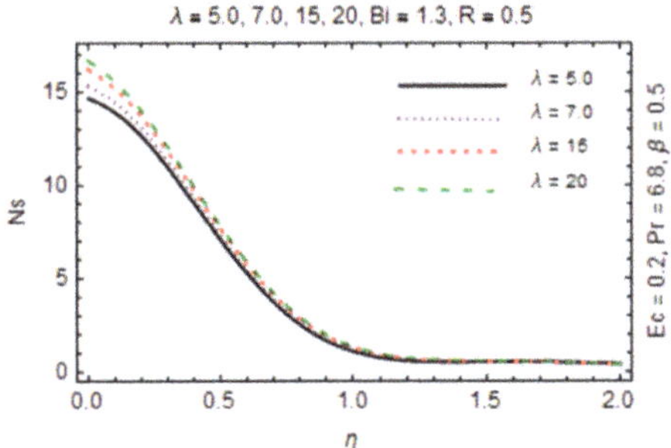

Figure 16. λ verses entropy generation number.

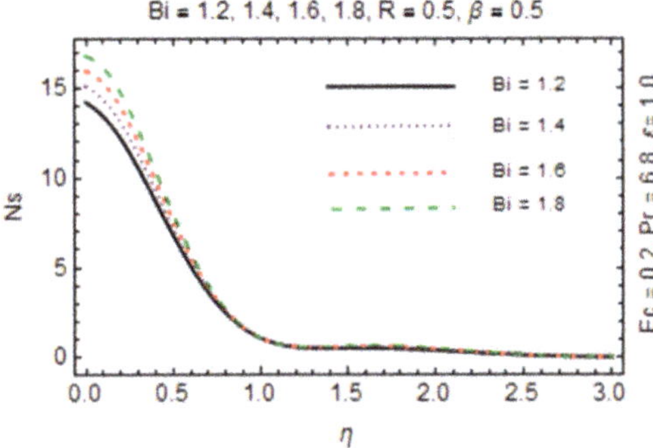

Figure 17. *Bi* verses entropy generation number.

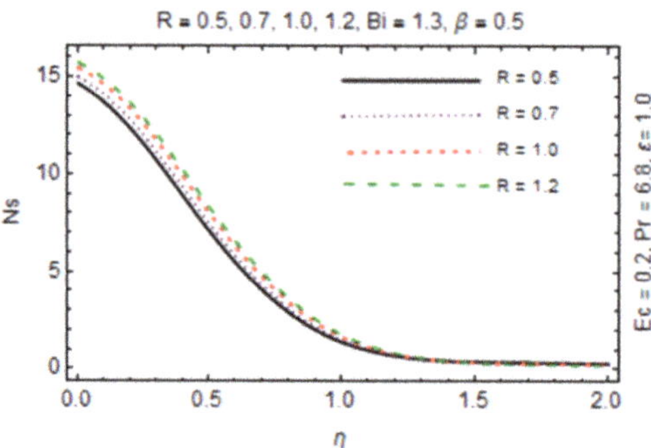

Figure 18. R verses entropy generation number.

Moreover, empirical formulas for the thermophysical properties of nanofluid and hybrid nanofluid are displayed in Table 2.

Table 2. Formulas for thermophysical properties of nanofluid and hybrid nanofluid [40].

Properties	Nanofluid	Hybrid Nanofluid
Density	$\rho_{nf} = \rho_f \left[(1-\varphi) + \varphi\left(\frac{\rho_s}{\rho_f}\right) \right]$	$\rho_{hnf} = \rho_f (1-\varphi_2)\left[(1-\varphi_1) + \varphi_1\left(\frac{\rho_{s_1}}{\rho_f}\right) \right] + \varphi_2 \rho_{s_2}$
Heat Capacity	$(\rho c_p)_{nf} = (\rho c_p)_f \left[(1-\varphi) + \varphi\left(\frac{(\rho c_p)_s}{(\rho c_p)_f}\right) \right]$	$(\rho c_p)_{hnf} =$ $(\rho c_p)_f (1-\varphi_2)\left[(1-\varphi_1) + \varphi_1\left(\frac{(\rho c_p)_{s_1}}{(\rho c_p)_f}\right) \right] + \varphi_2 (\rho c_p)_{s_2}$
Viscosity	$\mu_{nf} = \dfrac{\mu_f}{(1-\varphi)^{2.5}}$	$\mu_{hnf} = \dfrac{\mu_f}{(1-\varphi_1)^{2.5}(1-\varphi_2)^{2.5}}$
Thermal Conductivity	$\dfrac{\kappa_{nf}}{\kappa_f} = \dfrac{\kappa_s + (s-1)\kappa_f - (s-1)\varphi(\kappa_f - \kappa_s)}{\kappa_s + (s-1)\kappa_f + \varphi(\kappa_f - \kappa_s)}$	$\dfrac{\kappa_{hnf}}{\kappa_{bf}} = \dfrac{\kappa_{s_2} + (s-1)\kappa_{bf} - (s-1)\varphi_2(\kappa_{bf} - \kappa_{s_2})}{\kappa_{s_2} + (s-1)\kappa_{bf} + \varphi_2(\kappa_{bf} - \kappa_{s_2})}$, $where\ \dfrac{\kappa_{bf}}{\kappa_f} = \dfrac{\kappa_{s_1} + (s-1)\kappa_f - (s-1)\varphi_1(\kappa_f - \kappa_{s_1})}{\kappa_{s_1} + (s-1)\kappa_f + \varphi_1(\kappa_f - \kappa_{s_1})}$

The impacts of involving parameters on skin friction coefficient and heat transfer rate through Tables 3 and 4 as well as bar graphs in Figures 19–25 are explained in this section. Observations conclude that the surface velocity gradient is depressed as values of magnetic parameter β rises but enhances with elevation in λ. Moreover, an expansion in Nusselt number is noticed for rising values of the reciprocal of magnetic Prandtl number, Biot number, radiation parameter, and Eckert number whereas it is a decreasing function of magnetic parameter.

Table 3. Behavior of skin friction coefficient for Al_2O_3-Cu/H_2O against several parameters.

β	λ	$-C_f$
0.0	5.0	0.003467
0.5		0.002900
1.0		0.002331
1.5		0.001759
0.5	5.0	0.002900
	10	0.003185
	15	0.003280
	20	0.003327

Table 4. Behavior of Nusselt number for Al_2O_3-Cu/H_2O against several parameters.

β	λ	Bi	R	Ec	$-N_u$
0.0	5.0	1.5	2.5	0.2	0.026143
0.5					0.026134
1.0					0.026125
1.5					0.026117
0.5	5.0				0.026134
	10				0.026138
	15				0.026140
	20				0.026140
	5.0	0.1			0.001173
		0.5			0.011475
		1.0			0.021485
		1.5			0.026134
		1.5	0.5		0.004934
			1.0		0.00952
			2.0		0.022197
			2.5		0.026134
			2.5	0.2	0.031758
				0.4	0.033282
				0.6	0.034806
				0.8	0.034915

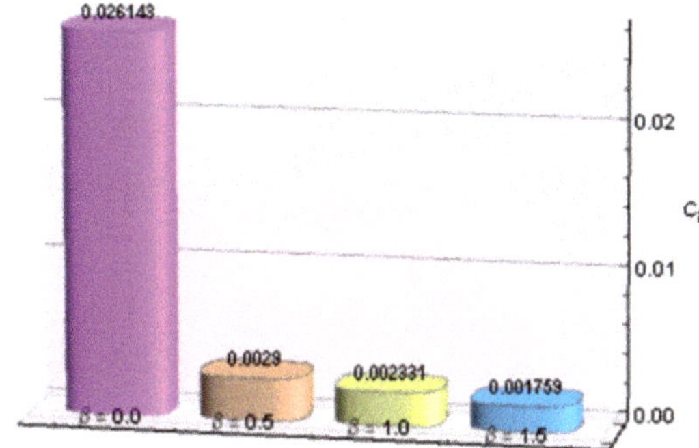

Figure 19. β verses C_f.

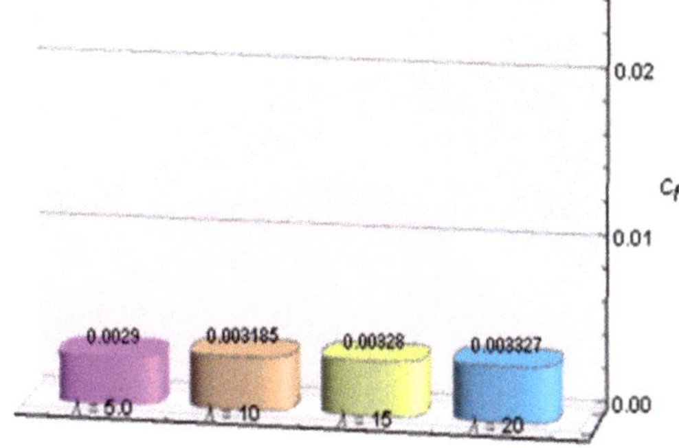

Figure 20. λ verses Cf.

Figure 21. β verses Nu.

Figure 22. λ verses Nu.

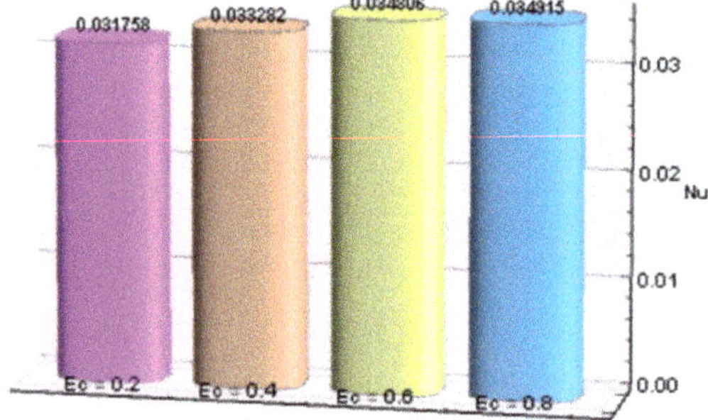

Figure 23. Ec verses Nu.

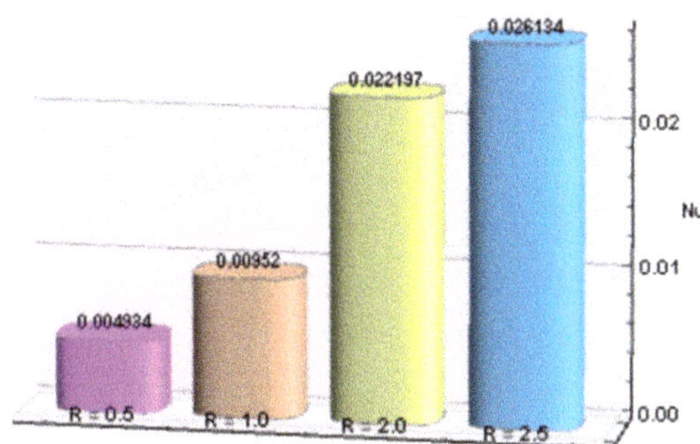

Figure 24. *R* verses *Nu*.

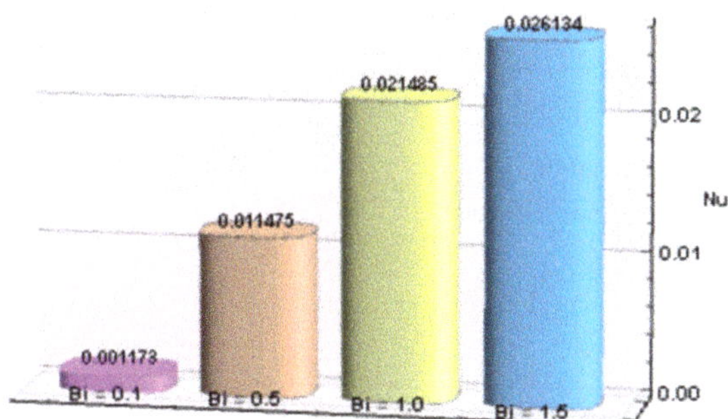

Figure 25. *Bi* verses *Nu*.

5. Concluding Remarks

A theoretical discussion of MHD viscous flow of Al_2O_3-Cu/H_2O hybrid nanofluid due to stretching of the disk is carried out. Some noteworthy influences of important emerging parameters on flow profiles and entropy generation are as follows:

1. $f'(\eta)$ and $g'(\eta)$ profiles are increasing functions of magnetic parameter while an opposite behavior is seen for enhancing values of λ.
2. Increment in β results in an enhancement in temperature, whereas it reduces against λ.
3. An increasing behavior for the temperature of the fluid is observed against rising values of parameters Bi and R at the surface. An opposite trend is depicted in the ambient region with a point of inflection in the field.
4. Entropy generation number enhances for enhancement in values of parameters Bi, R and λ but diminishes against β.
5. Flow profiles and entropy generation number are large near the surface of the disk and then decrease asymptotically far away from it. Also, these are more sensitive to fluctuate at the interface for several involved parameters.
6. At the interface, fluid temperature is significantly different from ambient temperature. Thus, convection at the wall corresponding to high values of convective heat coefficient leads to increased rate of heat transfer at the interface.
7. Values of skin friction and heat transfer coefficient can be optimized by choosing suitable values of involved parameters regarding different physical problems.

Author Contributions: Conceptualization, N.P.; Data curation, N.P.; Formal analysis, M.A. and M.Y.M.; Investigation, S.E.A.; Methodology, W.U.K.; Project administration, Y.H.; Resources, W.U.K. and Y.H.; Software, S.E.A. All authors have read and agreed to the published version of the manuscript.

Funding: This work was supported by theNational Natural Science Foundation of Chinaunder Grant No.51977153,51977161,51577046, State Key Program ofNational Natural Science Foundation of Chinaunder Grant No.51637004,National Key Research and Development Plan (China)"important scientific instruments and equipment development" Grant No.2016YFF010220,Equipment research project in advance (China)Grant No.41402040301, Wuhan Science and Technology Plan Project Grant No 20201G01.

Institutional Review Board Statement: Not applicable.

Informed Consent Statement: Not applicable.

Data Availability Statement: The data presented in this study is available in article.

Acknowledgments: All authors sincerely thank the editor and the referees for the useful suggestions regarding improvement of our work. M. Y. Malik present his appreciations to the Deanship of Scientific Research at King Khalid University, Abha, 61413, Saudi Arabia for support through R.G.P-1-223-42.

Conflicts of Interest: Authors have no conflict of interest.

Nomenclature

	Nomenclature		Greek Symbols
a	Constant	β	Magnetic parameter
Bi	Biot number	η	Dimensionless similarity variable
c_f	Skin friction coefficient	φ	Nanoparticles volume fraction
Ec	Eckert number	κ	Thermal conductivity
H_0	Uniform magnetic field	ρ	Density
h	Convective heat transfer coefficient	ρc_p	Heat capacity
k	Mean absorption coefficient	$\hat{\mu}$	Magnetic permeability
Nu	Nusselt number	μ_e	Magnetic diffusivity
Pr	Prandtl number	μ	Dynamic viscosity
q_r	Radiative heat flux	v	Kinematic viscosity
q_w	Heat transfer at wall	σ	Stefan-Boltzmann constant
R	Radiation parameter	θ	Dimensionless temperature
Re	Reynolds number	τ_w	Wall shear stress
s	Nanoparticles shape factor	ε	Temperature ratio parameter
T	Fluid temperature		**Subscripts**
T_w	Surface temperature	hnf	Hybrid nanofluid
T_∞	Ambient temperature	f	Base fluid
u, w	Velocity components in r-, z- directions.	w	Surface condition
U_w	Stretching velocity	∞	Ambient condition
U_∞	Ambient velocity	s_1, s_2	Shape factors of Copper andAlumina

References

1. Sakiadis, B.C. Boundary layer phenomenon. *J. AIChE* **1961**, *7*, 26–28. [CrossRef]
2. Crane, L.J. Flow past a stretching plate. *Z. Angew. Math. Phys.* **1970**, *21*, 645–647. [CrossRef]
3. Khan, A.S.; Nie, Y.; Shah, Z.; Dawar, A.; Khan, W.; Islam, S. Three-Dimensional Nanofluid Flow with Heat and Mass Transfer Analysis over a Linear Stretching Surface with Convective Boundary Conditions. *Appl. Sci.* **2018**, *8*, 2244. [CrossRef]
4. Gireesha, B.J.; Umeshaiah, M.; Prasannakumara, B.C.; Shashikumar, N.S.; Archana, N. Impact of nonlinear thermal radiation on magnetohydrodynamic three dimensional boundary layer flow of Jeffrey nanofluid over a nonlinearly permeable stretching sheet. *Phys. A Stat. Mech. Appl.* **2020**, *549*, 124051.
5. Hayat, T.; Khan, W.A.; Abbas, S.Z.; Nadeem, S.; Ahmad, S. Impact of induced magnetic field on second-grade nanofluid flow past a convectively heated stretching sheet. *Appl. Nanosci.* **2020**, *10*, 3001–3009. [CrossRef]
6. Rafiq, M.; Siddiqa, S.; Begum, N.; Al-Mdallal, Q.; Hossain, M.; Gorla, R.S.R. Non-linear radiation effect on dusty fluid flow near a rotating blunt-nosed body. *Proc. Inst. Mech. Eng. Part E J. Process. Mech. Eng.* **2021**. [CrossRef]
7. Khan, W.; Awais, M.; Parveen, N.; Ali, A.; Awan, S.; Malik, M.; He, Y. Analytical Assessment of (Al_2O_3–Ag/H_2O) Hybrid Nanofluid Influenced by Induced Magnetic Field for Second Law Analysis with Mixed Convection, Viscous Dissipation and Heat Generation. *Coatings* **2021**, *11*, 498. [CrossRef]

8. Adeniyan, A.; Mabood, F.; Okoya, S. Effect of heat radiating and generating second-grade mixed convection flow over a vertical slender cylinder with variable physical properties. *Int. Commun. Heat Mass Transf.* **2021**, *121*, 105110. [CrossRef]

9. Iyyappan, G.; Singh, A.K. MHD flows on irregular boundary over a diverging channel with viscous dissipation effect. *Int. J. Numer. Methods Heat Fluid Flow* **2021**, *31*, 2112–2127. [CrossRef]

10. Amjad, M.; Zehra, I.; Nadeem, S.; Abbas, N.; Saleem, A.; Issakhov, A. Influence of Lorentz force and induced magnetic field effects on Casson micropolar nanofluid flow over a permeable curved stretching/shrinking surface under the stagnation region. *Surf. Interfaces* **2020**, *21*, 100766. [CrossRef]

11. Xenos, M.A.; Petropoulou, E.N.; Siokis, A.; Mahabaleshwar, U.S. Solving the Nonlinear Boundary Layer Flow Equations with Pressure Gradient and Radiation. *Symmetry* **2020**, *12*, 710. [CrossRef]

12. Singh, A.K.; Roy, S. Analysis of mixed convection in water boundary layer flows over a moving vertical plate with variable viscosity and Prandtl number. *Int. J. Numer. Methods Heat Fluid Flow* **2019**, *29*, 602–616. [CrossRef]

13. Bilal, M.; Nazeer, M. Numerical analysis for the non-Newtonian flow over stratified stretching/shrinking inclined sheet with the aligned magnetic field and nonlinear convection. *Arch. Appl. Mech.* **2021**, *91*, 949–964. [CrossRef]

14. Agrawal, P.; Dadheech, P.K.; Jat, R.; Bohra, M.; Nisar, K.S.; Khan, I. Lie similarity analysis of MHD flow past a stretching surface embedded in porous medium along with imposed heat source/sink and variable viscosity. *J. Mater. Res. Technol.* **2020**, *9*, 10045–10053. [CrossRef]

15. Choi, S.U.S.; Eastman, J.A. *Enhancing Thermal Conductivity of Fluids with Nano Particles*; Argonne National Laboratory: Lemont, IL, USA, 1995; Volume 66, pp. 99–105.

16. Puneeth, V.; Manjunatha, S.; Gireesha, B.J.; Gorla, R.S.R. Magneto convective flow of Casson nanofluid due to Stefan blowing in the presence of bio-active mixers. *Proc. Inst. Mech. Eng. Part N J. Nanomater. Nanoeng. Nanosyst.* **2021**, *48*, 1137–1149. [CrossRef]

17. Awais, M.; Kumam, P.; Parveen, N.; Ali, A.; Shah, Z.; Thounthong, P. Slip and Hall Effects on Peristaltic Rheology of Copper-Water Nanomaterial through Generalized Complaint Walls With Variable Viscosity. *Front. Phys.* **2020**, *7*, 7. [CrossRef]

18. Liu, X.; Mohammed, H.I.; Ashkezari, A.Z.; Shahsavar, A.; Hussein, A.K.; Rostami, S. An experimental investigation on the rheological behavior of nanofluids made by suspending multi-walled carbon nanotubes in liquid paraffin. *J. Mol. Liq.* **2020**, *300*, 112269. [CrossRef]

19. Hu, Y.; He, Y.; Zhang, Z.; Wen, D. Enhanced heat capacity of binary nitrate eutectic salt-silica nanofluid for solar energy storage. *Sol. Energy Mater. Sol. Cells* **2019**, *192*, 94–102. [CrossRef]

20. Li, F.; Muhammad, N.; Abohamzeh, E.; Hakeem, A.A.; Hajizadeh, M.R.; Li, Z.; Bach, Q.-V. Finned unit solidification with use of nanoparticles improved PCM. *J. Mol. Liq.* **2020**, *314*, 113659. [CrossRef]

21. Bazdar, H.; Toghraie, D.; Pourfattah, F.; Akbari, O.A.; Nguyen, H.; Asadi, A. Numerical investigation of turbulent flow and heat transfer of nanofluid inside a wavy microchannel with different wavelengths. *J. Therm. Anal. Calorim.* **2020**, *139*, 2365–2380. [CrossRef]

22. Tarakaramu, N.; Narayana, P.S.; Venkateswarlu, B. Numerical simulation of variable thermal conductivity on 3D flow of nanofluid over a stretching sheet. *Nonlinear Eng.* **2020**, *9*, 233–243. [CrossRef]

23. Waini, I.; Ishak, A.; Pop, I. MHD flow and heat transfer of a hybrid nanofluid past a permeable stretching/shrinking wedge. *Appl. Math. Mech.* **2020**, *41*, 507–520. [CrossRef]

24. Parveen, N.; Awais, M.; Mumraz, S.; Ali, A.; Malik, M.Y. An estimation of pressure rise and heat transfer rate for hybrid nanofluid with endoscopic effects and induced magnetic field: Computational intelligence application. *Eur. Phys. J. Plus* **2020**, *135*, 1–41. [CrossRef]

25. Radhika, M.; Gowda, R.J.P.; Naveenkumar, R.; Siddabasappa; Prasannakumara, B.C. Heat transfer in dusty fluid with suspended hybrid nanoparticles over a melting surface. *Heat Transfer* **2021**, *50*, 2150–2167. [CrossRef]

26. Reddy, M.G.; Rani, M.V.V.N.L.S.; Kumar, K.G.; Prasannakumar, B.C.; Lokesh, H.J. Hybrid dusty fluid flow through a Cattaneo–Christov heat flux model. *Phys. A Stat. Mech. Its Appl.* **2020**, *551*, 123975.

27. Khan, A.M.; Jamil, M.; Mia, M.; He, N.; Zhao, W.; Gong, L. Sustainability-based performance evaluation of hybrid nanofluid assisted machining. *J. Clean. Prod.* **2020**, *257*, 120541. [CrossRef]

28. LLund, A.; Omar, Z.; Khan, I.; Sherif, E.S.M. Dual solutions and stability analysis of a hybrid nanofluid over a stretching/shrinking sheet executing MHD flow. *Symmetry* **2020**, *12*, 276. [CrossRef]

29. Qureshi, I.H.; Awais, M.; Awan, S.E.; Abrar, M.N.; Raja, M.A.Z.; Alharbi, S.O.; Khan, I. Influence of radially magnetic field properties in a peristaltic flow with internal heat generation: Numerical treatment. *Case Stud. Therm. Eng.* **2021**, *26*, 101019. [CrossRef]

30. Ellahi, R.; Hussain, F.; Abbas, S.A.; Sarafraz, M.M.; Goodarzi, M.; Shadloo, M.S. Study of Two-Phase Newtonian Nanofluid Flow Hybrid with Hafnium Particles under the Effects of Slip. *Inventions* **2020**, *5*, 6. [CrossRef]

31. Venkateswarlu, B.; Narayana, P.V.S. Cu-Al$_2$O$_3$/H$_2$O hybrid nanofluid flow past a porous stretching sheet due to temperatue-dependent viscosity and viscous dissipation. *Heat Transf.* **2021**, *50*, 432–449. [CrossRef]

32. Dogonchi, A.; Tayebi, T.; Karimi, N.; Chamkha, A.J.; Alhumade, H. Thermal-natural convection and entropy production behavior of hybrid nanoliquid flow under the effects of magnetic field through a porous wavy cavity embodies three circular cylinders. *J. Taiwan Inst. Chem. Eng.* **2021**, *124*, 162–173. [CrossRef]

33. Sahoo, A.; Nandkeolyar, R. Entropy generation and dissipative heat transfer analysis of mixed convective hydromagnetic flow of a Casson nanofluid with thermal radiation and Hall current. *Sci. Rep.* **2021**, *11*, 1–31. [CrossRef]

34. Shoaib, M.; Raja, M.A.Z.; Khan, M.A.R.; Farhat, I.; Awan, S.E. Neuro-Computing Networks for Entropy Generation under the Influence of MHD and Thermal Radiation. *Surf. Interfaces* **2021**, *25*, 101243.
35. Abrar, M.N.; Sagheer, M.; Hussain, S. Entropy analysis of Hall current and thermal radiation influenced by cilia with single- and multi-walled carbon nanotubes. *Bull. Mater. Sci.* **2019**, *42*, 250. [CrossRef]
36. Wang, J.; Muhammad, R.; Khan, M.I.; Khan, W.A.; Abbas, S.Z. Entropy optimized MHD nanomaterial flow subject to variable thicked surface. *Comput. Methods Programs Biomed.* **2020**, *189*, 105311. [CrossRef] [PubMed]
37. Dehghan, M.; Shirilord, A. The use of homotopy analysis method for solving generalized Sylvester matrix equation with applications. *Eng. Comput.* **2021**, 1–18. [CrossRef]
38. Odibat, Z. An improved optimal homotopy analysis algorithm for nonlinear differential equations. *J. Math. Anal. Appl.* **2020**, *488*, 124089. [CrossRef]
39. Naik, P.A.; Zu, J.; Ghoreishi, M. Estimating the approximate analytical solution of HIV viral dynamic model by using homotopy analysis method. *Chaos Solitons Fractals* **2020**, *131*, 109500. [CrossRef]
40. Iqbal, Z.; Azhar, E.; Maraj, E.N. Utilization of the computational technique to improve the thermophysical performance in the transportation of an electrically conducting Al_2O_3-Ag/H_2O hybrid nanofluid. *Eur. Phys. J. Plus* **2017**, *132*, 544. [CrossRef]

Article

Numerical Analysis of Thermal Radiative Maxwell Nanofluid Flow Over-Stretching Porous Rotating Disk

Shuang-Shuang Zhou [1], Muhammad Bilal [2,*], Muhammad Altaf Khan [3] and Taseer Muhammad [4,5]

[1] School of Science, Hunan City University, Yiyang 413000, China; zhoushuangshuang@hncu.edu.cn
[2] Department of Mathematics, City University of Science and Information Technology, Peshawar 25000, Pakistan
[3] Institute for Groundwater Studies, Faculty of Natural and Agricultural Sciences, University of Free State, Bloemfontein 9300, South Africa; altafdir@gmail.com
[4] Department of Mathematics, College of Sciences, King Khalid University, Abha 61413, Saudi Arabia; taseer_qau@yahoo.com
[5] Mathematical Modelling and Applied Computation Research Group (MMAC), Department of Mathematics, King Abdulaziz University, P.O. Box 80203, Jeddah 21589, Saudi Arabia
* Correspondence: bilalchd345@gmail.com; Tel.: +92-3429300825

Abstract: The fluid flow over a rotating disk is critically important due to its application in a broad spectrum of industries and engineering and scientific fields. In this article, the traditional swirling flow of Von Karman is optimized for Maxwell fluid over a porous spinning disc with a consistent suction/injection effect. Buongiorno's model, which incorporates the effect of both thermophoresis and Brownian motion, describes the Maxwell nanofluid nature. The dimensionless system of ordinary differential equations (ODEs) has been diminished from the system of modeled equations through a proper transformation framework. Which is numerically computed with the bvp4c method and for validity purposes, the results are compared with the RK4 technique. The effect of mathematical abstractions on velocity, energy, concentration, and magnetic power is sketched and debated. It is perceived that the mass transmission significantly rises with the thermophoresis parameter, while the velocities in angular and radial directions are reducing with enlarging of the viscosity parameter. Further, the influences of thermal radiation Rd and Brownian motion parameters are particularly more valuable to enhance fluid temperature. The fluid velocity is reduced by the action of suction effects. The suction effect grips the fluid particles towards the pores of the disk, which causes the momentum boundary layer reduction.

Keywords: bvp4c; RK4 technique; brownian motion; porous rotating disk; maxwell nanofluid; thermally radiative fluid; von karman transformation

Citation: Zhou, S.-S.; Bilal, M.; Khan, M.A.; Muhammad, T. Numerical Analysis of Thermal Radiative Maxwell Nanofluid Flow Over-Stretching Porous Rotating Disk. *Micromachines* **2021**, *12*, 540. https://doi.org/10.3390/mi12050540

Academic Editors: Lanju Mei and Shizhi Qian

Received: 21 March 2021
Accepted: 3 May 2021
Published: 10 May 2021

1. Introduction

The researchers have been interested in Maxwell nanofluid flow over a porous spinning disc because of its many uses in engineering and innovation. Non-Newtonian fluids are important in a variety of manufactured liquids, including plastics, polymers, pulps, toothpaste and fossil fluids. To simulate the analysis of these liquids, a variety of models have been suggested. Shear stress and shear rate are linked in non-Newtonian liquids because of their nonlinear existence. The momentum equation in these fluids involves dynamic nonlinear terms, making it difficult to solve. A variety of mathematical models exist in the literature to simulate the performance of these fluids.

In the present era, the role of nanotechnology to fulfill the increasing demand for energy and face energy challenges is remarkable. The usage of nanoparticles in ordinary base fluid (water, kerosene oil, etc.) effectively enhances the heat transfer and improves their thermal properties. The applications of nanoparticles in certain fields of engineering and industry are in the cooling systems of electronic devices and in cancer therapy, heat exchangers, transformer cooling, nuclear reactors, and space cooling systems. Due to the ability of oil wetting and dispersing, they are also used for cleaning purposes, in power generation, microfabrication, hyperthermia, and metallurgical purposes.

2. Literature Review

The rotating disk phenomena are widely used in centrifugal filtration, turbomachines, the braking system of vehicles, jet motors, sewing machines, turbine systems, heat exchangers and computer disk drives, etc. Von Karman [1] for the first time introduced similarity transformation. To solve Navier-Stokes equations, he studied fluid flow over an infinite rotating disk. Cochran [2] employed Von Karman's similarity transformation to incompressible fluid over a rotating disk and examined the asymptotic solution. Wagner [3] investigated the mechanism of heat transfer over the rotating disk, by considering Von Karman's velocity distribution, and analyzed convection in the non-turbulent flow. Turkyimazogl [4] looked at fluid movement over a spinning disc that was stretching under the influence of a static electric field. Liang et al. [5] reported a comparative study between semi-analytical model and experimental data to yield the best settlement. Millsaps and Pohlhausen [6] used Von Karman's similarity approach and analyzed heat distribution with the consequences of entropy generation over revolving disk. The three-dimensional (3D) magnetohydrodynamics (MHD) stagnation flow of ferrofluid, the numerical solution was revealed by Mustafa et al. [7]. Mustafa et al. [8], by taking MHD nanofluid over rotating surface with the effects of partial slips, observed that the boundary layer thickness and momentum transport are reduced due to slip effects. Rashidi et al. [9] have used a spinning disc to perform a viscous dissipation review for MHD nanofluid.

The people of the modern world are facing many challenges due to the increasing demand for energy by the latest technologies. Firstly Choi [10] presented the nanofluids terminology. The Brownian motion and thermophoresis mechanisms bring about a significant role in improving the thermal properties of base fluid presented by Buongiorno [11]. Turkyilmazolglu [12] analytically studied the energy and momentum equations of nanofluid flow, to deduce heat and flow transport. Pourmehran [13] considered Cu and Al_2O_3 nanoparticles to study heat and flow transfer in the microchannel. Hatami et al. [14] reported the heat transfer in nanofluid with the phenomena of natural convection. The Oldroyd-B fluid with nanoparticles over stretching sheet surface was reported by Nadeem et al. [15]. Aziz and Afify [16] have used the technique of the Lie group, to study non-Newtonian nanofluids. Yang et al. [17] studied the convective heat with Buongiorno Model's for nanofluid in the concentric annulus.

The study of a Newtonian fluid, due to its wide applicability in different fields of science and engineering, attracted the attention of scientists and researchers during the last few decades. Its major role is in geophysics, polymer solution, paper production, cosmetic processes, exotic lubricants, paints, suspensions, colloidal solutions, nuclear and chemical industries, pharmaceuticals, oil reservoirs, bioengineering, etc. [18]. Xiao et al. [19] attempted a fractal model for the capillary flow through a torturous capillary with the non-smooth surface in porous media. Attia [20] evaluated the Reiner-Rivlin numerical simulations for thermal convection over a porous spinning disc qualitatively. Griffiths [21] tested the Newtonian fluid Carreau viscosity model and high shear stresses on spinning discs. The numerical analysis of Reiner-Rivlin fluid flow for heat transfer and slip flow over a spinning disc is treated by Mustafa and Tabassum [22]. The micropolar fluid for thermophoretic diffusion generated by the rotation of the disk was examined by Doh and Muthtamilselvan [23].

Darcy's law is a mathematical equation that explains how fluid flows through a porous medium. Henry Darcy developed the law based on the effects of studies on the flow of water across sand beds, laying the groundwork for hydrogeology, a branch of earth sciences [24]. Fourier's law in heat conduction, Ohm's law in electrical networks, and Fick's law in diffusion theory are all examples of this law. Morris Muskat [25] improved Darcy's equation for a single-phase flow by incorporating viscosity into Darcy's single (fluid) phase equation. It is easy to see that viscous fluids have a harder time passing through a porous medium than less viscous fluids. Rasool et al. [26–28] numerically simulated the Darcy-Forchheimer effect on MHD nanoliquid flow between stretching non-linear sheets. Rasool et al. [29] scrutinized the consequences of thermal radiation, chemical reaction and Dufour-Soret on incompressible steady Darcy-Forchheimer flow of nanoliquid. Shafiq et al. [30] studied nanofluid flow under the influence of convective boundary conditions and thermal slip over a spinning frame. They found that the axial and transverse velocity fields all drop significantly due to the Forchheimer number's strong retardation. Skin friction is intensified by the Forchheimer number and porosity ratios, while skin friction is diminished by all slip parameters.

Viscoelastic fluids are a subclass of Newtonian fluid having memory effects. The intensity of energy discharged by these fluids is mainly accountable for recovery after the stress is removed. The Maxwell flow regime is the most basic viscoelastic fluid model, expressing memory effects by fluid relaxation time [31]. The attitude of the current model is very close to that of other geomaterials and polymers models. The aim of the present work is to provide a mathematical model for unsteady boundary layer flow of non-Newtonian Maxwell nanofluid with the heat transmission over a porous spinning disc. The present work has many industrial and engineering applications, which increases its worth. Using a resemblance method, the system of ODEs is limited to a structure of PDEs. A boundary value solver (bvp4c) technique is used to draw a numerical solution to the problem while RK4 method has been applied for validity.

3. Formulation of the Problem

Consider an unsteady hybrid nanoliquid flow over a stretching porous spinning disc. The magnetic force B_0 is introduced to the disc vertically. The disc rotates and stretches at different speeds $(u, v) = (cr, c\Omega)$, where c and w are the spinning and extending rates, respectively. The disk temperature is represented by τ_w. The formulation of the problems is conducted in (r, φ, z) cylindrical coordinates, where u, v, w is velocity component increasing in (r, φ, z) direction. At $z-$ the axis, the motion of the disk is assumed to be axisymmetric. The thermal radiation is significant in modeling the energy equation. The viscosity of a fluid is taken to be temperature-dependent $\mu(\tau) = \mu_0 e^{-\zeta(\tau - \tau_0)}$. The concentration and temperature are represented by (C_w, τ_w) and (C_∞, τ_∞) represent the concentration and temperature above the disk surface.

3.1. Governing Equations

Under the presuppositions stated above, the flow equations are as observes [31]:

$$\nabla.V = 0, \tag{1}$$

$$\rho_f(V.\nabla)V = \nabla p + \nabla.S + j \times B, \tag{2}$$

$$(V.\nabla)\tau = \alpha \nabla^2 \tau + \tau^* \left(D_B \nabla C.\nabla \tau + \frac{D\tau}{\tau_\infty} \nabla \tau.\nabla \tau \right) + \nabla.q_{rad}, \tag{3}$$

$$(V.\nabla)C = D_B \nabla^2 C + \frac{D\tau}{\tau_\infty} \nabla^2 \tau, \tag{4}$$

$$\nabla.B = 0, \tag{5}$$

$$\rho \frac{\partial B}{\partial t} = \rho \nabla + (V \times B) + \frac{\rho}{\sigma \mu_2} \nabla^2 V. \tag{6}$$

where D_B, D_T, ρ_f, V and α are the coefficients of Brownian motion, thermophoretic diffusion, fluid density, velocity, and thermal and diffusivity respectively. Equations (1)–(6) are simplified because of the boundary layer approximation concept:

$$\frac{\partial u}{\partial r} + \frac{u}{r} + \frac{\partial w}{\partial z} = 0, \tag{7}$$

$$\frac{\partial u}{\partial t} + u\frac{\partial u}{\partial r} + w\frac{\partial u}{\partial z} - \frac{v^2}{r} = \begin{bmatrix} \frac{1}{\rho}\left(\mu(\tau)\frac{\partial u}{\partial z}\right) - \lambda_1\left(u^2\frac{\partial^2 u}{\partial r^2} + w\frac{\partial^2 u}{\partial z^2} + 2uw\frac{\partial^2 u}{\partial r\partial z} - \frac{2uv}{r} \right. \\ \left. \frac{\partial v}{\partial r} - \frac{2vw}{r}\frac{\partial v}{\partial z} + \frac{uv^2}{r^2} + \frac{v^2}{r^2}\frac{\partial u}{\partial r}\right) - \frac{\sigma B_0^2}{\rho}\left(u + w\lambda_1\frac{\partial u}{\partial z}\right) \end{bmatrix}, \tag{8}$$

$$\frac{\partial v}{\partial t} + u\frac{\partial v}{\partial r} + w\frac{\partial v}{\partial z} - \frac{uv}{r} = \begin{bmatrix} \frac{1}{\rho}\left(\mu(\tau)\frac{\partial v}{\partial z}\right) - \lambda_1\left(u^2\frac{\partial^2 v}{\partial r^2} + w^2\frac{\partial^2 v}{\partial z^2} + 2uw\frac{\partial^2 v}{\partial r\partial z} + \frac{2uv}{r} \right. \\ \left. \frac{\partial u}{\partial r} + \frac{2vw}{r}\frac{\partial u}{\partial z} + \frac{2u^2 v}{r^2} - \frac{v^2}{r^2}\frac{\partial v}{\partial r}\right) - \frac{\sigma B_0^2}{\rho}\left(v + w\lambda_1\frac{\partial v}{\partial z}\right) \end{bmatrix}, \tag{9}$$

$$\frac{\partial \tau}{\partial t} + u\frac{\partial \tau}{\partial r} + w\frac{\partial \tau}{\partial z} = \frac{k}{\rho c_p}\left(\frac{\partial^2 \tau}{\partial z^2}\right) + \tau^*\left(D_B\frac{\partial \tau}{\partial z}\frac{\partial C}{\partial z} + \frac{D_\tau}{\tau_\infty}\left(\frac{\partial \tau}{\partial z}\right)^2\right) - \frac{1}{\rho c_p}\frac{\partial q_r}{\partial z}, \tag{10}$$

$$\frac{\partial C}{\partial t} + u\frac{\partial C}{\partial r} + w\frac{\partial C}{\partial z} = D_B\left(\frac{\partial^2 C}{\partial z^2}\right) + \frac{D_\tau}{\tau_\infty}\left(\frac{\partial^2 \tau}{\partial z^2}\right), \tag{11}$$

$$\frac{\partial B_r}{\partial t} = \left[-w\frac{\partial B_r}{\partial z} - B_r\frac{\partial w}{\partial z} + u\frac{\partial B_z}{\partial z} + B_z\frac{\partial u}{\partial z} + \frac{1}{\sigma\mu_2}\left(\frac{\partial^2 B_r}{\partial r^2} + \frac{\partial^2 B_r}{\partial z^2} + \frac{1}{r}\frac{\partial B_r}{\partial r} - \frac{B_r}{r^2}\right)\right], \tag{12}$$

$$\frac{\partial B_z}{\partial t} = \left[w\frac{\partial B_r}{\partial r} + B_r\frac{\partial w}{\partial r} + \frac{1}{r}wB_r - u\frac{\partial B_z}{\partial r} - B_z\frac{\partial u}{\partial r} - \frac{1}{r}uB_z + \frac{1}{\sigma\mu_2}\left(\frac{\partial^2 B_z}{\partial r^2} + \frac{\partial^2 B_z}{\partial z^2} + \frac{1}{r}\frac{\partial B_z}{\partial r}\right)\right]. \tag{13}$$

The temperature difference within a flow is assumed to be small, therefore higher-order in Taylor series and ignored at τ_∞, by using Rosseland approximation, the simple form of radiation heat flux is as follow [31]:

$$q_r = \frac{-4\sigma^*\partial\tau^4}{3k^*\partial z} = -\frac{16\sigma^*\tau^3}{3k^*}\frac{\partial\tau}{\partial z}, \tag{14}$$

using Equation (14) in (10), we get

$$\frac{\partial \tau}{\partial t} + u\frac{\partial \tau}{\partial r} + w\frac{\partial \tau}{\partial z} = \frac{k}{\rho c_p}\left(\frac{\partial^2 \tau}{\partial z^2}\right) + \tau^*\left(D_B\frac{\partial \tau}{\partial z}\frac{\partial C}{\partial z} + \frac{D_\tau}{\tau_\infty}\left(\frac{\partial \tau}{\partial z}\right)^2\right) - \frac{16\sigma^*\tau^3}{3k^*}\frac{\partial\tau}{\partial z}. \tag{15}$$

The boundary conditions are:

$$\begin{aligned} u = cr, v = \Omega r,\ w = W,\ C = C_w,\ T = T_w,\ B_r = 1, B_z = 1 \quad at\, z = 0 \\ u \to 0,\ v \to 0,\ w \to 0,\ C \to C_\infty,\ T \to T_\infty,\ B_r \to 0, B_z \to 0\ as\quad Z \to \infty. \end{aligned} \tag{16}$$

3.2. Similarity Transformation

Considering the following transformation, to reduce the system of PDEs to the system of ODEs:

$$\left.\begin{aligned} u = \frac{cr}{1-\alpha t}F(\eta), v = \frac{\Omega r}{1-\alpha t}G(\eta),\ w = \frac{\sqrt{cv}}{1-\alpha t}H(\eta), \eta = \sqrt{\frac{c}{v}}z,\ B_r = \frac{crM_0}{1-\alpha t}M\prime(\eta), \\ B_z = \frac{-M_0(2v_f c)^{\frac{1}{2}}}{1-\alpha t}N(\eta),\ T = (T_\infty) + \Theta(\eta)(T_W - T_\infty)C = (C_\infty) + \phi(\eta)(C_W - C_\infty). \end{aligned}\right\} \tag{17}$$

The following system of ODEs is obtained by using Equation (17) in Equations (7)–(13) and (15) and (16):

$$F'' = F'\Theta' + \frac{e^{\delta\Theta}}{\delta}\left\{S\left(\frac{F'\eta}{2} + F\right) + F^2 + HF' - G^2\right\} + \beta_1\frac{e^{\delta\Theta}}{\delta}\left\{HF' + 2FF'H - 2HGG'\right\} - M\frac{e^{\delta\Theta}}{\delta}(F + \beta_1 HF'), \tag{18}$$

$$G'' = \frac{\delta G'\Theta' + e^{\delta\Theta}\left\{S\left(\frac{G'-\eta}{2} + G\right) + 2FG + HG'\right\} + \beta_1 e^{\delta\Theta}\{2FHG' + 2FHG\} + Me^{\delta\Theta}(G + \beta_1 HG')}{1 + \beta_1 H^2 e^{\delta\Theta}}, \tag{19}$$

$$\Theta'' = \frac{-4Rd\Theta^2\Theta'^2(\Theta-1)^3 - 6\Theta'^2\Theta(Q_w-1)^2 - 3\Theta'^2(Q_w-1) - \Pr\left(-\frac{S}{2}(\Theta'\eta) - H\Theta' + Nb\Theta\phi' + Nt\Theta'^3\right)}{\left(1+\frac{4}{3}\right)Rd + \frac{4}{3}Rd\Theta^3(Q_w-1)^3 + 3\Theta^2(Q_w-1)^2 + 3\Theta(\Theta-1)}, \tag{20}$$

$$\phi'' = Sc\left(A\phi'\eta + H\phi'\right) - \frac{Nt}{Nb}\Theta'', \tag{21}$$

$$M''' = Bt\left[-HM'' + M'H' + FN' + NF' + S\left(\frac{M''\eta}{2} + M'\right)\right], \tag{22}$$

$$M'' = -Bt\left[2HM' + 2NF - \frac{S}{2}(N\eta + N)\right]. \tag{23}$$

the transforms conditions are:

$$F(0) = 1,\ G(0) = \omega,\ H = W_s,\ \Theta(0) = 1,\ f(0) = 1,\ M'(0) = 0,\ N(0) = 1,$$
$$F(\infty) = 0,\ G(\infty) = 0,\ H(0) = 0,\ \Theta(\infty) = 0,\ f(\infty) = 0,\ MI(\infty) = 0,\ N(\infty) = 0. \tag{24}$$

where $Pr = \nu/\alpha$ is the Prandtl number, $Sc = \nu/D_B$ is the Schmidt number, $\beta_1 = \lambda_1 c$ the Deborah number, while suction/injection parameter, Brownian motion, variable viscosity, thermophoresis parameter, magnetic parameter, temperature ratio, and thermal radiation are defined as [31]:

$$Ws = \frac{w}{\sqrt{c\nu}},\ Nb = \frac{\tau^* D_B(C_w - C_\infty)}{\nu},\ \delta = \zeta(\tau_w - \tau_\infty),\ Nt = \frac{\tau^* D_\tau(\tau_w - \tau_\infty)}{\tau_\infty \nu},\ M = \frac{\sigma B_0^2}{c_p},\ Rd = \frac{4\sigma^* T_\infty^3}{kk^*}. \tag{25}$$

The skin friction Cf_x, local Sherwood Sh_r and Nusselt number Nu_r are mathematically can be written as [32,33]:

$$Cf = \frac{\sqrt{\tau_{zr}^2 + \tau_{z\varphi}^2}}{\rho(\Omega r)^2}, \tag{26}$$

$$Sh_r = -\frac{r}{(C_w - C_\infty)}\left(\frac{\partial\tau}{\partial C}\right)\bigg|_{z=0}, \tag{27}$$

$$Nu_r = \frac{r}{(\tau_w - \tau_\infty)}\left[1 + \frac{16\sigma^*\tau^3}{3kk^*}\right]\left(\frac{\partial\tau}{\partial z}\right)\bigg|_{z=0}. \tag{28}$$

The skin friction, Sherwood and Nusselt numbers have a non-dimensional structure as:

$$Cf\mathrm{Re}_r{}^{1/2} = \sqrt{f''^2(0) + g'^2(0)}, \tag{29}$$

$$Re^{-\frac{1}{2}}Sh_r = -\phi(0), \tag{30}$$

$$Re^{-\frac{1}{2}}Nu_r = -(1 + \frac{4}{3}Rd\{1 + (\Theta_w - 1)\Theta(0)\}^3)\Theta'(0). \tag{31}$$

Here $Re = \frac{ru}{\nu}$ is the local Reynold number.

4. Solution Procedures

The higher-order model equation is brought down to first order by choosing variables:

$$\left.\begin{array}{l}\chi_1 = H,\ \chi_2 = F,\ \chi_3 = FI,\ \chi_4 = G,\ \chi_5 = GI,\ \chi_6 = \theta,\ \chi_7 = \theta I, \\ \chi_8 = \phi,\ \chi_9 = \phi I,\ \chi_{10} = M,\ \chi_{11} = MI,\ \chi_{13} = N,\ \chi_{14} = NI.\end{array}\right\} \tag{32}$$

$$\chi_1' = -2\chi_2, \chi_1' = \chi_3,$$
$$\chi_3' = \chi_7\chi_3 + \frac{e^{\delta\theta}}{\delta}\left\{S(\frac{\eta\chi_3}{2} + \chi_2) + \chi_2^2 + \chi_1\chi_2^2 - \chi_4^2\right\} + \beta_1\frac{e^{\delta\theta}}{\delta}\{\chi_1\chi_3 + 2\chi_1\chi_2\chi_3 - 2\chi_1\chi_4\chi_5\}$$
$$M_0\frac{e^{\delta\theta}}{\delta}(\chi_2 + \beta_1\chi_1\chi_3), \chi_4' = \chi_5,$$
$$\chi_5' = \frac{\delta\chi_5\chi_7 + e^{\delta\theta}\left\{S(\frac{\chi_5-\eta}{2}) + 2\chi_2\chi_4 + \chi_1\chi_5\right\} + \beta_1\{2\chi_1\chi_2\chi_5 - 2\chi_1\chi_4\chi_2\} + M_0 e^{\delta\theta}(\chi_4 + \beta_1\chi_1\chi_5)}{1 + \beta_1 e^{\delta\theta}\chi_1^2},$$
$$\chi_6' = \chi_7, \chi_7' = \frac{\frac{4}{3}Rd\chi_7^2(\theta_w - 1)\left\{3\chi_6^2(\theta_w - 1)^2 + 6\chi_6(\theta_w - 1) - 3\right\} - Pr\left\{Nb\chi_7\chi_9 + Nt\chi_7^2 - A\chi_7\chi_{10} - \chi_1\chi_7\right\}}{1 - \frac{4}{3}Rd - \frac{4}{3}Rd\chi_6(\theta_w - 1)\left\{(\theta_w - 1)\chi_6^2 - 3\chi_6(\theta_w - 1) + 3\right\}},$$
$$\chi_8' = \chi_9,$$

$$\chi_9' = \frac{Nt}{Nb}\left(\frac{\frac{4}{3}Rd\chi_7^2(\theta_w-1)\left\{3\chi_6^2(\theta_w-1)^2+6\chi_6(\theta_w-1)-3\right\}-Pr\left\{Nb\chi_7\chi_9+Nt\chi_7^2-A\chi_7\chi_{10}-\chi_1\chi_7\right\}}{1-\frac{4}{3}Rd-\frac{4}{3}Rd\chi_6(\theta_w-1)\left\{(\theta_w-1)\chi_6^2-3\chi_6(\theta_w-1)+3\right\}}\right)$$
$$Sc(A\chi_9\chi_{10}+\chi_{91}\chi_9),$$

$$\chi_{10}' = \chi_{11}, \chi_{11}' = Bt[-2\chi_{11}\chi_2 - \chi_1\chi_{11} + \chi_2\chi_{14} + \chi_{13}\chi_2 + S(\tfrac{\eta\chi_{12}}{2} + \chi_{11})],$$
$$\chi_{12}' = \chi_{13}, \chi_{13}' = Bt[2\chi_{11}\chi_1 + 2\chi_{13}\chi_2 + \tfrac{S}{2}(\eta\chi_{14} + \chi_{13})] \tag{33}$$

The boundary conditions are:

$$\chi_1(0) = 1, \chi_2(0) = 0, \chi_4(0) = 1, \chi_6(0) = 1, \chi_8(0) = 1, \chi_{11}(0) = 0, \chi_{13}(0) = 1,$$
$$\chi_2(\infty) = 0, \chi_4(\infty) = 0, \chi_6(\infty) = 0, \chi_8(\infty) = 0, \chi_{11}(\infty) = 1, \chi_{13}(\infty) = 0. \tag{34}$$

5. Results and Discussion

The discussion section is devoted to understanding better the graphical and physical description. The system of non-linear Equations (18)–(23) along with their boundary conditions, Equation (24), are solved through numerical method bvp4c. The configuration of the problem is described in Figure 1. The velocities, energy profile, concentration distribution $\phi(\eta)$, magnetic strength in the radial direction $M(\eta)$, and azimuthal magnetic strength $N(\eta)$ are explored graphically through different physical constraints Figures 2–10. While keeping $Pr = 6.7, \omega = 1.0, \delta = 0.5, \beta = 0.1, M = 1.1, Nb = 0.5, Nt = 0.7, Q_w = 1.2, Sc = 2.0$, and $Rd = 0.5$.

Figure 2a–c depicts the behavior velocities profiles against the variation of Deborah number β. All three velocity shows decreasing behavior for incremented of β. Deborah number is the measure of the content evaluation period to content recreational time, so having optimum stress relaxation or eliminating observation time increases the value of β. It reflects the fluid's solid-like reaction. The hydrodynamic boundary layer thins out, and the velocity experiences more resistance.

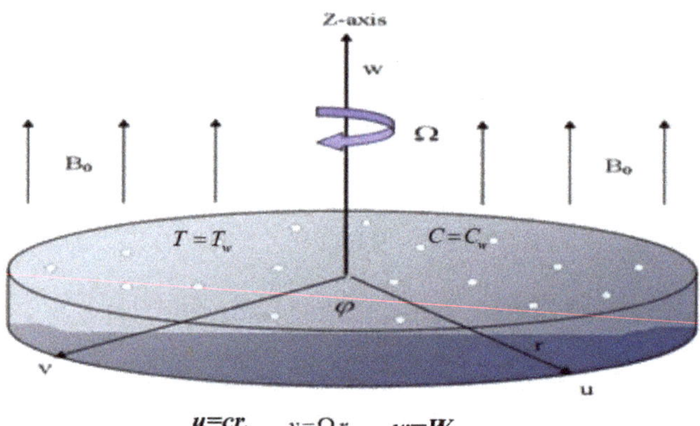

Figure 1. Stretchable porous rotating disk.

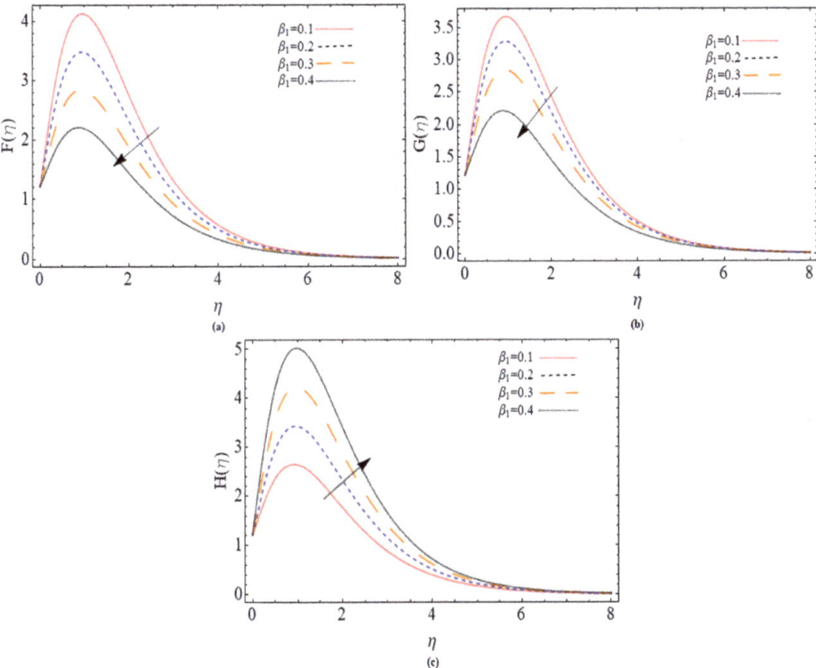

Figure 2. The out-turn of Deborah number β on the axial, radial and azimuthal velocity profiles, respectively.

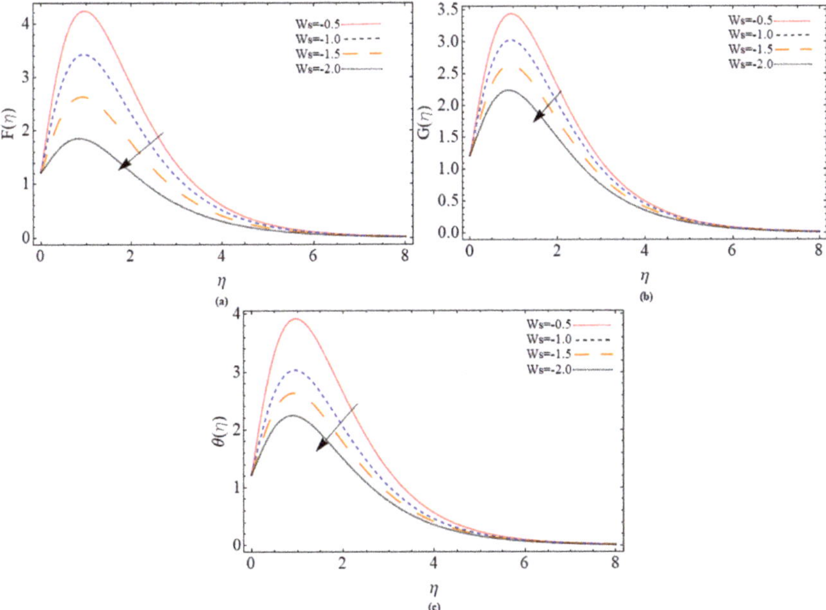

Figure 3. The out-turn of the suction parameter ($Ws < 0$) on the axial velocity, radial velocity and temperature distribution profiles, respectively.

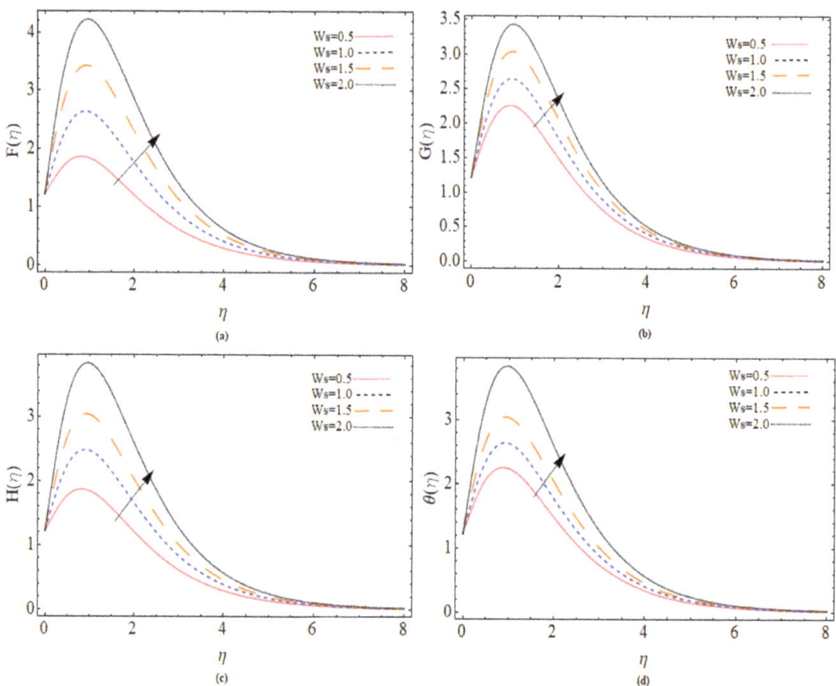

Figure 4. The out-turn of injection parameter $(Ws > 0)$ on the axial, radial and azimuthal velocity profiles and temperature distribution, respectively.

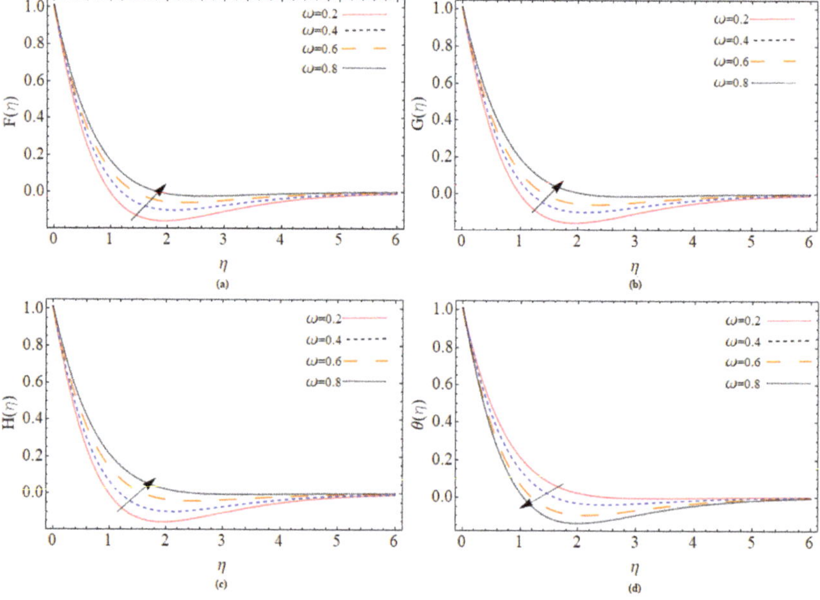

Figure 5. The out-turn of rotation parameter ω on the axial, radial and azimuthal velocity and temperature profiles, respectively.

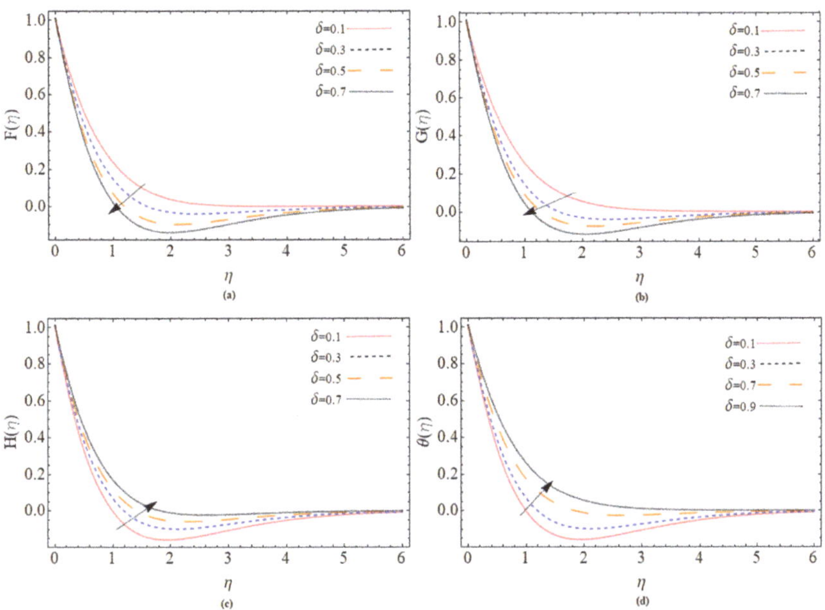

Figure 6. The out-turn of viscosity parameter δ on the axial, radial and azimuthal velocity and temperature profiles, respectively.

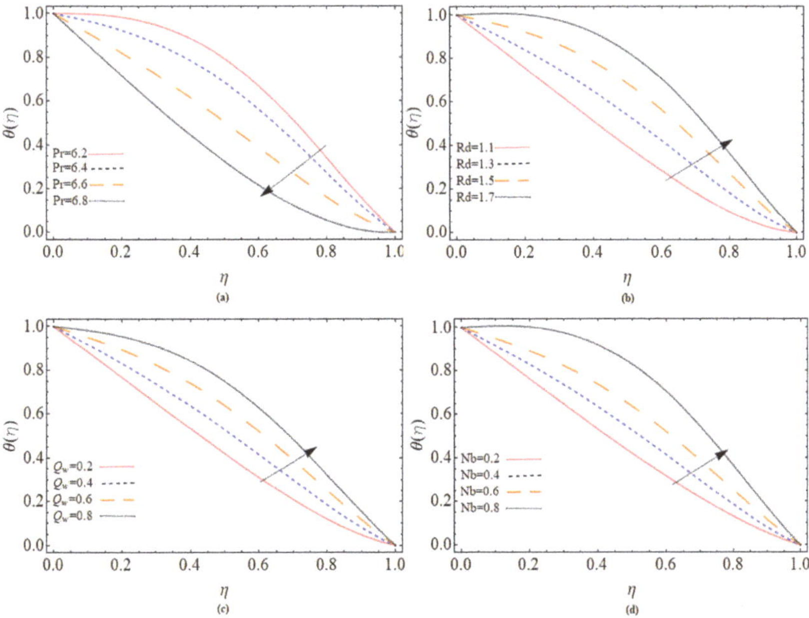

Figure 7. The out-turn of different parameters (Pr, Rd, Θ, Nb) on the temperature distribution profile.

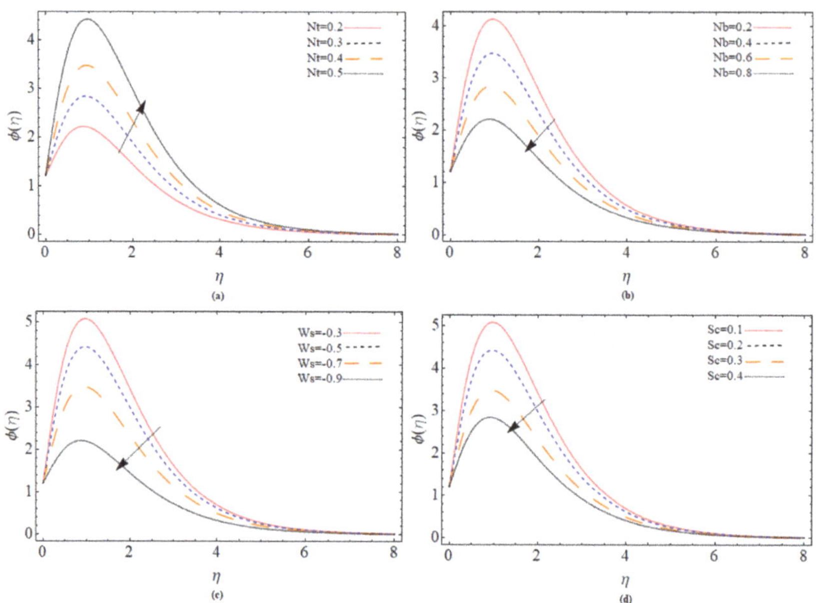

Figure 8. The out-turn of different parameters (Nt, Nb, Ws, Sc) on the concentration profile.

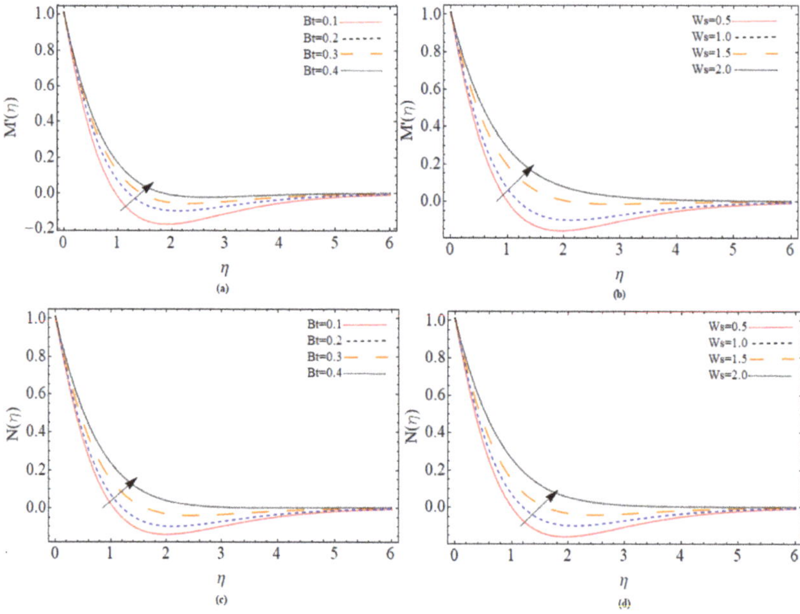

Figure 9. The out-turn of different parameters (Bt, Ws) on the axial and radial magnetic strength profiles, respectively.

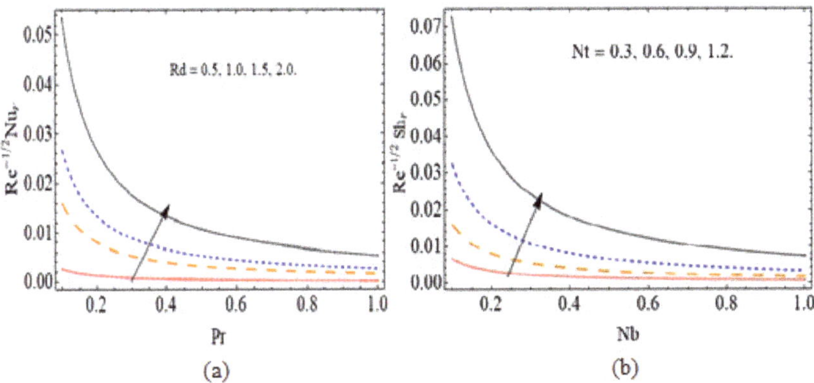

Figure 10. The out-turn of radiation parameter and Brownian coefficient versus the heat and mass transmission profiles, respectively.

The implications of the suction factor $(Ws < 0)$ on the velocities profile and heat spectrum on the disc surface are depicted in Figure 3a–c. It seems to be that as the suction velocity rises, the velocity decreases. Because the suction velocity draws the fluid particles towards the pores in the disk, which causes the momentum boundary layer reduction. The enhancement of suction velocity also decreases the fluid temperature.

Figure 4a–d are drawn to depict the effects of injection velocity parameter $(Ws > 0)$ on the radial, azimuthal, tangential velocity profile and temperature distribution. It can be seen that the enhancing of injection velocity increases the velocity and fluid temperature.

The radial, azimuthal, tangential velocity profile and temperature distribution variation are illustrated in Figure 5a–d. Figure 5a depicts the dominant behavior of radial velocity with ω. Physically, fluid particles are moved in a radial direction owing to centrifugal force with the enhancement of parameters ω. The increasing value of ω shows that the rotation parameter becomes greater than extending. The radial velocity exceeds the disc stretching velocity too close to the disc stretching surface. It does, however, gradually fade away from the disc. It is concluded that the impact of centrifugal force is limited and dominant in the vicinity of the disk's surface. Besides, angular velocity $G(\eta)$ near the disk flourished shown in Figure 5b. The axial velocity increasing against the strengthening of ω is illustrated in Figure 5c. Figure 5d demonstrates the decreases of temperature $\theta(\eta)$ with ω. Physically, a faster spinning disc reduces the width of the thermal boundary, which play a significant role in the cooling of the system.

Figure 6a–d are rough sketches that show the effects of the viscosity factor on the velocity and temperature spectrum on the disk's surface. The viscosity parameter causes radial velocity, tangential velocity, and temperature profile to increase, while azimuthal velocity reduces.

Figure 7a,b indicates how the heat transfer changes when the Prandtl number Pr and the thermal radiation factor change. The thermal dispersion of a strong Prandtl fluid is low, while the thermal diffusivity of a low Prandtl fluid is high. Figure 7b is drawn to explore the influence of radiation parameters Rd on the thermal mechanism of fluid. With a higher value of the radiation parameter, an increasing trend in heat is analyzed. Because, with enhancing of thermal radiation the fluid absorbs more heat, and as a result increment occurs in boundary layer thickness and fluid temperature. Figure 7c is sketched to illustrate temperature $\theta(\eta)$ variation versus temperature ratio Q_w. While Figure 7d indicates the consequence of molecular diffusion on the thermal performance of a Maxwell nanofluid. The Brownian motion produces random movement between fluid particles, which generate more heat, and as a result, the fluid temperature increases.

Figure 8a,b accordingly represents the action of the concentration field as a function of the thermophoresis term and Brownian. It can be shown that with Nt, the nanofluid concentration field raises while with Nb, the concentration profile drops. The suction parameter $(Ws) < 0$ and Schmidt number Sc effect are illustrated in Figures 8c and 8d respectively. The concentration field reduces with an increase of both suction velocity and Schmidt number. The consequences of different parameters (Bt, Ws) on the axial and radial magnetic strength profiles are illustrated through Figure 9a–d respectively. It can observe that both parameters Batchlor number Bt and injection parameter Ws positively effect the magnetic strength profile along axial and radial direction.

Figure 10a,b express the nature of heat $-\theta'(0)$ and mass transfer $-\phi'(0)$ against radiation parameter Rd and Brownian motion parameter Nb, respectively. Because of the improving effect of radiation, the fluid temperature also rises, which enhances the heat transmission rate. Table 1 displays the numerical outcomes for skin friction and compared it with the published literature. Table 2 illustrates the comparison of bvp4c and RK4 techniques for the numerical outcomes. The numerical outputs for Sherwood number Sh_r and Nusselt number Nu_r are plotted in Table 3.

Table 1. The numerical outcomes for skin friction $F'(0)$.

ω	Mustafa et al. [7]	Ahmed et al. [31]	Present Paper
0	−1.1737	−1.1379	−1.1380
1	−0.9483	−0.9485	−0.9487
2	−0.3262	−0.3264	−0.3266
5	3.1937	3.11937	3.11939
10	12.7209	12.7209	12.7811
20	40.9057	40.9057	40.9058

Table 2. Comparison between Runge Kutta order four and bvp4c method.

η	RK4	bvp4c	Absolute Error
1.0	1.000000	1.000000	8.146310×10^{-13}
1.2	1.199831	1.199831	3.387821×10^{-9}
1.4	0.988459	0.988459	2.845561×10^{-9}
1.6	0.879189	0.879189	2.813281×10^{-9}
1.8	0.539393	0.539393	3.287961×10^{-9}

Table 3. The comparison of $RK4$ and Bvp4c for Sherwood Sh_r and Nusselt number Nu_r, while keeping $\omega = 1.0$, $\delta = 0.5$, $\beta = 0.1$, $M = 1.1$, $Nb = 0.5$, $Nt = 0.7$, $Q_w = 1.2$ and $Rd = 0.5$.

Sh_r			Nu_r		
Pr	Bvp4c	RK4	Sc	Bvp4c	RK4
3.0	0.7718284	0.7718285	1.0	1.457986	1.457995
4.0	0.6999885	0.6999885	1.5	1.531211	1.531220
5.0	0.6290089	0.6290088	2.0	1.596949	1.596949
6.0	0.5632372	0.5632370	2.5	1.685639	1.685639

6. Conclusions

In the present mathematical model, the Maxwell nanoliquid flow over a porous spinning disk with suction/injection effects has been examined. The flow is studied in the context of magnetization and radiation. The numerical results are found through bvp4c, while for comparison purposes, the computation is carried out via the RK4 technique. From the above studies the following conclusions have been drawn:

- The centrifugal force is more effective and dominant near the disk surface, so causes the maximum velocity in the neighborhood of the surface of the disk.
- The fluid velocity reduces by the action of suction effects. Because the suction effect gripping the fluid particles towards the pores of the disk, which causes the momentum boundary layer reduction.

- The axial, radial and tangential velocities are boosts with the increases of injection parameter.
- The enhancement in rotation parameter also boosts the azimuthal and radial flows.
- The thermal energy profile enhances by the consequence of Brownian motion parameter Nb. The Brownian motion produces random movement between fluid particles, which generate more heat, as a result, the fluid temperature increases.
- The fluid temperature decline with Prandtl number Pr, while incline with thermal radiation parameter, Rd.
- The nanofluid concentration field enhances with Nt, while Nb causes the reduction of concentration profile.

Author Contributions: M.B. wrote the original manuscript and performed the numerical simulations. M.A.K. reviewed the mathematical results and S.-S.Z., T.M. and M.A.K. restructured the manuscript and revised the paper. All authors have read and agreed to the published version of the manuscript.

Funding: This research was funded by the Deanship of Scientific Research at King Khalid University, Abha, Saudi Arabia for funding this work through research groups program under grant number R.G.P-1/142/42 and the APC was funded by S.S.Z.

Acknowledgments: The authors extend their appreciation to the Deanship of Scientific Research at King Khalid University, Abha, Saudi Arabia for funding this work through research groups program under grant number R.G.P-1/142/42.

Conflicts of Interest: The authors declare no conflict of interest.

References

1. Kármán, T.V. Über laminare und turbulente Reibung. *ZAMM J. Appl. Math. Mech. Z. Für Angew. Math. Mech.* **1921**, *1*, 233–252. [CrossRef]
2. Cochran, W. The flow due to a rotating disc. *Math. Proc. Camb. Philos. Soc.* **1934**, *30*, 365–375. [CrossRef]
3. Wagner, C. Heat transfer from a rotating disk to ambient air. *J. Appl. Phys.* **1948**, *19*, 837–839. [CrossRef]
4. Turkyilmazoglu, M. MHD fluid flow and heat transfer due to a stretching rotating disk. *Int. J. Therm. Sci.* **2012**, *51*, 195–201. [CrossRef]
5. Liang, M.; Fu, C.; Xiao, B.; Luo, L.; Wang, Z. A fractal study for the effective electrolyte diffusion through charged porous media. *Int. J. Heat Mass Transf.* **2019**, *137*, 365–371. [CrossRef]
6. Millsaps, K.; Pohlhausen, K. Heat transfer by laminar flow from a rotating plate. *J. Aeronaut. Sci.* **1952**, *19*, 120–126. [CrossRef]
7. Mustafa, I.; Javed, T.; Ghaffari, A. Heat transfer in MHD stagnation point flow of a ferrofluid over a stretchable rotating disk. *J. Mol. Liq.* **2016**, *219*, 526–532. [CrossRef]
8. Mustafa, M. MHD nanofluid flow over a rotating disk with partial slip effects: Buongiorno model. *Int. J. Heat Mass Transf.* **2017**, *108*, 1910–1916. [CrossRef]
9. Rashidi, M.; Abelman, S.; Mehr, N.F. Entropy generation in steady MHD flow due to a rotating porous disk in a nanofluid. *Int. J. Heat Mass Transf.* **2013**, *62*, 515–525. [CrossRef]
10. Choi, S.U.; Eastman, J.A. *Enhancing Thermal Conductivity of Fluids with Nanoparticles*; Argonne National Lab.: Lemont, IL, USA, 1995.
11. Buongiorno, J. Convective transport in nanofluids. *J. Heat Transf.* **2006**, *128*, 240–250. [CrossRef]
12. Turkyilmazoglu, M. Analytical solutions of single and multi-phase models for the condensation of nanofluid film flow and heat transfer. *Eur. J. Mech. B Fluids* **2015**, *53*, 272–277. [CrossRef]
13. Pourmehran, O.; Rahimi-Gorji, M.; Hatami, M.; Sahebi, S.; Domairry, G. Numerical optimization of microchannel heat sink (MCHS) performance cooled by KKL based nanofluids in saturated porous medium. *J. Taiwan Inst. Chem. Eng.* **2015**, *55*, 49–68. [CrossRef]
14. Hatami, M.; Song, D.; Jing, D. Optimization of a circular-wavy cavity filled by nanofluid under the natural convection heat transfer condition. *Int. J. Heat Mass Transf.* **2016**, *98*, 758–767. [CrossRef]
15. Nadeem, S.; Haq, R.U.; Akbar, N.S.; Lee, C.; Khan, Z.H. Numerical study of boundary layer flow and heat transfer of Oldroyd-B nanofluid towards a stretching sheet. *PLoS ONE* **2013**, *8*, e69811. [CrossRef]
16. Afify, A.A.; Abd El-Aziz, M. Lie group analysis of flow and heat transfer of non-Newtonian nanofluid over a stretching surface with convective boundary condition. *Pramana* **2017**, *88*, 31. [CrossRef]
17. Yang, C.; Li, W.; Nakayama, A. Convective heat transfer of nanofluids in a concentric annulus. *Int. J. Therm. Sci.* **2013**, *71*, 249–257. [CrossRef]
18. Hayat, T.; Qayyum, S.; Alsaedi, A.; Asghar, S. Radiation effects on the mixed convection flow induced by an inclined stretching cylinder with non-uniform heat source/sink. *PLoS ONE* **2017**, *12*, e0175584. [CrossRef] [PubMed]

19. Xiao, B.; Huang, Q.; Chen, H.; Chen, X.; Long, G. A fractal model for capillary flow through a single tortuous capillary with roughened surfaces in fibrous porous media. *Fractals* **2021**, *29*, 2150017.
20. Attia, H.A. Numerical study of flow and heat transfer of a non-Newtonian fluid on a rotating porous disk. *Appl. Math. Comput.* **2005**, *163*, 327–342. [CrossRef]
21. Griffiths, P.T. Flow of a generalised Newtonian fluid due to a rotating disk. *J. Non-Newton. Fluid Mech.* **2015**, *221*, 9–17. [CrossRef]
22. Tabassum, M.; Mustafa, M. A numerical treatment for partial slip flow and heat transfer of non-Newtonian Reiner-Rivlin fluid due to rotating disk. *Int. J. Heat Mass Transf.* **2018**, *123*, 979–987. [CrossRef]
23. Doh, D.; Muthtamilselvan, M. Thermophoretic particle deposition on magnetohydrodynamic flow of micropolar fluid due to a rotating disk. *Int. J. Mech. Sci.* **2017**, *130*, 350–359. [CrossRef]
24. Whitaker, S. Flow in porous media I: A theoretical derivation of Darcy's law. *Transp. Porous Media* **1986**, *1*, 3–25. [CrossRef]
25. Muskat, M. The flow of homogeneous fluids through porous media. *Soil Sci.* **1938**, *46*, 169. [CrossRef]
26. Rasool, G.; Shafiq, A.; Alqarni, M.S.; Wakif, A.; Khan, I.; Bhutta, M.S. Numerical Scrutinization of Darcy-Forchheimer Relation in Convective Magnetohydrodynamic Nanofluid Flow Bounded by Nonlinear Stretching Surface in the Perspective of Heat and Mass Transfer. *Micromachines* **2021**, *12*, 374. [CrossRef]
27. Rasool, G.; Shafiq, A. Numerical exploration of the features of thermally enhanced chemically reactive radiative Powell–Eyring nanofluid flow via Darcy medium over non-linearly stretching surface affected by a transverse magnetic field and convective boundary conditions. *Appl. Nanosci.* **2020**, 1–18. [CrossRef]
28. Rasool, G.; Zhang, T. Darcy-Forchheimer nanofluidic flow manifested with Cattaneo-Christov theory of heat and mass flux over non-linearly stretching surface. *PLoS ONE* **2019**, *14*, e0221302. [CrossRef]
29. Rasool, G.; Shafiq, A.; Baleanu, D. Consequences of Soret–Dufour effects, thermal radiation, and binary chemical reaction on Darcy Forchheimer flow of nanofluids. *Symmetry* **2020**, *12*, 1421. [CrossRef]
30. Shafiq, A.; Rasool, G.; Khalique, C.M. Significance of thermal slip and convective boundary conditions in three-dimensional rotating Darcy-Forchheimer nanofluid flow. *Symmetry* **2020**, *12*, 741. [CrossRef]
31. Ahmed, J.; Khan, M.; Ahmad, L. Stagnation point flow of Maxwell nanofluid over a permeable rotating disk with heat source/sink. *J. Mol. Liq.* **2019**, *287*, 110853. [CrossRef]
32. Ferdows, M.; Shamshuddin, M.; Zaimi, K. Dissipative-Radiative Micropolar Fluid Transport in a NonDarcy Porous Medium with Cross-Diffusion Effects. *CFD Lett.* **2020**, *12*, 70–89. [CrossRef]
33. Devi, S.S.U.; Mabood, F. Entropy anatomization on Marangoni Maxwell fluid over a rotating disk with nonlinear radiative flux and Arrhenius activation energy. *Int. Commun. Heat Mass Transf.* **2020**, *118*, 104857. [CrossRef]

micromachines

MDPI

Article

Calculation of Effective Thermal Conductivity for Human Skin Using the Fractal Monte Carlo Method

Guillermo Rojas-Altamirano [1], René O. Vargas [1,*], Juan P. Escandón [1], Rubén Mil-Martínez [2] and Alan Rojas-Montero [1]

[1] Departamento de Termofluidos, Instituto Politécnico Nacional, SEPI-ESIME Azcapotzalco, Av. de las Granjas No. 682, Col. Santa Catarina, Alcaldía Azcapotzalco, Ciudad de México 02250, Mexico; grojasa1400@alumno.ipn.mx (G.R.-A.); jescandon@ipn.mx (J.P.E.); alan_rojas102@hotmail.com (A.R.-M.)

[2] Escuela Militar de Ingenieros, Universidad del Ejército y Fuerza Aérea, Av. Industria Militar No. 261, Col. Lomas de San Isidro, Naucalpan de Juárez 53960, Mexico; rbnm2@hotmail.com

* Correspondence: rvargasa@ipn.mx; Tel.: +52-55-5729-6000 (ext. 64511)

Abstract: In this work, an effective thermal conductivity (ETC) for living tissues, which directly affects the energy transport process, is determined. The fractal scaling and Monte Carlo methods are used to describe the tissue as a porous medium, and blood is considered a Newtonian and non-Newtonian fluid for comparative and analytical purposes. The effect of the principal variables—such as fractal dimensions D_T and D_f, porosity, and the power-law index, n—on the temperature profiles as a function of time and tissue depth, for one- and three-layer tissues, besides temperature distribution, are presented. ETC was improved by considering high tissue porosity, low tortuosity, and shear-thinning fluids. In three-layer tissues with different porosities, perfusion with a non-Newtonian fluid contributes to the understanding of the heat transfer process in some parts of the human body.

Keywords: effective thermal conductivity; fractal scaling; Monte Carlo; porous media; non-Newtonian fluid; power-law model; bioheat equation; human body

Citation: Rojas-Altamirano, G.; Vargas, R.O.; Escandón, J.P.; Mil-Martínez, R.; Rojas-Montero, A. Calculation of Effective Thermal Conductivity for Human Skin Using the Fractal Monte Carlo Method. *Micromachines* **2022**, *13*, 424. https://doi.org/10.3390/mi13030424

Academic Editor: Lanju Mei and Shizhi Qian

Received: 4 February 2022
Accepted: 6 March 2022
Published: 10 March 2022

Publisher's Note: MDPI stays neutral with regard to jurisdictional claims in published maps and institutional affiliations.

1. Introduction

The skin is the largest single organ of the body, enabling protection from the surrounding environment. It consists of several layers and plays an important role in thermoregulation, sensory, and host defense functions [1–3]. The skin is generally described by a three-layer tissue: epidermis, dermis, and hypodermis (also called subcutaneous) [4–6]. The thickness of these layers varies depending on the location of the skin. The epidermis is the outer layer (75–150 µm), this layer plays a barrier role between environment and organism [7,8]. The dermis is much thicker than the epidermis, in this, there are blood vessels, nerves, lymph vessels, and skin appendages. Dermis performances important functions in thermoregulation and supports the vascular network to supply the non-vascularized epidermis with nutrients. This layer is formed by an irregular network with wavy and unaligned collagen fiber bundles, allowing considerable deformations in all directions. The hypodermis is composed of loose fatty connective tissue. It is not part of the skin, but appears as a deep extension of the dermis, and depends on the age, sex, race, endocrine, and nutritional status of the individual [7–10]. The thermoregulation function of skin is realized mainly by modifying the blood flow, which is located in a microcirculatory bed, composed of arterioles, arterial and vein capillaries, and venules (blood perfusion). Blood perfusion has great effect on the heat transfer process in living tissues [8,11,12]. Heat transfer in human tissues takes place through different mechanisms, such as heat conduction, blood perfusion, metabolic heat generation, and external interactions [13,14]. One of the earliest models of heat transfer in biological tissues was developed by Pennes in 1948 [15], who proposed a model to describe the effects of metabolism and blood perfusion on the energy balance within tissue, this model is based on the classical Fourier's law [15]. Wulff [16] questioned the assumptions of the Pennes model and provided an alternative analysis. He assumed that heat transfer between blood flow and tissue should be modeled proportionally to the temperature difference between these two media and not between the two temperatures of the blood flow. Klinger [17] consider the convective heat transfer caused by the blood flow inside the tissue, since this term was neglected by Pennes. Chen and Holmes [18] assumed that the total tissue control volume is composed of the solid-tissue subvolume and blood subvolume. They determined an effective thermal conductivity using the tissue porosity and the local mean tissue temperature, together with a simplified volume-averaging technique for the solid and tissue spaces. Weinbaum et al. [19,20] determined an effective thermal conductivity (ETC) based on the hypothesis that small arteries and veins are parallel and the flow direction is countercurrent, which is a function of the blood flow rate and vascular geometry. The model included a perfusion bleed-off term that apparently resembles the Pennes perfusion term. Weinbaum and Jiji [11] derived a simplified equation to study the influence of the blood flow on the tissue temperature distribution defining an ETC.

In biological tissues, many body parts reveal anisotropy in heat transport that can not be explained by the Fourier's law [6,8]. This leads to formulation of thermal wave bioheat model based on two main approaches: the Maxwell–Cattaneo approach with heat flux time lag (also known as the single-phase approach), and the double-phase-lag (DPL) approach with relaxations in both the heat flux and temperature gradient propagation [21–23]. Various researchers have contributed in this area with analytical and experimental work. Hobiny and Abbas [24] presented an analytical solution of the hyperbolic bioheat equation under intense moving heat source. Alzahrani and Abbas [25] also presented an analytical approach, experimental temperature data, and a time sequential concept to obtain the thermal damage and temperature in a living tissue due to laser irradiation. Hobiny and Abbas [26] provided a method to determine numerical solutions for thermal damage of cylindrical living tissues using hyperbolic bioheat model. Hobiny et al. [27] proposed a new interpretation to study thermal damage in a skin tissue caused by laser irradiation, using the fractional order bioheat model. Hobiny et al. [28] presented an analytical method and experimental verification, to estimate thermal damage and temperature due to laser irradiation, using skin surface measurement data. Kumari and Singh [29] generated a space-fractional mathematical model of bioheat transfer to graphically analyze thermal behavior within living tissue, using a three-phase-lag constitutive relation. Li et al. [30] developed a generalized model of bioheat transfer to explore heat transport properties involving different thermal phase lagging effect. Important reviews and articles that the reader can consult additionally in this context are the following: classical mathematical models of bioheat [13]; developments in modeling heat transfer in blood perfused tissues [9,31]; bioheat models based on the porous media theory [32]; general heat transfer review [33,34]; concepts, derivation, and experimental versus porous media modeling [35]; and modeling and scaling of the bioheat equation [3].

Alternatively to continuum models, concepts that consider the tissue matrix, arteries, veins, and capillary vessels in a porous medium with specific porosity variations, ETC, and heat dispersion by blood flow have been developed [3,12,13,32,36]. Porous medium is defined as a material volume consisting of solid matrix with an interconnected pores. It is characterized by porosity, ratio of the pore space to the total volume of the medium, permeability, and tortuosity [32]. Khanafer and Vafai [36] remarked that the most appropriate treatment for heat transfer in biological tissues is the porous media theory because of fewer assumptions as compared with other models. Roetzel and Xuan [37] introduced a two-equation bioheat model in which the biological system is a porous media. It is divided into two different regions, the vascular and extravascular, without considering local thermal equilibrium between the two phases, introducing an equivalent ETC in the energy equations of blood and tissue [37]. Nakayama and Kuwahara [38] developed a model that consists of two energy equations based on the volume average theory (VAT), these equations are correct for all cases of thermal non-equilibrium.

The ETC is one of the most important thermo-physical properties for quantifying conductive heat transfer of porous media with gas, liquid, and solid phases [39]. The prediction of the ETC of porous media is essential to many engineering applications, such as thermally enhanced oil recovery, geothermal energy, and chemical and biological engineering [40]. With the development of computer technology, many numerical methods, such as Monte Carlo [41] and lattice Boltzmann [42], have been proposed to study the conductive heat transfer and evaluate the ETC of porous media. In addition to conventional methods based on Euclidean geometry, fractal geometry has been shown with evident advantages for addressing the complexity and multiple scales of porous media [43]. Hence, the fractal geometry has been successfully applied to characterize structures of transport processes in porous media [44,45]. Kou et al. proposed a fractal model for ETC of porous media based on fractal scaling law for water and gas phases in the pores [46]. Extensive studies have shown that most natural porous media and some synthetic porous media possess self-similar fractal scaling laws over multiple scales [45]. Therefore, two kinds of fractal models based on pore and solid phases have been proposed [47]. Fractal scaling laws can be applied to characterize the geometrical and morphological structures for pore and solid phases in porous media, respectively.

The fractal Monte Carlo method has been applied in different areas to model a porous media. Yu et al. [48] performed Monte Carlo simulations to predict the permeability of fractal porous media, their results were verified by comparison with the analytical solution for the permeability of bi-dispersed porous media. Zou et al. [49] used the Monte Carlo simulation technique to model the surface topography in a scale-invariant manner with the fractal nature of rough surfaces. Yu [44] presented a review article summarizing the theories, methods, mathematical models, achievements, and open questions in the area of flow in fractal porous media by applying the theory and technique of fractal geometry. Feng et al. [50] combined the Monte Carlo technique with fractal geometry theory to predict the thermal conductivity of nanofluids. Xu et al. [51] performed Monte Carlo simulations of radial seepage flow in the fractured porous medium, where the fractal probability model was applied to characterize the fracture size distribution. Vadapalli et al. [52] proposed a permeability estimation method for a sandstone reservoir, which considers the fractal behavior of pore size distribution and tortuosity of capillary pathways using Monte Carlo simulations. Xu et al. [53] used fractal Monte Carlo simulations to predict the effective thermal conductivity of porous media. Xiao et al. [54] employed the fractal Monte Carlo to simulate the Kozeny–Carman constant of fibrous porous media with the micropore size characterized by the fractal scaling law. Yang et al. [55] performed Monte Carlo simulations based on the fractal probability law to understand gas flow mechanisms and predict the apparent gas permeability of shale reservoirs.

In this work, the calculation of ETC for human skin using the fractal scaling and Monte Carlo methods is presented. This ETC involves a bundle of tortuous capillaries whose size distribution follow fractal scaling laws. The power-law model was chosen because of its simplicity in describing different non-Newtonian fluids by modifying a single parameter, in the case of blood as a shear-thinning fluid. The heat transfer process in the perfused tissue is analyzed, and a heat source is applied on the tissue surface for a period of time without reaching the degradation temperature. The tissue is considered as a uniform porous medium of one and three layers, which can be assigned different porosity and conductivity. The temperature profiles, as well as their distribution when modifying the main variables of the model, are presented. To the authors' knowledge, there are no studies that determine an ETC for heat transfer in biological tissues, considering the techniques mentioned above and especially for non-Newtonian fluids.

2. Heat Transfer in Human Skin

Figure 1 presents the human skin structure, considering three layers; epidermis, dermis and hypodermis. These layers differ in having their own physical properties such as density, specific heat, thermal conductivity and porosity.

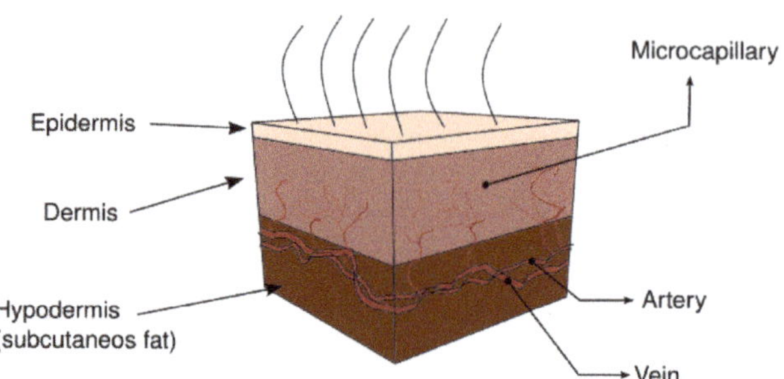

Figure 1. Human skin structure.

The first equation that described heat transfer in human tissue and included the effects of blood flow on tissue temperature on a continuum basis was presented by Pennes [15]. He presented the heat transfer analysis in the human forearm, considering the metabolic heat rate in the tissue and the perfusion heat source term. This last term has been the focus of attention since its inception and many researches have found alternative representations of the effect of blood perfusion on tissue heat transfer. The Pennes equation is given by:

$$\rho_t c_t \frac{\partial T_t}{\partial t} = \nabla \cdot (k_t \nabla T_t) + \rho_b c_b \omega_b (T_a - T_t) + q_m, \tag{1}$$

where ρ, c, T, t, k, q_m, and ω_b are density, specific heat, temperature, time, thermal conductivity, metabolic heat production, and blood perfusion rate per unit volume of tissue, respectively. The subscripts t, b, and a refer to tissue, blood, and artery, respectively [15] .

However, some inconsistencies in the Pennes model include the following: the thermal equilibrium take place in arteries and veins (not in the capillaries, as it assumes); it does not take into account any vascular architecture; and the most critical assumption is on the blood perfusion term, which is not a global term—it is local along the capillary and depends on direction.

3. Mathematical Modeling

Hyperthermia treatment consists of applying heat in a specific area of the human body. In this work, an external heat source is applied to an area of the forearm, as shown in Figure 2a. In addition, in Figure 2b, it is described that H is the total thickness of the tissue and the thicknesses of each layer are $H_E = 0.04H$, $H_D = 0.48H$, and $H_H = 0.48H$. Initially, the tissue is at a constant temperature, $T_c \sim 37\,°\text{C}$, subsequently, a heat source is applied to an area of the tissue. The area around the heat application area is open to the surroundings, and generally is a temperature lower than the human body. For this work, the surrounding temperature is considered to be $T_\infty \sim 25\,°\text{C}$. Furthermore, Figure 2b shows a mathematical representation of the hyperthermia treatment, note that the deepest internal temperature is maintained at the body temperature. According to Weinbaum and Jiji [11], the simplified two-dimensional governing equation of heat transport in biological tissue is given by:

$$(\rho C_p)_t \frac{\partial T}{\partial t} = \frac{\partial}{\partial x}\left(k_{eff}\frac{\partial T}{\partial x}\right) + \frac{\partial}{\partial y}\left(k_{eff}\frac{\partial T}{\partial y}\right) + q_m, \tag{2}$$

where k_{eff} is the effective thermal conductivity (ETC) and q_m is the metabolic heat.

According to physical model showed in Figure 2, Equation (2) is subject to the following initial and boundary conditions:

$$T(x, y, 0) = T_c,\tag{3}$$

$$\frac{\partial}{\partial x} T(0, y, t) = 0,\tag{4}$$

$$\frac{\partial}{\partial x} T(W, y, t) = 0,\tag{5}$$

$$T(x, 0, t) = T_c,\tag{6}$$

$$-k_t \frac{\partial}{\partial y} T(x, H, t) = h[T_\infty - T(x, H, t)], \quad \frac{2}{5}W \geq x \geq \frac{3}{5}W,\tag{7}$$

$$-k_t \frac{\partial}{\partial y} T(x, H, t) = f, \quad \frac{2}{5}W < x > \frac{3}{5}W,\tag{8}$$

$$f = \begin{cases} q_{\text{app}} = 200\left[\frac{\text{W}}{\text{m}^2}\right], & t \leq t_{\text{app}}, \\ h[T_\infty - T(x, H, t)], & t > t_{\text{app}}, \end{cases}$$

where W and H are the width and the height of the domain, respectively. The h, q_{app}, and t_{app} are the heat transfer coefficient, applied external heat, and application time, respectively.

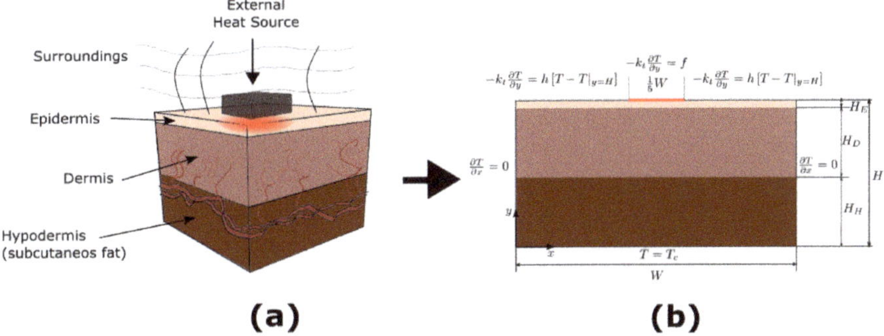

Figure 2. (**a**) Hyperthermia treatment; (**b**) human skin three-layer model, and the corresponding boundary conditions.

ETC is an important parameter in Equation (2). Weinbaum and Jiji [11] proposed a vascular function $V(y)$ that can be constructed knowing the vascular data, which is the distribution of the arteries, veins, and capillaries. According to Weinbaum and Jiji [11], the vascular function increases with tissue depth [14]. In this work, the *representative elementary volume* (REV), is defined and the fractal scaling method is used to depict the vascular tissue structure.

3.1. Fractal Scaling Method

We consider a cubic REV as shown in Figure 3, with defined length side L_0. All REV capillaries extend throughout the volume from one side to the other as is showed in Figure 3.

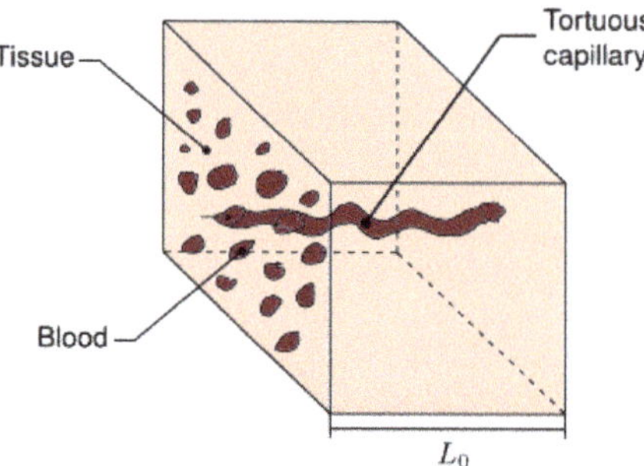

Figure 3. Representative elementary volume of human skin.

The fractal scaling method establishes the relationship between the number and pore size in the porous medium. The fundamental fractal scaling law is applied to REV cross-section, as follows [56]:

$$N(>\lambda) = \left(\frac{\lambda_{\max}}{\lambda}\right)^{D_f},\tag{9}$$

where N, D_f, λ, and $\lambda_{\max}$ are the number of capillaries, the fractal dimension, the equivalent diameter, and the maximum equivalent diameter of the capillaries in the REV cross-section, respectively. The number of capillaries with equivalent diameter between $\lambda + d\lambda$ in the REV cross-section is:

$$-dN(\lambda) = D_f \lambda_{\max}^{D_f} \lambda^{-D_f-1} d\lambda.\tag{10}$$

The total number of pores in the range from $\lambda_{\min}$ to $\lambda_{\max}$ is obtained using Equation (9), as follows:

$$N_T(\leq \lambda_{\min}) = \left(\frac{\lambda_{\max}}{\lambda_{\max}}\right)^{D_f},\tag{11}$$

where N_T is the total pores. Dividing Equation (10) by Equation (11), we obtain:

$$-\frac{dN}{N_T} = D_f \lambda_{\min}^{D_f} \lambda^{-(D_f+1)} d\lambda = f(\lambda)d\lambda,\tag{12}$$

where $f(\lambda)$ is the probability density function, which satisfies that $f(\lambda) \geq 0$. According to probability theory, $f(\lambda)$ must satisfy the following normalization relation or total cumulative probability:

$$-\int_{\lambda_{\min}}^{\lambda_{\max}} \frac{dN}{N_T} = \int_{\lambda_{\min}}^{\lambda_{\max}} f(\lambda)d\lambda = 1 - \left(\frac{\lambda_{\min}}{\lambda_{\max}}\right)^{D_f} \equiv 1.\tag{13}$$

The integration result of Equation (13) shows that it holds if—and only if—the following holds:

$$\left(\frac{\lambda_{\min}}{\lambda_{\max}}\right)^{D_f} \cong 0.\tag{14}$$

The above equation implies that $\lambda_{\min} \ll \lambda_{\max}$, it must be satisfied for fractal analysis of a porous media. According to Yu et al. [48], Equation (14) can be considered as a criterion whether a porous medium can be characterized by fractal theory and technique. If Equation (13) is expressed as:

$$R(\lambda) = \int_{\lambda_{\min}}^{\lambda} f(\lambda)d\lambda = 1 - \left(\frac{\lambda_{\min}}{\lambda}\right)^{D_f},\tag{15}$$

then the pore diameter λ can be found as:

$$\lambda = \frac{\lambda_{min}}{(1-R)^{1/D_f}} = \left(\frac{\lambda_{min}}{\lambda_{max}}\right)\frac{\lambda_{max}}{(1-R)^{1/D_f}}. \tag{16}$$

On the other hand, the fractal path of the tortuous capillary can be described as follows [57]:

$$L(\lambda) = L_0^{D_T}\lambda^{1-D_T}, \tag{17}$$

where $L(\lambda)$, L_0, and D_T are capillary tortuous length, characteristic length of a straight capillary (REV side length), and fractal dimension describing the capillary tortuous length, respectively. An important parameter involved in the ETC is the total pore area, A_c, determined by Wu and Yu [58]:

$$A_c = -\int_{\lambda_{min}}^{\lambda_{max}} \frac{\pi\lambda^2}{4}dN(\lambda) = \frac{\pi D_f \lambda_{max}^{D_f}\left(\lambda_{max}^{2-D_f} - \lambda_{min}^{2-D_f}\right)}{4\left(2-D_f\right)}, \tag{18}$$

the total cross area, A_T, is:

$$A_T = L_0^2 = \frac{A_c}{\phi}, \tag{19}$$

where ϕ is the porosity.

3.2. The Fractal ETC of the REV

In order to obtain the ETC and according to Fourier's law, the total heat flux in the REV is given by:

$$Q_T = k_{eff}A_T\frac{\Delta T}{L_0}, \tag{20}$$

where k_{eff} is the ETC including the tissue conductivity and convective blood flow in the capillaries, and ΔT is the difference temperature between two faces in the REV. The heat flux in a single tortuous capillary of the REV is:

$$Q_c = k_b\frac{\pi\lambda^2}{4}\frac{\Delta T}{L(\lambda)} = k_b\frac{\pi\lambda^2}{4}\frac{\Delta T}{L_0^{D_T}\lambda^{1-D_T}}, \tag{21}$$

where k_b is the blood thermal conductivity. The heat flux corresponding only to the tissue is expressed by:

$$Q_t = k_t A_t\frac{\Delta T}{L_0} = k_t(1-\phi)A_T\frac{\Delta T}{L_0}, \tag{22}$$

where the subscript t refers to the tissue. According to superposition theorem, the total heat flux Q_T is the addition of the fluxes as follows:

$$Q_T = Q_t + Q_c. \tag{23}$$

Substituting Equations (21) and (22) into Equation (23) the following equation is obtained:

$$k_{eff} = k_t(1-\phi) + k_b\frac{\phi\left(2-D_f\right)\lambda^{D_T+1}}{L_0^{D_T-1}D_f\lambda_{max}^{D_f}\left(\lambda_{max}^{2-D_f} - \lambda_{min}^{2-D_f}\right)}, \tag{24}$$

where L_0 is determined from Equation (19). According to the work presented by Weinbaum and Jiji [11], from an energy conservation balance in a countercurrent artery and mean value theory, derived the effective enhancement of the tissue conductivity. Therefore, by comparing the Weinbaum and Jiji ETC to Equation (24), it can be seen that the two equations have a similar form, proposing the following the expression:

$$k_{eff} = k_t(1 + \Gamma(\mathbf{r})\psi(\mathbf{r})), \tag{25}$$

where Γ is a parameter that depends on conduction or convective heat transport [59], ψ is related to skin structure (also called dimensionless vascular geometry function) [14], and $\mathbf{r}$ is the position vector. This work focuses on the convective transport of blood. Hence, the dimensionless ETC for this work is given by:

$$k_{eff}^* = \frac{k_{eff}}{k_t} = \left[(1 - \phi) + Pe^2 \left(\frac{k_b}{k_t} \right)^2 \frac{\phi \left(2 - D_f \right) \lambda^{D_T + 1}}{L_0^{D_T - 1} D_f \lambda_{\max}^{D_f} \left(\lambda_{\max}^{2 - D_f} - \lambda_{\min}^{2 - D_f} \right)} \right], \tag{26}$$

where k_{eff}^* is the dimensionless ETC and Pe is the Peclet number defined as $Pe = PrRe = \rho_b C_b \lambda u / k_b$; where ρ_b, C_b, and u are the density, specific heat, and average blood velocity in the capillary, respectively.

3.3. Fractal Dimensions, D_f and D_T

Equation (26) is a function of porosity, fractal dimensions, maximum and minimum diameters, conductivity ratio, and Peclet number, the latter being a function of physical properties, velocity, and diameter.

There is a relationship between porosity and fractal dimension, D_f, according to Yu and Li [56], is given by:

$$\phi = \left(\frac{\lambda_{\min}}{\lambda_{\max}} \right)^{D_E - D_f}, \tag{27}$$

where D_E is the Euclidean dimension ($D_E = 2$ and 3, for two- and three-dimensional space, respectively). Another important aspect of Equation (27), from experimentation, it has been found that the ratio of minimum and maximum diameter, $\lambda_{\min}/\lambda_{\max}$, in several natural porous media, is the order of $10^{-2} \sim 10^{-4}$, [60]. Feng et al. derived a generalized model covering a wide range of porosities, for the effective thermal conductivity, based on the fact that statistical self-similarity exists in porous media [61].

In this work, D_T is established manually, taking into account whether $D_T > 1$ the tortuosity is present and $D_T = 1$ are straight capillaries.

3.4. Non-Newtonian Fractal Velocity

The Peclet number in Equation (26) is defined by:

$$Pe = PrRe = \frac{\rho_b C_b}{k_b} \lambda u, \tag{28}$$

where the velocity u is a function of the microcapillary diameter, is obtained from the non-Newtonian fluid flow in a single microcapillary, as presented by Zhang [62], as follows:

$$q_c = \left[\frac{dp}{dL_0} \frac{L_0^{1 - D_T} 2^{D_T - 1}}{\mu_b (D_T + 1) D_T} \right]^{\frac{1}{n}} \frac{n\pi}{D_T + 3n} \left(\frac{\lambda}{2} \right)^{\frac{D_T}{n} + 3}, \tag{29}$$

where q_c, p, and μ_b are the flow rate in a single microcapillary, pressure, and dynamic blood viscosity, respectively. By considering $q_c = u A_{sc}$, where A_{sc} is the area of a single capillary and the velocity can be determined as:

$$u = \left[\frac{dp}{dL_0} \frac{L_0^{1 - D_T} 2^{D_T - 1}}{\mu (D_T + 1) D_T} \right]^{\frac{1}{n}} \frac{n}{D_T + 3n} \left(\frac{\lambda}{2} \right)^{\frac{D_T}{n} + 1}, \tag{30}$$

where n is the power-law index of the constitutive equation, when $n = 1$ the Newtonian velocity is recovered. For $n < 1$ and $n > 1$ describes shear-thinning and shear-thickening fluid behavior, respectively. Figure 4 presents the average velocity and Peclet number as a function of the capillary diameter for different values of the power-law index, n, considering two values of porosity, $\phi = 0.1$ and $\phi = 0.5$. By increasing the pore diameters, the flow through the tissue is greater for all cases. This effect is magnified when the power-law index, n, decreases, which indicates a lower resistance to flow, because the viscosity decreases, which is characteristic of shear-thinning fluids. The opposite case can be seen in this figure for shear-thickening fluids ($n > 1$). The number of Peclet is proportional to the velocity, therefore they have the same tendency. On the other hand, by increasing the porosity, both u and Pe have a slight increase because the pores are very small. Figure 5 shows the average velocity and Peclet number as a function of the fractal tortuosity, D_T, for different values of porosity, ϕ, considering two values of power-law index, $n = 1$ and $n = 0.6$. For this case, the maximum pore diameter was taken into account. Keeping constant porosity and increasing tortuosity, D_T, the velocity and Peclet number both decrease. This is correct for complex vascular architectures where the flow experiences higher resistance as well as being very small. For straight or slightly tortuous capillaries—where the highest velocities and Peclet numbers are found—there is a large increase as the power-law index decrease, as shown in Figure 5.

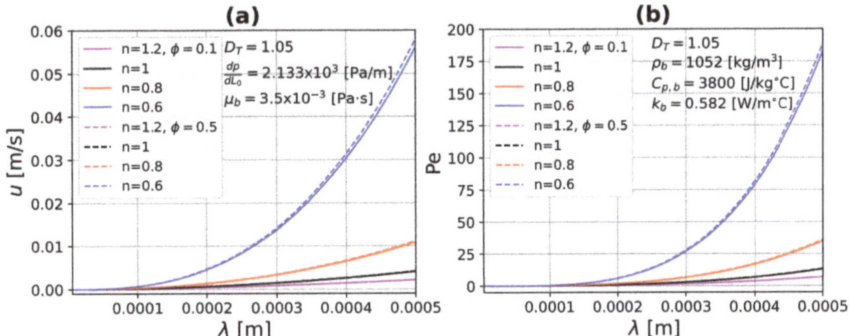

Figure 4. (**a**) Average velocity and (**b**) Peclet number as a function of capillary diameter for different values of the power-law index n. In a single microcapillary—continuous lines ($\phi = 0.1$), and dashed lines ($\phi = 0.5$).

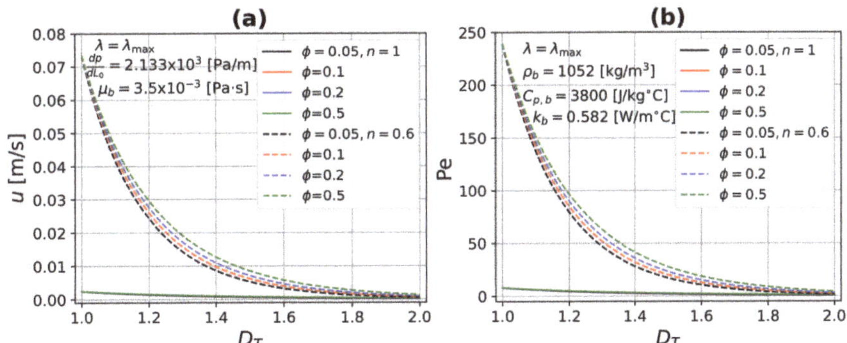

Figure 5. (**a**) Average velocity and (**b**) Peclet number as a function of fractal tortuosity D_T for different porosity values. For both in a single microcapillary, continuous lines (Newtonian fluid), and dashed lines (non-Newtonian fluid).

3.5. Monte Carlo Method

The Monte Carlo method is used to establish the pore size and the porous medium distribution, as is shown in Figure 6. Figure 6a,b present one- and three-layer tissue with different porosities, with $\phi = 0.1$ (one layer) and $\phi = 0$, $\phi = 0.5$, and $\phi = 0.05$ for epidermis, dermis, and hypodermis, respectively. Yu et al. [48] was the first to propose the fractal Monte Carlo methodology to simulate the transports in fractal porous media. In Figure 6c for one-layer tissue and Figure 6d for three-layer tissue show the Monte Carlo simulations, which are performed in the range of $\lambda_{\min}$–$\lambda_{\max}$. Figure 6d shows the pore sizes variation for dermis and hypodermis layers, since the porosity of the epidermal layer is not considered. The pore size variation is determined using the Monte Carlo method in Equation (16), as follows:

$$\lambda_i = \frac{\lambda_{\min}}{(1 - R_i)^{1/D_f}} = \left(\frac{\lambda_{\min}}{\lambda_{\max}}\right)\frac{\lambda_{\max}}{(1 - R_i)^{1/D_f}}, \tag{31}$$

where λ_i is the variation of the pore diameter and R_i are the random numbers between $0 - 1$. On the other hand, the random numbers also help us to construct the porous medium presented in Figure 6. Equation (31) is derived from Equation (9), which implies that there is only one largest pore in the REV cross-section [44]. This is consistent to the pore size distribution shown in Figure 6.

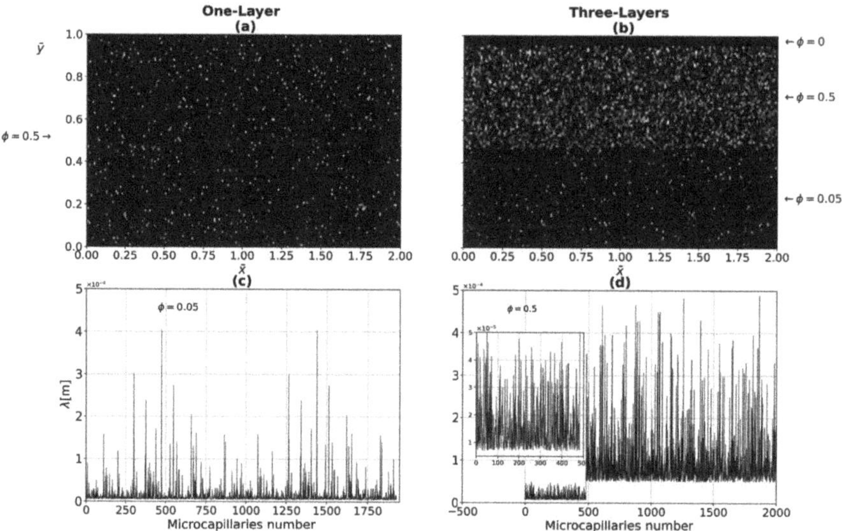

Figure 6. Two-dimensional pore size distribution for (**a**) one-layer tissue with $\phi = 0.1$, (**b**) three-layer tissue with $\phi = 0, 0.5$, and 0.05 for epidermis, dermis, and hypodermis, respectively. (**c,d**) The simulated pore sizes by the Monte-Carlo technique in the range of $\lambda_{\min} = 5 \times 10^{-6}$ to $\lambda_{\max} = 5 \times 10^{-4}$.

3.6. Dimensionless Governing Equation

The dimensionless variables are:

$$\bar{x} = \frac{x}{H}; \qquad \bar{y} = \frac{y}{H}; \qquad \tau = t\frac{\alpha_t}{H^2}; \qquad \theta = \frac{T - T_\infty}{T_c - T_\infty}, \tag{32}$$

where α_t is the tissue thermal diffusivity. Substituting dimensionless variables of Equation (32) into Equation (2), the dimensionless governing equation is:

$$\frac{\partial \theta}{\partial \tau} = \frac{\partial}{\partial \bar{x}}\left(k^*_{eff}\frac{\partial \theta}{\partial \bar{x}}\right) + \frac{\partial}{\partial \bar{y}}\left(k^*_{eff}\frac{\partial \theta}{\partial \bar{y}}\right) + \Phi_m, \tag{33}$$

where $k^*_{eff} = k_{eff}/k_t$ is defined in Equation (26) and $\Phi_m = q_m H^2/k_t(T_c - T_\infty)$ is the dimensionless metabolic heat generation. Respectively, the dimensionless initial and boundary conditions are given by:

$$\theta(\bar{x}, \bar{y}, 0) = 1, \tag{34}$$

$$\frac{\partial}{\partial \bar{x}}\theta(0, \bar{y}, \tau) = 0, \tag{35}$$

$$\frac{\partial}{\partial \bar{x}}\theta(2, \bar{y}, \tau) = 0, \tag{36}$$

$$\theta(\bar{x}, 0, \tau) = 1, \tag{37}$$

$$\frac{\partial}{\partial \bar{y}}\theta(\bar{x}, 1, \tau) = Bi[\theta_\infty - \theta(\bar{x}, 1, \tau)], \quad \frac{2}{5} \geq \bar{x} \geq \frac{3}{5}, \tag{38}$$

$$\frac{\partial}{\partial \bar{y}}\theta(\bar{x}, 1, \tau) = f^*, \quad \frac{2}{5} < \bar{x} > \frac{3}{5}, \tag{39}$$

$$f^* = \begin{cases} \frac{H}{k_t \Delta T} q_{app}, & \tau \leq \tau_{app}, \\ Bi[\theta_\infty - \theta(\bar{x}, 1, \tau)], & \tau > \tau_{app}, \end{cases}$$

where Bi is the Biot number defined as $Bi = hH/k_t$, θ_∞ is the dimensionless temperature of the environment, and f^* takes the boundary condition according to the dimensionless time of application τ_{app}.

4. Results

The numerical results presented in this section were generated from a numerical code developed in the Fortran programming language. Equation (33) subject to boundary conditions Equations (34)–(39) was solved using an explicit finite difference method.

The values used for the tissue physical properties are: $\rho_t = 1200 \text{ kg/m}^3$, $C_{p,t} = 3600 \text{ J/kg·°C}$, $k_t = 0.293 \text{ W/m·°C}$. When considering three-layer tissue, their corresponding thermal conductivities are: $k_e = 0.25 \text{ W/m·°C}$, $k_d = 0.45 \text{ W/m·°C}$, and $k_h = 0.2 \text{ W/m·°C}$, for epidermis, dermis, and hypodermis, respectively. The blood physical properties are: $\rho_b = 1052 \text{ kg/m}^3$, $C_{p,b} = 3800 \text{ J/kg·°C}$, and $k_b = 0.582 \text{ W/m·°C}$ [2,63]. The metabolic heat is $q_m = 368.1 \text{ W/m}^3$ [63]. The heat transfer coefficient due to convection and radiation in surroundings is: $h = 5 \text{ W/m}^2\text{·°C}$. The environmental temperature was chosen as $T_\infty = 25 \text{ °C}$. For the three-layer model, the main changes in porosity are considered to be in the dermis, due to the greater interaction between the tissue and the vascular network [8]. This work does not take into account thermal degradation (thermal damage) in tissue, which occurs at 44 °C and higher [64].

4.1. ETC Analysis

According to Equation (26), ETC depends on the conductivity ratio, porosity, ϕ, fractal coefficients, D_f, D_T, and Peclet number. In this work, both conductivities of tissue and blood are constant; therefore, the conductivity ratio is also constant. Once the porosity value is assigned manually, D_f can be determined using Equation (27). The Peclet number depends on the physical properties of the blood, capillary diameter, and velocity, see Equation (28). An important contribution of this work is that it allows the analysis of blood as a Newtonian and non-Newtonian fluid. This is possible through the power-law model, which is immersed in the velocity calculation Equation (30). This equation also depends on the fractal coefficients, in addition to the viscosity and the power-law index, n. The latter makes it possible to consider a Newtonian fluid ($n = 1$) and non-Newtonian fluids such as shear-thinning fluid ($n < 1$) and shear-thickening $n < 1$. The blood is a shear-thinning fluid described by a power-law index value of $n = 0.6$, according to Johnston et al. [65].

Figure 7a presents the ETC as a function of the fractal coefficient D_T for different porosity values, in addition for two values of the power-law index $n = 1$ and $n = 0.6$. D_T describes the capillary tortuous length, the influence of D_T on the ETC is in the range of 1–1.3, which is in accordance with that reported by Yu and Cheng [57]. By increasing the porosity, the ETC also increases, due to higher blood flow through the tissue having a higher thermal conductivity. An important increase is found in the ETC when considering blood as a non-Newtonian fluid, since it increases with porosity as mentioned above, but especially for a shear-thinning fluid such as blood ($n = 0.6$).

Figure 7b shows the fractal dimension as a function of porosity. When porosity tends to unity, the fractal dimension tends to two. That for a surface indicates that it is totally occupied by pores, which corresponds to a fractal dimension of two, it is consistent according to Yu et al. [66].

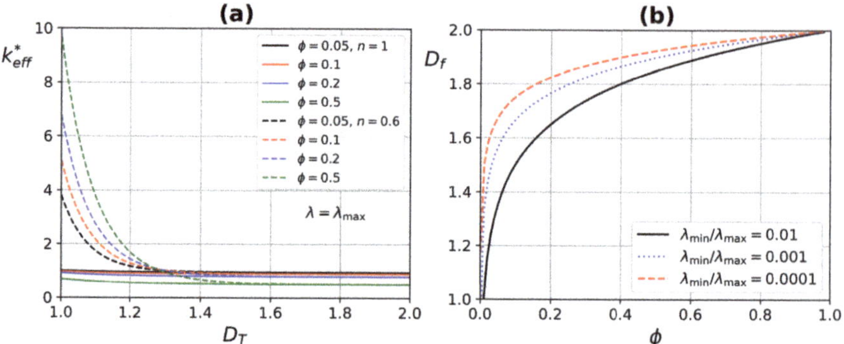

Figure 7. (**a**) Effective thermal conductivity as a function of fractal dimension D_T. Continuous lines (Newtonian fluid) and dashed lines (non-Newtonian fluid). (**b**) Fractal dimension D_f as a function of porosity, for different ratios $\lambda_{min}/\lambda_{max}$.

Figure 8 exhibits the ETC as a function of fractal dimension D_f, and the porosity ϕ; for two pore diameter values, $\lambda = 4 \times 10^{-4}$ and $\lambda_{max} = 5 \times 10^{-4}$, and by considering straight capillaries $D_T = 1$ for different values of the power-law index. k_{eff}^{*} has an increase by augmenting the pore diameter, but the main improvement is for shear-thinning fluids, due to the lower resistance that the fluid experiences in large pores, increasing the velocity and generating a higher energy dissipation.

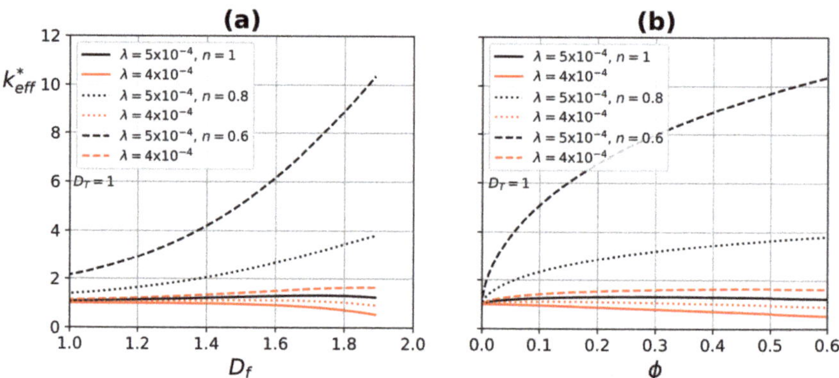

Figure 8. Effective thermal conductivity as a function of (**a**) fractal dimension D_f, and (**b**) porosity ϕ. Continuous lines (Newtonian fluid) and dashed lines (non-Newtonian fluid).

4.2. Code Validation

To validate the ETC, the solution of the Weinbaum and Jiji [11] (WJ) is compared with the present work using a one-layer tissue with porosity $\phi = 0.05$, straight capillaries

$D_T = 1$, $k_b/k_t = 1.9863$, D_f is determined by Equation (27) and Newtonian fluid $n = 1$. The Peclet number depends on the pore diameter λ and the velocity u, WJ determined $Pe = 20$, which is in accordance with this work of $Pe = 16.3$ for maximum pore size. Figure 9 shows the comparison of the temperature profile as a function on time and a function of deep-tissue layer (at $\tau = 0.09$) between WJ model and the present work. The ETC obtained by Weinbaum and Jiji [11] is as follows:

$$\frac{k_{eff,WJ}}{k_t} = \left[1 + Pe_0^2 V(\bar{y})\right],\tag{40}$$

where $Pe_0 = 2\rho_b c_b a_0 u_0/k_b = 20$, $V(\bar{y}) = A + B\bar{y} + C\bar{y}^2$ with $A = 6.32 \times 10^{-5}$, $B = -15.9 \times 10^{-5}$ and $C = 10 \times 10^{-5}$. The metabolic heat $\Phi_m = q_m L^2/k_t(T_c - T_\infty) = 0.094$.

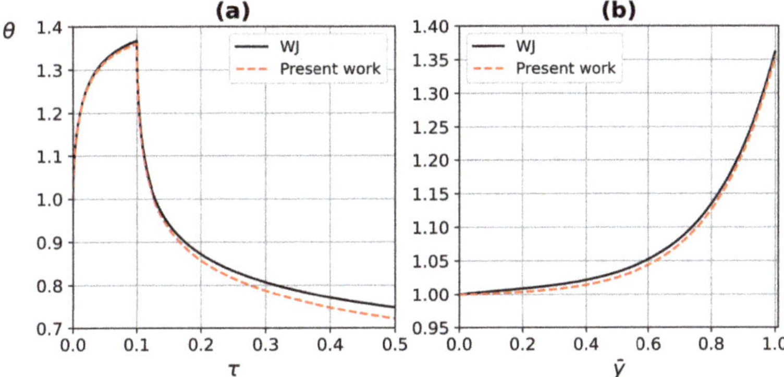

Figure 9. Comparison of temperature profile between WJ and the present work as a function of (**a**) time, and (**b**) tissue layer depth.

The comparison between WJ and the present work exhibited in Figure 9, for the temperature as a function of (a) time and (b) deep of the tissue layer, both fit in a good agreement. The comparison between the temperature distribution of the WJ model (40) and the present work (26) is shown in Figure 10. The heat source is applied on the central part of the tissue surface, as shown in Figure 2. The temperature distribution corresponds to a heat source application dimensionless time of $\tau = 0.09$. The temperature distribution of the WJ model shows a higher evolution, indicating a higher heat transfer in the tissue. The difference between models according to temperature contours is less than 2%.

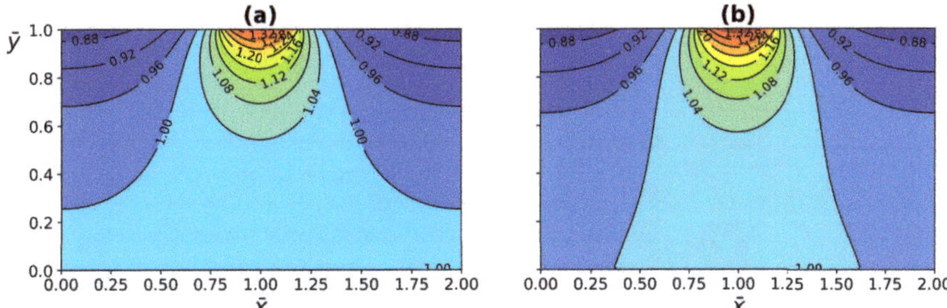

Figure 10. Comparison of temperature distribution between (**a**) WJ model, and (**b**) present work at dimensionless time. Just before the heat source was removed.

4.3. Dynamical Test Simulations

4.3.1. One-Layer Tissue Analysis

To evaluate the different parameters involved in the ETC, a dynamic test is performed, which consists of applying a heat source on the tissue surface for a period of time, removing the source when the thermal relaxation process begins, the scheme and conditions are shown in Figure 2. This test does not reach the tissue degradation temperature, $T < 44\,^\circ\text{C}$ [8,64].

Figure 11 presents the temperature profile as a function of time and tissue layer depth for one-layer (at $\tau = 0.09$), by modifying the fractal coefficient D_T in the range of 1–1.1, with porosity $\phi = 0.1$ and Newtonian fluid $n = 1$. There are no significant changes when the heat source is applied, at the end of the relaxation process is when there is a difference between the WJ model and this work. This difference is due to the fact that there is a better energy transfer process in the present model, but there is no appreciable effect of D_T due to the low tissue porosity.

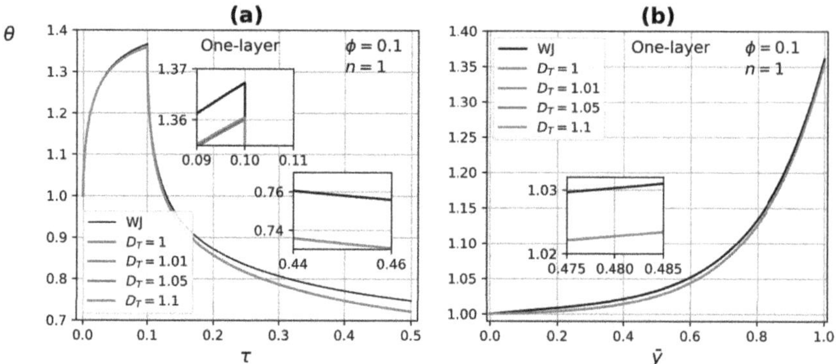

Figure 11. Temperature profiles as a function of (**a**) time and (**b**) tissue layer depth, for different D_T values at $\tau = 0.09$.

Figure 12 shows the temperature profile as a function of time and tissue layer depth (at $\tau = 0.09$) for one-layer tissue by modifying the porosity ϕ in the range of 0.01 to 0.5, with $D_T = 1.05$ and Newtonian fluid $n = 1$. As the porosity increases, higher temperatures are reached when the heat source is applied, and lower temperatures are reached at the end of the relaxation process. This is due to increased perfusion in the tissue, on the other hand, blood has a higher thermal conductivity which improves heat transfer in the porous medium.

Figures 13 and 14 exhibit the temperature profile as a function of time and tissue layer depth (at $\tau = 0.09$) one layer for two porosity values $\phi = 0.1$ and $\phi = 0.5$, respectively. Modifying the power-law index n in the range of 0.6 to 1 and $D_T = 1.05$. In the case of porosity $\phi = 0.1$, there are no significant changes when modifying the power-law index, due to low perfusion through the tissue. For porosity $\phi = 0.5$, blood perfusion is increased and the effect of the power-law index is reflected both in the application of the heat source and in the thermal relaxation process. For a shear-thinning fluid $n = 0.6$, there is a greater dissipation of energy and therefore it reaches lower temperatures compared with a Newtonian fluid, in addition to experiencing a higher thermal relaxation due to a greater influence of the convective condition of the surface, as shown in Figure 14a. The effect of higher porosity is also reflected in the temperature profile as a function of tissue depth, as it presents distortions, mainly for the non-Newtonian fluid with $n = 0.6$, see Figure 14b.

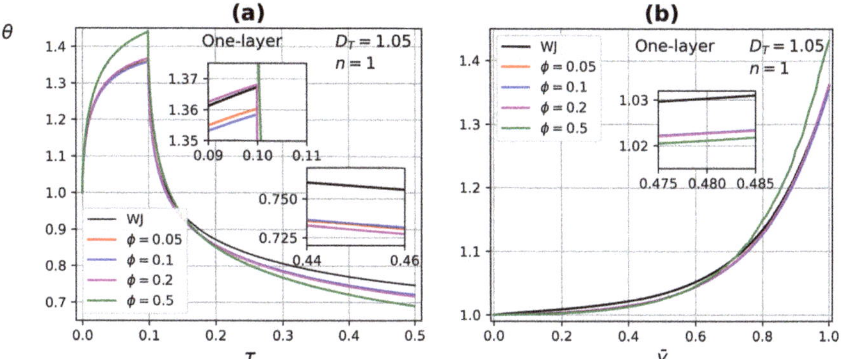

Figure 12. Temperature profiles as a function of (**a**) time and (**b**) tissue layer depth, for different porosity ϕ values at $\tau = 0.09$.

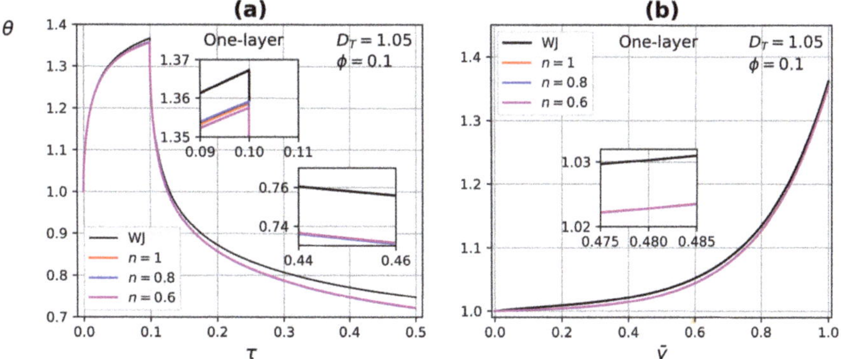

Figure 13. Temperature profiles as a function of (**a**) time and (**b**) tissue layer depth for different power-law index n at $\tau = 0.09$ and porosity $\phi = 0.1$.

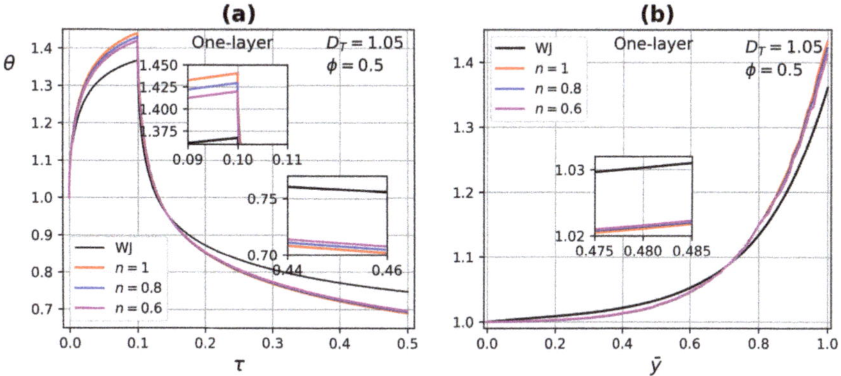

Figure 14. Temperature profiles as a function of (**a**) time and (**b**) tissue layer depth, for different power-law index n at $\tau = 0.09$ and porosity $\phi = 0.5$.

4.3.2. Three-Layer Tissue Analysis

The structure of the skin is very complex and is generally considered in three layers—epidermis, dermis, and hypodermis—as is shown in Figure 1. The thickness of these layers varies depending on the location of the skin and many factors, such as age, sex, race, endocrine, and nutritional status of the individual [7–10].

Figure 15 shows the temperature profile as a function of time and tissue three-layer depth (at $\tau = 0.09$), each layer with its own conductivity, $k_e = 0.25\,\mathrm{W/m\cdot{}^\circ C}$, $k_d = 0.45\,\mathrm{W/m\cdot{}^\circ C}$ and $k_h = 0.2\,\mathrm{W/m\cdot{}^\circ C}$, for epidermis, dermis, and hypodermis, respectively, [8]. The porosity ϕ varies uniformly throughout the tissue in the range of 0.01–0.5, with $D_T = 1.05$ and non-Newtonian fluid $n = 0.6$. The shear-thinning effect increases the heat transfer in tissue, which is reflected in the temperatures reached when applying the heat source and in the thermal relaxation process. The effect of having three different conductivities can be seen clearly in the temperature profile as a function of tissue depth, where changes in the slope of the curve at the layer interfaces are presented, as is shown in Figure 15b. This behavior has been reported by Xu et al. [2].

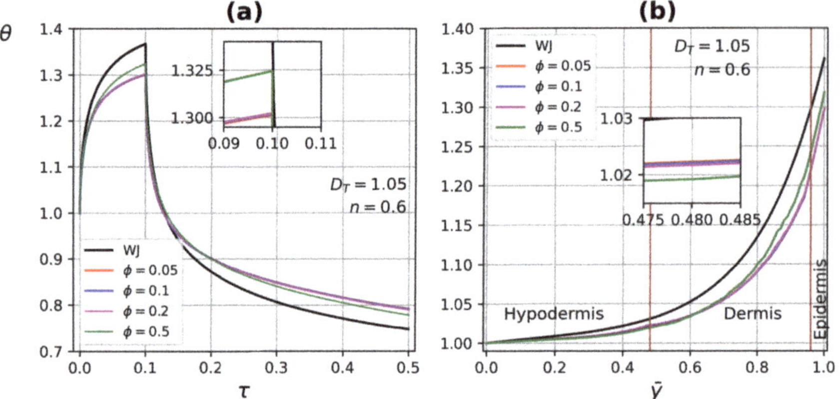

Figure 15. Temperature profiles as a function of (**a**) time and (**b**) tissue three-layer depth with different porosities, $\phi = 0$, 0.5, and 0.05 for epidermis, dermis, and hypodermis, respectively.

Figure 16 shows the temperature profile comparison between one- and three-layer tissue at different depths as a function of time and tissue depth, the properties of the three-layer tissue are those used in Figure 15. In the case of a one-layer tissue, the uniform porosity is $\phi = 0.5$, for the case of three-layer tissue the following porosities are assigned $\phi = 0$, 0.5, and 0.05 for epidermis, dermis, and hypodermis, respectively. These values were assigned according to the fact that the epidermis has very low porosity, the dermis contains the highest density of capillaries, and the hypodermis is composed of loose fatty connective tissue. The other variable values are $D_T = 1.05$ and $n = 0.6$ [8]. Considering blood as a non-Newtonian shear-thinning fluid, in addition to different tissue properties, generates important modifications in the temperature profiles both in time and depth. In contrast, uniform properties throughout the biological tissue, as shown in Figure 16a,b.

Figure 16a shows red dots on temperature profiles used to indicate the time corresponding to the instantaneous temperature distribution presented in Figure 17. This figure shows the comparison of temperature distribution between the one- and three-layer tissues at different times. In the case of three-layer tissues, the effect of the different tissue properties on the temperature contours, which show small variations due to porosity depending on the area. In the case of one-layer tissues, the heat transfer is more similar to homogeneous solids, which have higher thermal conductivity at the top and bottom of the tissue compared with the three-layer model. Finally, the right half is overlapped on the left half of the temperature distribution for one- and three-layer tissue to demonstrate the asymmetry generated by the global effect of all the variables involved in the model, as shown in Figure 18.

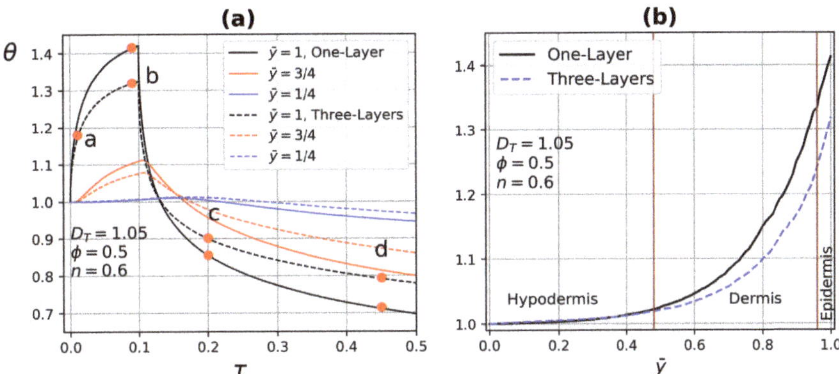

Figure 16. Temperature profile comparison between one- and three-layer tissue as a function of (**a**) time at three different depths of the tissue and (**b**) depth of the three-layer tissue with different porosities $\phi = 0$, 0.5, and 0.05 for epidermis, dermis, and hypodermis, respectively.

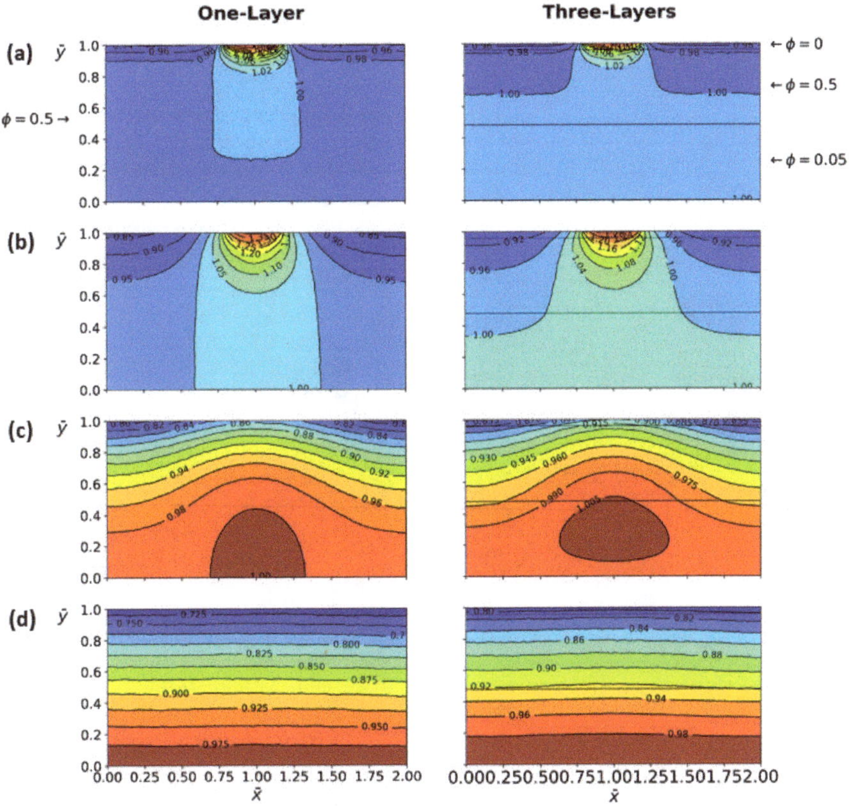

Figure 17. Comparison of the temperature distribution for one-layer tissue with $\phi = 0.5$ and three-layer tissue with $\phi = 0$, 0.5, and 0.05 for epidermis, dermis, and hypodermis, respectively. For different times from top to bottom (**a**) $\tau = 0.01$, (**b**) $\tau = 0.09$, (**c**) $\tau = 0.2$, and (**d**) $\tau = 0.45$. For a non-Newtonian fluid, $n = 0.6$.

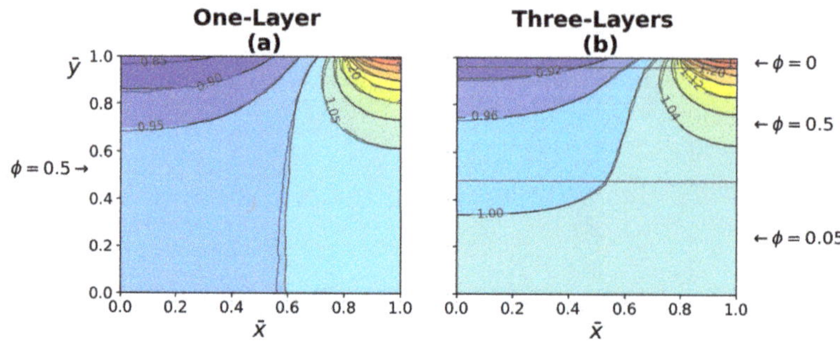

Figure 18. The right half is overlapped on the left half of the temperature distribution for the one- and three-layer tissue. Temperature distribution was taken from Figure 17 at time $\tau = 0.09$.

5. Conclusions

In this work, the ETC for human skin using the fractal scaling and Monte Carlo methods was obtained. These methods were used to describe the tissue as a porous medium, the blood was considered as Newtonian and non-Newtonian shear-thinning fluid. The numerical code developed was validated by comparing the ETC obtained by Weinbaum and Jiji [11] and the present work using one-layer tissue with the same porosity, straight capillaries, and Newtonian fluid. The difference between models according to temperature contours is less than 2%.

The ETC involves various parameters, such as fractal dimensions D_T and D_f, porosity, and the power-law index n; in order to evaluate these parameters, dynamical tests were performed, which consisted of applying a heat source on the tissue surface for a period of time—when the heat source was removed, the thermal relaxation process began. The effect of the main parameters on the temperature profiles as a function of time and tissue depth, for one- and three-layer tissue, besides temperature distribution, were presented. The main findings of this work are the following:

- The effect of fractal dimension D_T on the ETC was mainly in the range of 1–1.3.
- Higher porosity improves ETC, due to increased blood flow through the tissue, having a higher thermal conductivity.
- In one-layer tissues of low porosity, no significant changes in ETC were found. Increasing porosity, the effect of the power-law index is reflected in both heating and relaxation processes.
- The Peclet number increases substantially due to the combination of large pore diameters and shear thinning fluids.
- In three-layer tissues with different porosities, perfusion with a shear-thinning fluid contributes to the understanding of the heat transfer process in some parts of the human body.
- The ETC involves the main variables of the heat transfer process in human skin; moreover, it is easy to implement for other case studies.

Author Contributions: Conceptualization, R.O.V. and G.R.-A.; methodology, R.O.V., G.R.-A., J.P.E. and R.M.-M.; software, R.O.V., G.R.-A., J.P.E. and A.R.-M.; validation, R.O.V. and G.R.-A.; formal analysis, R.O.V. and G.R.-A.; investigation, R.O.V., G.R.-A., J.P.E., A.R.-M. and R.M.-M.; resources, R.O.V., G.R.-A., J.P.E., A.R.-M. and R.M.-M.; data curation, G.R.-A.; writing—original draft preparation, R.O.V., G.R.-A. and J.P.E.; writing—review and editing, R.O.V., G.R.-A., J.P.E., A.R.-M. and R.M.-M.; visualization, R.O.V., G.R.-A. and J.P.E.; supervision, R.O.V.; project administration, R.O.V.; funding acquisition, R.O.V. and J.P.E. All authors have read and agreed to the published version of the manuscript.

Funding: The authors R.O.V. and J.P.E. would like to acknowledge the financial support SIP-20221783 and SIP-20221572.

Acknowledgments: Guillermo Rojas-Altamirano gratefully acknowledges support by the Instituto Politécnico Nacional of Mexico for the PhD scholarship.

Conflicts of Interest: The authors declare no conflict of interest.

Abbreviations

The following abbreviations are used in this manuscript:

ETC	effective thermal conductivity
WJ	Weinbaum and Jiji
REV	representative elementary volume

References

1. Xu, F.; Lu, T.J.; Seffen, K.A. Skin thermal pain modeling—A holistic method. *J. Therm. Biol.* **2008**, *33*, 223–237. [CrossRef]
2. Xu, F.; Seffen, K.; Lu, T. Non-Fourier analysis of skin biothermomechanics. *Int. J. Heat Mass Transf.* **2008**, *51*, 2237–2259. [CrossRef]
3. Hristov, J. Bio-heat models revisited: Concepts, derivations, nondimensalization and fractionalization approaches. *Front. Phys.* **2019**, *7*, 1–36. [CrossRef]
4. Dai, T.; Pikkula, B.M.; Wang, L.V.; Anvari, B. Comparison of human skin opto-thermal response to near-infrared and visible laser irradiations: A theoretical investigation. *Phys. Med. Biol.* **2004**, *49*, 4861–4877. [CrossRef] [PubMed]
5. Dai, W.; Wang, H.; Jordan, P.M.; Mickens, R.E.; Bejan, A. A mathematical model for skin burn injury induced by radiation heating. *Int. J. Heat Mass Transf.* **2008**, *51*, 5497–5510. [CrossRef]
6. Xu, F.; Wen, T.; Seffen, K.; Lu, T. Modeling of skin thermal pain: A preliminary study. *Appl. Math. Comput.* **2008**, *205*, 37–46. [CrossRef]
7. Xu, F.; Lu, T.J.; Seffen, K.A.; Ng, E.Y.K. Mathematical modeling of skin bioheat transfer. *Appl. Mech. Rev.* **2009**, *62*, 1–35. [CrossRef]
8. Xu, F.; Lu, T. *Introduction to Skin Biothermomechanics and Thermal Pain*, 1st ed.; Springer: Berlin/Heidelberg, Germany, 2011.
9. Zolfaghari, A.; Maerefat, M. Bioheat transfer. *Dev. Heat Transf.* **2011**, 153–170. [CrossRef]
10. Liu, K.C.; Wang, Y.N.; Chen, Y.S. Investigation on the bio-heat transfer with the Dual-Phase-Lag effect. *Int. J. Therm. Sci.* **2012**, *58*, 29–35. [CrossRef]
11. Weinbaum, S.; Jiji, L.M. A new simplified bioheat equation for the effect of blood flow on local average tissue temperature. *J. Biomech. Eng.* **1985**, *107*, 131–139. [CrossRef]
12. Nakayama, A.; Sano, Y.; Yoshikawa, K. A rigorous derivation of the bioheat equation for local tissue heat transfer based on a volume averaging theory. *Heat Mass Transf.* **2010**, *46*, 739–746. [CrossRef]
13. Charny, C.K. Mathematical models of bioheat transfer. *Adv. Heat Transf.* **1992**, *22*, 19–155.
14. Jiji, L.M. *Heat Conduction*, 3rd ed.; Springer: Berlin/Heidelberg, Germany, 2009.
15. Pennes, H.H. Analysis of tissue and arterial blood temperatures in the resting human forearm. *J. Appl. Physiol.* **1948**, *1*, 93–122. [CrossRef] [PubMed]
16. Wulff, W. The energy conservation equation for living tissues. *IEEE Trans. Biomed. Eng.* **1974**, *21*, 494–495. [CrossRef]
17. Klinger, H.G. Heat transfer in perfused biological tissue I: General theory. *Bull. Math. Biol.* **1974**, *36*, 403–415. [PubMed]
18. Chen, M.M.; Holmes, K.R. Microvascular contributions in tissue heat transfer. *Ann. N. Y. Acad. Sci.* **1980**, *335*, 137–150. [CrossRef]
19. Weinbaum, S.; Jiji, L.M.; Lemons, D.E. Theory and experiment for the effect of vascular microstructure on surface tissue heat transfer: Part I: Anatomical foundation and model conceptualization. *J. Biomech. Eng.* **1984**, *106*, 321–330. [CrossRef] [PubMed]
20. Weinbaum, S.; Jiji, L.M.; Lemons, D.E. Theory and experiment for the effect of vascular microstructure on surface tissue heat transfer: Part II: Anatomical foundation and model conceptualization. *J. Biomech. Eng.* **1984**, *106*, 331–341. [CrossRef] [PubMed]
21. Yang, W.H. Thermal (heat) shock biothermomechanical viewpoint. *J. Biomech. Eng.* **1993**, *115*, 617–621. [CrossRef] [PubMed]
22. Liu, J.; Chen, X.; Xu, L.X. New thermal wave aspects on burn evaluation of skin subjected to instantaneous heating. *IEEE Trans. Biomed. Eng.* **1999**, *46*, 420–428.
23. Tzou, D.Y. *Macro- to Microscale Heat Transfer: The Lagging Behaviour*, 1st ed., Taylor and Francis: Abingdon, UK, 1997.
24. Hobiny, A.D.; Abbas, I.A. Theoretical analysis of thermal damages in skin tissue induced by intense moving heat source. *Int. J. Heat Mass Transf.* **2018**, *124*, 1011–1014. [CrossRef]
25. Alzahrani, F.S.; Abbas, I.A. Analytical estimations of temperature in a living tissue generated by laser irradiation using experimental data. *J. Therm. Biol.* **2019**, *85*, 1–5. [CrossRef] [PubMed]
26. Hobiny, A.; Abbas, I. Thermal response of cylindrical tissue induced by laser irradiation with experimental study. *Int. J. Numer. Methods Heat Fluid Flow* **2019**, *30*, 4013–4023. [CrossRef]
27. Hobiny, A.; Alzahrani, F.; Abbas, I.; Marin, M. The effect of fractional time derivative of bioheat model in skin tissue induced to laser irradiation. *Symmetry* **2020**, *12*, 602. [CrossRef]
28. Hobiny, A.; Alzahrani, F.; Abbas, I. Analytical estimation of temperature in living tissues using the tpl bioheat model with experimental verification. *Mathematics* **2020**, *8*, 1188. [CrossRef]
29. Kumari, T.; Singh, S.K. A numerical study of space-fractional three-phase-lag bioheat transfer model during thermal therapy. *Heat Transf.* **2022**, *51*, 470–489. [CrossRef]

30. Li, M.; Wang, Y.; Liu, D. Generalized bio-heat transfer model combining with the relaxation mechanism and nonequilibrium heat transfer. *J. Heat Transf.* **2021**, *144*, 031209. [CrossRef]

31. Arkin, H.; Xu, L.X.; Holmes, K.R. Recent developments in modeling heat transfer in blood perfused tissues. *IEEE Trans. Biomed. Eng.* **1994**, *41*, 97–107. [CrossRef]

32. Khaled, A.R.; Vafai, K. The role of porous media in modeling flow and heat transfer in biological tissues. *Int. J. Heat Mass Transf.* **2003**, *46*, 4989–5003. [CrossRef]

33. Goldstein, R.; Ibele, W.; Patankar, S.; Simon, T.; Kuehn, T.; Strykowski, P.; Tamma, K.; Heberlein, J.; Davidson, J.; Bischof, J.; et al. Heat transfer—A review of 2003 literature. *Int. J. Heat Mass Transf.* **2006**, *49*, 451–534. [CrossRef]

34. Goldstein, R.J.; Ibele, W.E.; Patankar, S.V.; Simon, T.W.; Kuehn, T.H.; Strykowski, P.J.; Tamma, K.K.; Heberlein, J.V.R.; Davidson, J.H.; Bischof, J.; et al. Heat transfer—A review of 2005 literature. *Int. J. Heat Mass Transf.* **2010**, *53*, 4397–4447. [CrossRef]

35. Yang, X.; Liang, Y.; Chen, W. A spatial fractional seepage model for the flow of non-Newtonian fluid in fractal porous medium. *Commun. Nonlinear Sci. Numer. Simul.* **2018**, *65*, 70–78. [CrossRef]

36. Khanafer, K.; Vafai, K. Synthesis of mathematical models representing bioheat transport. *Adv. Numer. Heat Transf.* **2009**, *3*, 1–28.

37. Roetzel, W.; Xuan, Y. Bioheat equation of the human thermal system. *Chem. Eng. Technol.* **1997**, *20*, 268–276.

38. Nakayama, A.; Kuwahara, F. A general bioheat transfer model based on the theory of porous media. *Int. J. Heat Mass Transf.* **2008**, *51*, 3190–3199. [CrossRef]

39. Shen, Y.; Xu, P.; Qiu, S.; Rao, B.; Yu, B. A generalized thermal conductivity model for unsaturated porous media with fractal geometry. *Int. J. Heat Mass Transf.* **2020**, *152*, 119540. [CrossRef]

40. Zhao, C. Review on thermal transport in high porosity cellular metal foams with open cells. *Int. J. Heat Mass Transf.* **2012**, *55*, 3618–3632. [CrossRef]

41. Belova, I.V.; Murch, G.E. Monte Carlo simulation of the effective thermal conductivity in two-phase material. *J. Mater. Process. Technol.* **2004**, *153*, 741–745. [CrossRef]

42. Song, Y.; Youn, J. Evaluation of effective thermal conductivity for carbon nanotube/polymer composites using control volume finite element method. *J. Carbon* **2006**, *44*, 710–717. [CrossRef]

43. Mandelbrot, B. *The Fractal Geometry of Nature*; Freeman: New York, NY, USA, 1982.

44. Yu, B. Analysis of flow in fractal porous media. *Appl. Mech. Rev.* **2008**, *61*, 050801. [CrossRef]

45. Xu, P.; Mujumdar, A.S.; Sasmito, A.P.; Yu, B.M. *Multiscale Modeling of Porous Media*, 1st ed.; Taylor and Francis: Abingdon, UK, 2019.

46. Kou, J.; Liu, Y.; Wu, F.; Fan, J.; Lu, H.; Xu, Y. Fractal analysis of effective thermal conductivity for three-Phase (unsaturated) porous media. *J. Appl. Phys.* **2009**, *106*, 054905. [CrossRef]

47. Xu, P. A discussion on fractal models for transport physics of porous media. *Fractals* **2015**, *23*, 1530001. [CrossRef]

48. Yu, B.; Zou, M.; Feng, Y. Permeability of fractal porous media by Monte Carlo simulations. *Int. J. Heat Mass Transf.* **2005**, *48*, 2787–2794. [CrossRef]

49. Zou, M.; Yu, B.; Feng, Y.; Xu, P. A Monte Carlo method for simulating fractal surfaces. *Physica A* **2007**, *386*, 176–186. [CrossRef]

50. Feng, Y.; Yu, B.; Feng, K.; Xu, P.; Zou, M. Thermal conductivity of nanofluids and size distribution of nanoparticles by Monte Carlo simulations. *J. Nanopart. Res.* **2008**, *10*, 1319–1328. [CrossRef]

51. Xu, P.; Yu, B.; Qiao, X.; Qiu, S.; Jiang, Z. Radial permeability of fractured porous media by Monte Carlo simulations. *Int. J. Heat Mass Transf.* **2013**, *57*, 369–374. [CrossRef]

52. Vadapalli, U.; Srivastava, R.P.; Vedanti, N.; Dimri, V.P. Estimation of permeability of a sandstone reservoir by a fractal and Monte Carlo simulation approach: A case study. *Nonlinear Process. Geophys.* **2014**, *21*, 9–18. [CrossRef]

53. Xu, Y.; Zheng, Y.; Kou, J. Prediction of effective thermal conductivity of porous media with fractal-Monte Carlo simulations. *Fractals* **2014**, *22*, 1440004. [CrossRef]

54. Xiao, B.; Zhang, X.; Jiang, G.; Long, G.; Wang, W.; Zhang, Y.; Liu, G. Kozeny–Carman constant for gas flow through fibrous porous media by fractal-Monte Carlo simulations. *Fractals* **2019**, *27*, 1950062. [CrossRef]

55. Yang, J.; Wang, M.; Wu, L.; Liu, Y.; Qiu, S.; Xu, P. A novel Monte Carlo simulation on gas flow in fractal shale reservoir. *Energy* **2021**, *236*, 121513. [CrossRef]

56. Yu, B.; Li, J. Some fractal characters of porous media. *Fractals* **2001**, *9*, 365–372. [CrossRef]

57. Yu, B.; Cheng, P. A fractal permeability model for bi-dispersed porous media. *Int. J. Heat Mass Transf.* **2002**, *45*, 2983–2993. [CrossRef]

58. Wu, J.; Yu, B. A fractal resistance model for flow through porous media. *Int. J. Heat Mass Transf.* **2007**, *50*, 3925–3932. [CrossRef]

59. Shen, H.; Ye, Q.; Meng, G. Anisotropic fractal model for the effective thermal conductivity of random metal fiber porous media with high porosity. *Phys. Lett. A* **2017**, *381*, 3193–3196. [CrossRef]

60. Qin, X.; Cai, J.; Xu, P.; Dai, S.; Gan, Q. A fractal model of effective thermal conductivity for porous media with various liquid saturation. *Int. J. Heat Mass Transf.* **2019**, *128*, 1149–1156. [CrossRef]

61. Feng, Y.; Yu, B.; Zou, M.; Zhang, D. A generalized model for the effective thermal conductivity of porous media based on self-similarity. *J. Phys. D Appl. Phys.* **2004**, *37*, 3030–3040. [CrossRef]

62. Zhang, B.; Yu, B.; Wang, H.; Yun, M. A fractal analysis of permeability for power-law fluids in porous media. *Fractals* **2006**, *14*, 171–177. [CrossRef]

63. Xu, F.; Wen, T.; Lu, T.; Seffen, K.A. Skin biothermomechanics for medical treatments. *J. Mech. Behav. Biomed. Mater.* **2008**, *1*, 172–187. [CrossRef]
64. Kumar, S.; Damor, R.S.; Shukla, A.K. Numerical study on thermal therapy of triple layer skin tissue using fractional bioheat model. *Int. J. Biomath.* **2018**, *11*, 1850052. [CrossRef]
65. Johnston, B.M.; Johnston, P.R.; Corney, S.; Kilpatrick, D. Non–Newtonian blood flow in human right coronary arteries: Steady state simulations. *J. Biomech.* **2004**, *37*, 709–720. [CrossRef] [PubMed]
66. Yu, B.; Cai, J.; Zou, M. On the physical properties of apparent two-phase fractal porous media. *Vadose Zone J. Fractals* **2009**, *8*, 177–186. [CrossRef]

 *micromachines*

Article

Viscoelastic Particle Focusing and Separation in a Spiral Channel

Haidong Feng [1,†], Alexander R. Jafek [1,†], Bonan Wang [2], Hayden Brady [1], Jules J. Magda [2] and Bruce K. Gale [1,*]

[1] Department of Mechanical Engineering, University of Utah, Salt Lake City, UT 84112, USA; haidong.feng@utah.edu (H.F.); u1058324@utah.edu (A.R.J.); u0949394@utah.edu (H.B.)
[2] Department of Chemical Engineering, University of Utah, Salt Lake City, UT 84112, USA; bonan.wang@utah.edu (B.W.); magda@chemeng.utah.edu (J.J.M.)
* Correspondence: bruce.gale@utah.edu
† These authors contributed equally to this work.

Abstract: As one type of non-Newtonian fluid, viscoelastic fluids exhibit unique properties that contribute to particle lateral migration in confined microfluidic channels, leading to opportunities for particle manipulation and separation. In this paper, particle focusing in viscoelastic flow is studied in a wide range of polyethylene glycol (PEO) concentrations in aqueous solutions. Polystyrene beads with diameters from 3 to 20 μm are tested, and the variation of particle focusing position is explained by the coeffects of inertial flow, viscoelastic flow, and Dean flow. We showed that particle focusing position can be predicted by analyzing the force balance in the microchannel, and that particle separation resolution can be improved in viscoelastic flows.

Keywords: particle separation; viscoelastic flow; inertial focusing; spiral channel

Citation: Feng, H.; Jafek, A.R.; Wang, B.; Brady, H.; Magda, J.J.; Gale, B.K. Viscoelastic Particle Focusing and Separation in a Spiral Channel. *Micromachines* **2022**, *13*, 361. https://doi.org/10.3390/mi13030361

Academic Editors: Lanju Mei and Shizhi Qian

Received: 1 February 2022
Accepted: 22 February 2022
Published: 25 February 2022

Publisher's Note: MDPI stays neutral with regard to jurisdictional claims in published maps and institutional affiliations.

1. Introduction

Many biological and sensing assays require the ability to isolate particular types of particles from a mixture of particle types [1]. Existing label-free protocols select unique particle populations by relying on differences in particle size, shape, density, or surface properties [2]. While the optimal separation technique varies by target particle and application, generally the best separation method is one that is fast, robust, has a high throughput, and does not apply excessive force to the particles [3].

The field of microfluidics has been particularly well suited for applications involving particle separation and isolation [4,5]. Microfluidic devices manipulate fluids through channels with dimensions between 1–100 μm [6]. At this scale, channel features are commensurate with cells and physical phenomena arise which are unique to the microscale. In a microfluidic channel, particles can experience different types of forces that contribute to particle movement, retention, and separation. Force fields, such as electrical fields [7–9], surface acoustic waves [10], and optical forces [11], have been applied for particle manipulation and trapping in microfluidic channels. Furthermore, particles can experience forces with controllable magnitude and direction in a dielectrophoretic force field based on their size and polarizability, leading to particle separation with high resolution [12–14]. In optical trapping devices, particles experience the force due to incident photon scattering, and single particle manipulation is realized using optical forces [11]. One particularly useful hydrodynamic force for particle selection and separation on the microscale is inertial focusing [15–17]. Inertial focusing refers to the tendency of identical particles to align themselves to certain locations within a channel cross-section as they travel along a channel. Since the forces which drive this behavior are dependent on the particle geometry, it is possible to separate particles of different shapes or sizes in microfluidic channels [18,19]. In many applications, the efficiency of this separation can be increased through the additional incorporation of a Dean flow. In curved channels, a Dean flow arises due to the radial component of the fluid velocity and is generally expressed as counter-rotating vortices in the direction tangent to the fluid flow [20,21]. The Dean flow in the channel cross-section provides extra drag force leading to lateral particle movement. The inertial focusing force drives particles towards the channel top and bottom wall, while the Dean flow transports particle focusing positions to the channel inner sidewall. In inertial-Dean coupled flow, size-based separation can be realized since large particles have a focusing position closer to the inner sidewall when compared with small particles [22]. In addition, Dean flow can be induced in microfluidic channels with expansion/contraction structure [23] or pillar structures [24]. The Dean flow vortices are combined with inertial flow, and leads to a change of particle focusing position and particle trapping in the channel. A detailed description of inertial flow and Dean flow can be found in recent published review articles [25–27].

Viscoelastic fluid presents one type of non-Newtonian solution that exhibits both viscous and elastic properties under shear stress. In viscoelastic dominated flows, particle lateral movement and focusing stream phenomena have been observed and used for particle manipulation and separation in microchannels [28]. Polymer solutions, such as high molecular weight poly(ethylene) oxide (PEO) solutions, exhibit non-linear elastic stress distribution in the channel cross-section, leading to particles focusing towards the channel center [29,30]. In a confined channel, the particle focusing position is affected by the force balance between the viscoelastic forces and inertial forces [31]. In a straight channel, the interaction of the inertial focusing force and the viscoelastic focusing force leads to a change of particle focusing stream location. It is found that large particles bear split focusing streams near the channel sidewalls, while small particles focus to the channel center [32]. In a curved channel, the Dean flow leads to a change of particle focusing positions towards the channel outer sidewall, which effect has been applied for size-based particle separation [33].

In this paper, we demonstrate a particle separation regime showing the co-effects of inertial flow, viscoelastic flow, and Dean flow. With an increase of PEO concentration in the fluid, the force balance between the inertial flow and viscoelastic flow changes, leading to a change of particle focusing positions for different sized particles. The Dean flow, which provide a transverse drag force, improves the separation efficiency by changing particle focusing position in the horizontal direction. As we explore this separation regime, we principally focus on three main areas: (1) Contributions of the PEO solution viscoelastic properties under ultra-high shear rates in a microfluidic channel; (2) Identification of particle focusing positions and focusing regimes over a wide range of PEO concentrations; and (3) Evaluation and improvement of particle separation resolution in the viscoelastic flow. We expect that these results will find application in both sensing and biological assays.

2. Theory

Inertial microfluidics depends on the flow lift forces that drive suspended particles to focused locations within a confined microfluidic channel. These focused positions arise as basins of attraction [34] where the inertial forces on particles are balanced. In a straight channel with a Newtonian flow, the shear gradient lift force drives particles up the shear gradient towards channel walls. When particles approach the channel wall, the wall repulsive force pushes particles away from the wall. The balance of shear gradient lift force and wall repulsive force form particle focusing positions with zero net forces [35,36]. In a channel with a rectangular cross-section as is shown in Figure 1a, a higher shear stress gradient appears in the channel height direction since the channel height is smaller than the channel width. As a result, particles have two focusing positions in the vertical center of channel top and bottom walls. In the inertial flow, the inertial lift force is determined by the flow velocity gradient and particle dimension. It can be calculated by Equation (1) [35].

$$F_L = f_L \rho U_{max}^2 a^4 / h^2 \tag{1}$$

where f_L is the lift coefficient, ρ is the fluid density, U_{max} is the maximum flow velocity, a is particle diameter, and h is channel height.

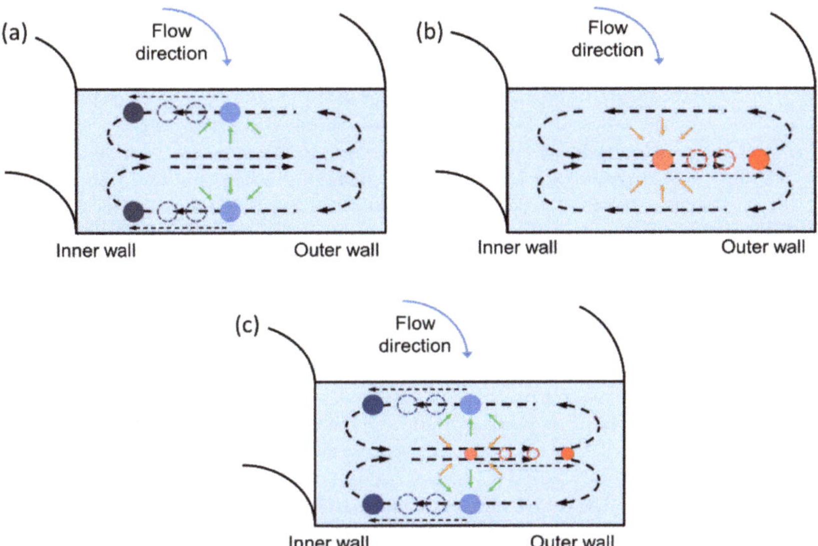

Figure 1. Representation of particle focusing in curved microfluidic channels. (**a**) In Newtonian fluids, particles are driven towards the inner wall. (**b**) In the viscoelastic regime, particles are driven towards the outer wall. (**c**) In the intermediate regime, size-based separations can be achieved.

In viscoelastic flow, the non-linear shear stress distribution induces particle viscoelastic focusing. This elastic force acts inwards from all of the walls driving the particles towards the channel centerline as shown in Figure 1b. The viscoelastic focusing force is determined by the first normal stress difference and particle dimension. It can be calculated by Equation (2) [37].

$$F_E = 8a^3\lambda(U_{max}/h)^3 \tag{2}$$

where λ is the flow relaxation time.

The ratio of F_E and F_L is used to evaluate the dominant force on the particle. With an increase of particle diameter, F_L dominates particle movement. With the increase of λ and U, F_E dominates particle movement. Based on Equation (3), it is possible to guide large particles to inertial focusing positions and small particles to viscoelastic focusing positions, and the change of particle focusing mechanism leads to a significant change of particle focusing position.

$$F_E/F_L = 8\lambda U_{max}/f_L\rho ah \tag{3}$$

In a curved channel, a Dean flow is generated due to the imbalance of pressure and velocity gradient [38]. Dean flow vortices appear in the channel cross-section and flow towards the channel outer sidewall in the channel center and circulate back to the channel inner sidewall near the channel top/bottom sidewall. The influence of Dean flow on particle movement is characterized by the Dean drag force F_D [37]:

$$F_D = 5.4 \times 10^{-4}\pi\mu De^{1.63}a \tag{4}$$

where De is the dimensionless Dean number. F_D has lower magnitude when compared with F_E and F_L, while it determines particle focusing position in the horizontal direction when F_E is balanced with F_L. Particle lateral migration in the horizontal direction due to

Dean flow is used to increase the distance between particle focusing streams [39]. As a result, an enhanced separation resolution is realized in a curved microchannel.

3. Methods

3.1. Microfluidic Device Design and Fabrication

In this work, we fabricate a device composed of a spiral channel with two inlets and two outlets (Figure 2). Our chip contains a channel with a cross-section of 200 μm × 50 μm (width × height). The channel forms a spiral with a radius that varies between 7–9 mm. There are 3 loops and the total channel length is 172 mm. The channel geometry is fabricated in polydimethylsiloxane (PDMS), according to traditional soft lithography protocols that we have explained previously [39]. Briefly, negative photoresist SU-8 (SU-8 2050, MicroChem, Westborough, MA, USA) is used for the mold fabrication. SU-8 is spin coated on a clean 100 mm (4 inch) wafer, and baked at 95 °C for 15 min. The pattern of the spiral channel is transferred on the photoresist layer using a UV mask aligner, and a post-baking process is performed to stabilize the cross-linked SU-8. Then, the mold is developed using SU-8 developer and the mold quality is checked under a microscope. Uncured PDMS (Sylgard 184, Dow, Midland, MI, USA) is mixed with a curing agent at a ratio of 10:1. Air bubbles generated from the mixing process are removed using a vacuum chamber. About 40 mL degassed PDMS mixture is poured on the SU-8 mold, which leads to a PDMS device with a thickness of 8 mm. PDMS is cured in an oven at 80 °C for 6 h. The cured PDMS is sealed with a glass slide. The PDMS and glass slide surfaces are treated by oxygen plasma to generate permanent bonding. The oxygen plasma treatment is carried out using a Technics PEII-A plasma system, with 200 W power, and 50 sccm O_2 flow for 2 min. After the bonding process, a post-baking process is performed by putting the spiral channel device on a hotplate at 120 °C for 15 min.

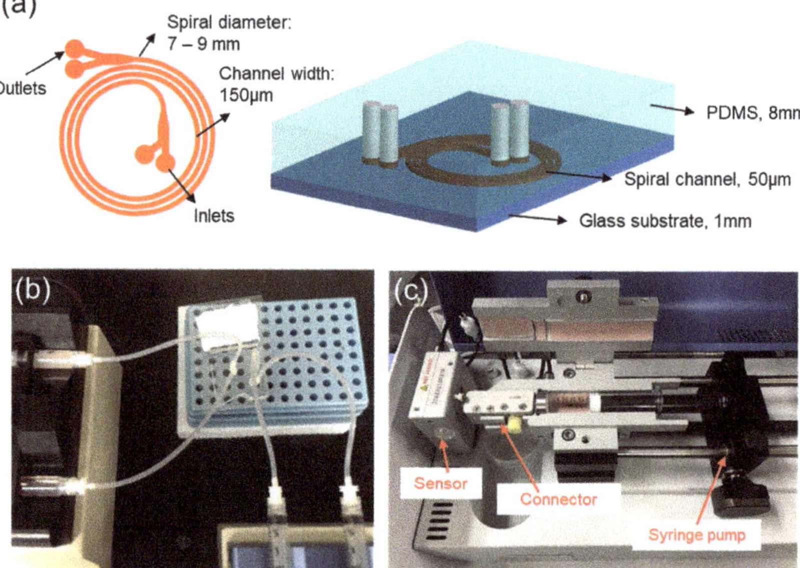

Figure 2. Picture of experimental device set up. (**a**) Schematic view of spiral channel device. (**b**) Microfluidic chip channel connection for particle loading and flow rate control. (**c**) Device structure of m-VROC rheometer.

3.2. Experimental Set-Up

Two sets of data are collected: one to establish an understanding of fluid elasticity and flow rate and one to explore the application of these findings in detection and separation protocols. The details of the experimental variables are introduced in Table 1. For both sets of experiments, images are acquired of fluorescent beads suspended in the liquid flowing through the channel at a controlled flow rate. Fluorescent beads with diameters of 3 µm(18861-1), 10 µm(19103-2), and 20 µm(19096-2) were purchased from Polysciences, Inc., (Warrington, PA, USA), and beads with diameter of 8µm(FSFR007) were purchased from Bangs Laboratories, Inc., (Fishers, IN, USA). Fluorescent beads are initially suspended in water, and PEO concentrations of 0.001%, 0.003%, 0.005%, 0.025%, 0.05%, 0.1%, 0.2%, and 0.4% by weight are created with serial dilutions of water. The flow rate is controlled with a syringe pump connected to the inlet of the device as is shown in Figure 2. The chip is imaged through the glass near the channel outlet.

Table 1. Summary of experiment parameters in the spiral channel particle separation test.

Variables	Value
Channel geometry	150 µm width; 50 µm height; 172 mm length
PEO concentration	0.001%, 0.003%, 0.005%, 0.025%, 0.05%, 0.1%, 0.2%, and 0.4%
Particle diameters	3 µm, 8 µm, 10 µm, and 20 µm
Flow rate	0.05 mL/min, 0.1 mL/min, 0.15 mL/min, 0.2 mL/min, 0.25 mL/min, 0.3 mL/min, and 0.35 mL/min

3.3. Characterization Experiments

Fluorescent beads of 3 and 8 µm are suspended in three different PEO-water dilutions and infused into the spiral channel at seven different flow rates. The bead concentration is kept low (0.1 M/mL) to limit particle–particle interactions. Images are acquired with a high-speed fluorescent camera and about 100 images collected over 1 s are compiled to form the images presented. The spiral channel device is rinsed with water between successive experiments, and experiments with a single chip are acquired starting from the lowest concentrations of PEO to mitigate the possibility of cross-contamination between experiments.

3.4. PEO Solution Viscosity Characterization

For viscoelastic aqueous solutions with various PEO concentrations, viscosity is directly associated with peak-to-peak separation distance and resolution. A pressure sensing microfluidics rheometer m-VROC (RheoSense, LLC, San Ramon, CA, USA) is employed in this work that combines a microfluidic channel and a pressure sensor array embedded along the microchannel to measure dynamic viscosity [40]. The principles of VROC are based on measuring the pressure drop by using the Hagen–Poiseuille law, where fluid flows through a given enclosed rectangular slit microfluidic channel. The apparent viscosity depends on the applied shear rate for dilute PEO solutions. The flow rate is proportional to the shear rate; a syringe pump is applied to control the flow rate of an optimal dilute PEO solution. The microfluidics channel etched into a silicon chip contains a depth of 20 um, a width of the 3 mm, and a length of 10 mm. The shallow depth of the microfluidics channel allows one to investigate high shear rates without occurrence of turbulence.

All PEO concentrations from 0.001% to 1% have a measured viscosity at an apparent shear rate between 1.7×10^4 and 1×10^5/s, a constant flow rate from 142.6 µL/min to 823 µL/min through the measuring channel where the pressure sensors monitor the pressure drop at room temperature. The PEO solution is pre-filtered with 0.2 µm PTFE filters and degassed to avoid fibers and bubbles at high shear rate measurements. (The relaxation time is measured by using the Zimm module at 810 µs. Hydrodynamic radius R_H is 68.8 nm, calculated from the Einstein module.)

$$\lambda_{Zimm} = \frac{\eta_s [\eta] M_w}{RT} \tag{5}$$

$$R_H = \left(\frac{3[\eta] M_w}{10 \, \pi N_A} \right)^{1/3}$$

(6)

where η_s is the solvent viscosity, $[\eta]$ is PEO intrinsic viscosity, M_w is PEO molecular weight, R is the molar gas constant $R = 8.314 \, \text{J}/(\text{K mol})$, and T is the temperature (K).

3.5. Detection and Separation Experiments

Fluorescent beads of 3, 10, and 20 µm are suspended in four different PEO-water dilutions and infused into the spiral channel at three different flow rates using a syringe pump KDS200 (KD Scientific, Holliston, MA, USA). Separation device outlets are connected with another syringe pump (Chemyx F200X, Stafford, TX, USA) for flow withdrawal during sample loading. The bead concentration is much higher ($10\text{--}100 \times 10^6/\text{mL}$) to replicate concentrations common to biological samples. Images are acquired with a fluorescent microscope and camera (Nikon A1R). The 4X objective lens provides a field of view of 4 mm $\times$ 4 mm to observe the spiral channel outlet region. Fluorescent imaging lasers, including a 405 nm diode laser, a 488 nm Argon gas laser, and a 638 nm diode laser, are used for the observation of fluorescent beads with different colors. Filter cube sets are selected for the optimization with DAPI, FITC, and TRITC signal detection. The spiral channel device is rinsed with water between successive experiments, and experiments with a single chip are acquired starting from the lowest concentrations of PEO to mitigate the possibility of cross-contamination between experiments.

3.6. Data Processing

Quantitative data is compiled from images by a Matlab program that reads and interprets the pixel intensity across the channel width. The program averages values across the columns of a selected area, which is manually selected. For comparison and presentation, the data is then reformed to a length of 100 entries, and normalized by the total value of all intensity measurements.

The peak-to-peak separation distance is calculated as a distance between the maximum values of the two distributions. The separation resolution is calculated as:

$$R = \frac{Separation\ Distance}{2(W_1 + W_2)}$$

(7)

where W_1 and W_2 represent the width at half maximum calculated from the distributions [41].

4. Results & Discussion

4.1. PEO Solution Characterization

Capillary viscometers or cone plate rheometers are commonly used for the precise description of fluid viscosity and viscoelastic behavior. In our application, the shear rate in the microfluidic channel is in the range of 10,000 to 60,000/s, which is much higher than the measurement limit of these measurement methods. The rheometer m-VROC is used to evaluate PEO solution viscoelastic property at a wide range of shear rates. Figure 3 shows the viscosity of the PEO solution with the shear rate of 10,000 to 100,000/s. The ultra-low concentration of the dilute PEO solutions, PEO concentrations of 0.001%, 0.003%, 0.005%, 0.025%, and 0.05% exhibit Newtonian behavior as their viscosities do not significantly change in a lower shear rate range. The viscosity values can be considered as constant. However, higher PEO concentrations greater than 0.05% show non-Newtonian shear-thinning behavior as their viscosities reduce while the shear rate increases. The true wall shear rate is corrected by applying the rigorous Weissenberg–Rabinowitsch correction [42].

$$\dot{\gamma}_{true} = \frac{\dot{\gamma}_{apparent}}{3} \left(2 + \frac{d \ln \dot{\gamma}_{apparent}}{d \ln \tau} \right)$$

(8)

where, $\dot{\gamma}_{true}$ is the corrected true shear rate, $\dot{\gamma}_{apparent}$ is the apparent shear rate measured by using m-VROC, and τ is the wall shear stress in the rectangle slit microchannel.

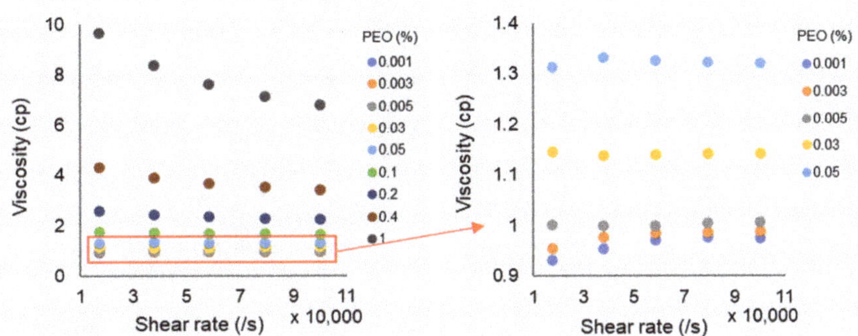

Figure 3. PEO solution viscosity under high shear rates. High concentration PEO solutions are shear-thinning with an increase of flow rate. Low concentration PEO solutions (details are shown on the right figure) exhibit constant shear rate.

4.2. Particle Movement in Low Concentration PEO Solutions

Figure 4 shows the distribution of 3 and 8 µm particles near the outlet of the spiral channel device in low concentrations PEO solutions. As seen from the fluorescent images of particle focusing, the focused positions of particles within the channel width are highly dependent on the concentration of PEO in the fluid. With an increase of PEO concentration, particle streams shift from the inertial focusing regime to the viscoelastic focusing regime. Particles with different sizes have different transitional PEO concentrations, leading to the change of particle separation behavior.

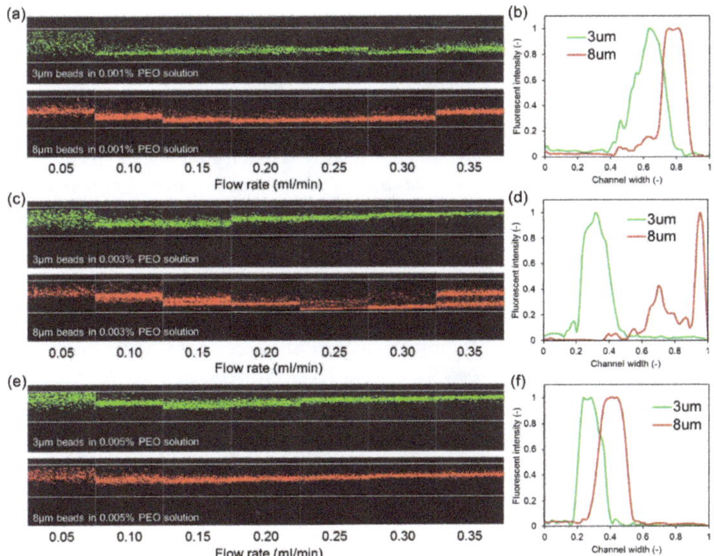

Figure 4. Distribution of 3 µm beads (green) and 8 µm beads (red) near the outlet of the spiral channel device. (**a,c,e**) represents particle focusing stream under different flow rates in PEO solutions with concentration of 0.001%, 0.003%, and 0.005%. (**b,d,f**) represent the corresponding particle focusing stream distributions at 0.3 mL/min.

When the concentration of PEO measures 0.001%, the viscoelastic focusing force is low, and the inertial focusing force dominates particle movement as shown in Figure 1a. Both the 3 and 8 μm particles are focused towards the inner side of the channel under all tested flow rates. The diagram of stream distribution indicates that 8 μm particles have a focusing position closer to the channel inner sidewall due to Dean flow. However, the focused peaks of the two particles are very close to each other with significant overlap. This proximity could inhibit the detection of two different types of particles and renders separation impossible.

In comparison, when the concentration of PEO is 0.005%, the viscoelastic force dominates particle movement (Figure 1b). Both the 3 and 8 μm particles are focused towards the outer side of the channel, and the 3 μm particles have focusing positions closer to the sidewall. The focused bands are tightest at 0.30 mL/min, representing approximately 10% of the channel width for both particle types. As shown in the right column of Figure 4, the focused peaks of the two particles are still very close to each other with significantly overlapping peaks. Again, this proximity inhibits the detection of two different types of particles and would make complete separation impossible. The focusing stream width in the viscoelastic focusing regime is smaller than that in the inertial focusing regime. It could be explained by the fact that particles have two focusing streams in the vertical direction in inertial focusing, while there is only one focusing stream in the viscoelastic focusing.

Between the two concentrations, when the concentration of PEO is 0.003%, particles with different sizes are dominated by different focusing regimes as shown in Figure 1c. The movement of 3 μm particles is dominated by the viscoelastic focusing. Particles get focused to the channel centerline and pushed to the outer sidewall by Dean flow. The movement of 8 μm particles is dominated by inertial focusing. Particles become focused to the channel top and bottom wall and are pushed to the inner sidewall by Dean flow. With the beads focused towards different edges of the spiral channel, we observe the complete separation of the two focused peaks which are also spread by about half of the channel's width. This greatly enhanced separation is beneficial for both detection and separation as will be explored in the following section.

Compared with the particle dimension or PEO concentration, the flow rate does not have a significant influence on the particle focusing regime. The increase of flow rate promotes inertial focusing and viscoelastic focusing and leads to the increase of the Dean flow velocity. Particle focusing stream width decreases with the increase of flow rate and the tightest focusing stream occurs at a flow rate between 0.20 and 0.30 mL/min for all cases. In addition, it is observed that particle focusing position is close to the channel sidewall since the Dean flow velocity increases. At a flow rate of 0.35 mL/min, particles dominated by the inertial focusing regime (3 and 8 μm particles in 0.001% PEO, 8 μm particles in 0.003% PEO) have focusing streams shift towards the channel center. In this case, the increase of flow rate leads to the increase of viscoelastic force, and particle focusing positions change.

4.3. Particle Movement in Medium Concentration PEO Solution

With the increase of PEO concentration, the viscoelastic force increases, and the balance between F_E, F_L and F_D need to be configured. Small-sized particles have viscoelastic dominated movement in low concentration PEO solution, while large-sized particles experience a high inertial effect even in medium concentration PEO solution. In the experiment, focusing behaviors of 3, 10, and 20 μm particles at flow rates of 0.1, 0.2, and 0.4 mL/min are observed in three different PEO concentrations. Plots of the data set are shown in Figure 5. In Figure 5 we can observe many of the same trends that were present in our initial data set. At the lowest PEO concentration, particles of all sizes are focused to the inner side of the channel, with the tightest focusing achieved by larger particles and focusing improving at the higher flow rates. Particle focusing position is highly affected by particle dimension and PEO concentration. 3 μm beads have a focusing position close to the outer sidewall in all PEO solutions. 10 μm beads focus to the channel center in 0.005% PEO solution, and the focusing position migrates to the channel outer sidewall in 0.025% PEO solution. 20 μm beads changed focusing position in 0.1% PEO solution.

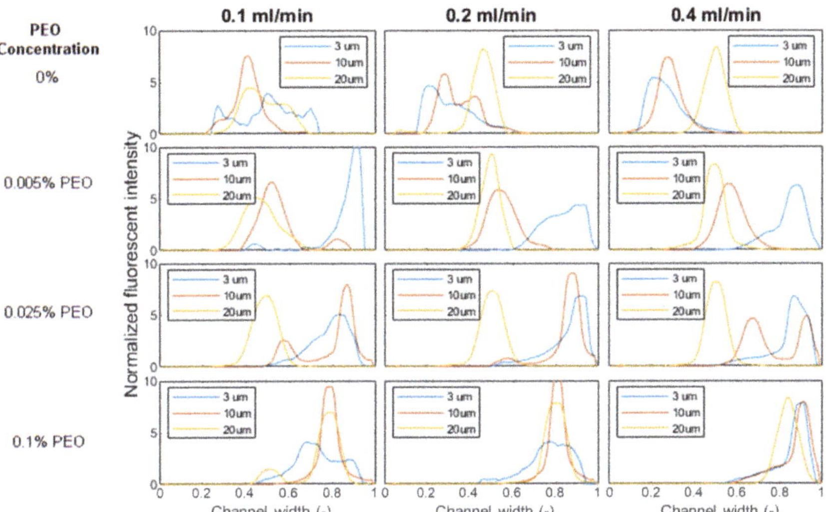

Figure 5. Plots showing focusing of 3, 10, and 20 μm beads under varied flow conditions. Across the rows of the figure, the flow rate increases from left to right while the PEO concentration is increased down each of the rows. The x-axis shown in each figure is the normalized channel width from channel inner side wall to outer side wall.

The increase of PEO concentration leads to the increase of F_E, and viscoelastic force bears a high influence on particle movement. With the increase of particle size, F_L grows faster than F_E, and the inertial effect dominates particle movement. The balance between F_E and F_L determines particle focusing position. Figure 6 shows the magnitude of F_L, F_E, and F_D in the current data set. All forces increase with flow rate, while F_E has a higher increase rate compared with F_L. It should be noticed that F_D has a much low magnitude compared with F_E and F_L for large-sized particles, and the effect of F_D decreases with the increase of PEO concentration.

For large-sized particles, a focusing position in the channel center is observed when $F_L > F_E$. Large-sized particles occupy a large volume in the confined channel. 3, 10, and 20 μm particles have particle to channel height ratios a/h of 0.06, 0.2, and 0.4, respectively. When F_L dominates large particle movement, particles tend to migrate towards the channel top/bottom wall, while this movement is balanced by the wall repulsive force. As a result, particles have focusing positions in the middle between the channel top/bottom wall and channel center. At such a focused position, the Dean flow is counter-balanced on the particle, and will not drive particle towards the channel sidewalls in the horizontal direction. Particles have focused positions in the channel center. Figure 7 illustrates the ratio of F_E and F_L in different PEO concentrations, which can be used to explain particle movement behavior. For 3 μm beads, F_E/F_L is above one in all three PEO concentrations. F_E dominates particle movement and 3 μm beads have a focus position near the channel outer sidewall. For 10 μm beads, F_E/F_L is above one in 0.025% PEO solution, and the transition of focusing positions occurs. In this case, $F_E/F_L = 1.58$ in 0.025% PEO solution. The magnitudes of F_E and F_L are very close to each other, leading to the double peak focusing phenomenon since particles may have stable focusing positions under two focusing regimes. For 20 μm beads, the particle changed focusing position in 0.1% PEO solution. It should be noticed that $F_E/F_L = 0.28$ in 0.005% PEO, and the inertial force dominates 20 μm bead movement. However, the 20 μm beads are notably large compared with the channel height (50 μm), the inertial focusing position is close to the channel center, and Dean flow is still counter-balanced in this case.

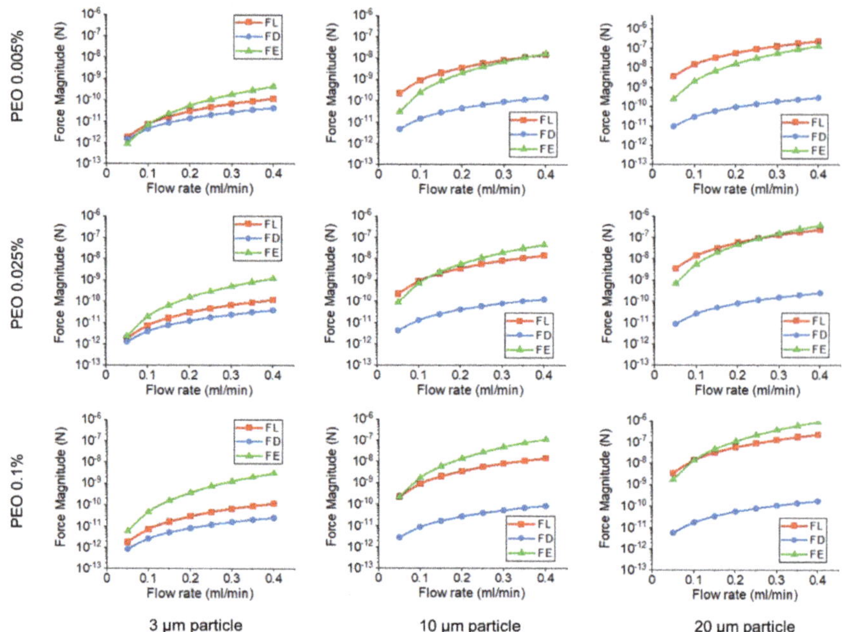

Figure 6. Force magnitude of F_L, F_E, and F_D in medium concentration PEO flow.

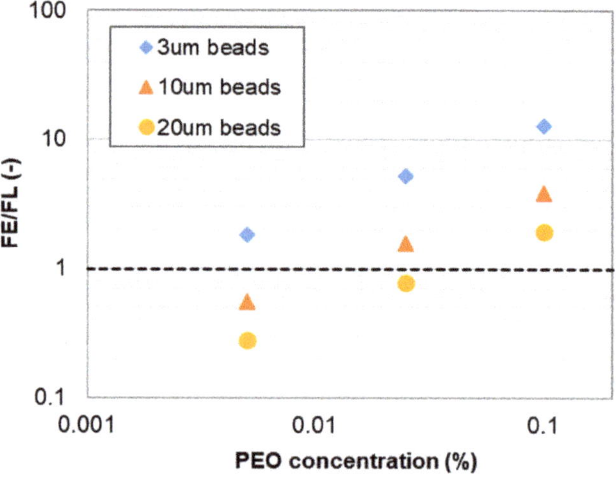

Figure 7. The ratio of F_E/F_L under different PEO concentrations at 0.2 mL/min.

4.4. Particle Movement in High Concentration PEO Solution

When the PEO concentration increases above 0.1%, the PEO solution is a semi-dilute solution, in which the polymer molecules start to interact with each other, leading to a change of polymer solution properties. In a semi-dilute PEO solution, a viscoelastic secondary flow appears in the channel cross-section, leading to a multiple stream focusing (MSF) phenomenon. The mechanism and application of MSF in high viscoelastic flow is described in detail in our previous publication [43]. MSF is induced by a viscoelastic secondary flow drag force and exhibits a larger influence on small-sized particles. Large-sized particles are less affected by MSF and maintain a single focusing stream, while the focusing stream position changes in high concentration PEO solution. Figure 8a shows the 10 μm particle focusing stream position in different PEO solutions. It is found that the focusing stream moves from the channel outer sidewall towards the channel center in high concentration PEO solutions. With the continuous increase of PEO concentration, the solution viscosity and relaxation time increase dramatically, and the flow is purely viscoelastic-dominated. Particles are focused to the channel center due to the viscoelastic force and are driven by Dean flow to the channel outer sidewall. With the increase of PEO concentration, the Dean flow has less effect on particle focusing positions. Figure 8b shows the ratio of F_E to F_D in high concentration PEO solution. The increase of solution viscosity reduces the Dean flow drag force, and F_E/F_D increases from 500 to 4500 in high concentration PEO solution. With the combination of these factors, particle movement is less affected by the Dean force, and the focusing position shift towards the channel centerline.

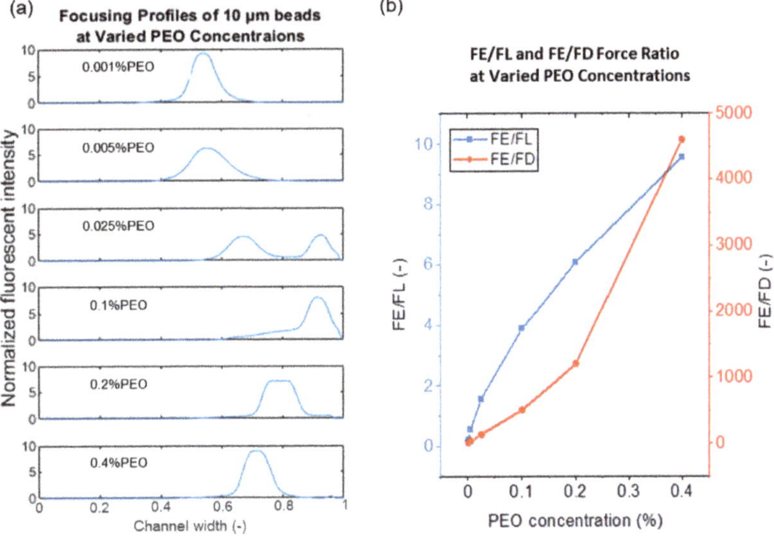

Figure 8. (**a**) Effect of PEO concentration on the focusing of 10 μm beads at a constant flow rate of 0.2 mL/min. The PEO increases in each figure from top to bottom with the absolute value shown in the legend. The x-axis shows the normalized channel width from the channel inner side wall to the outer side wall. (**b**) Force ratio of F_E/F_L and F_E/F_D for 10 μm beads.

4.5. Detection Resolution

As discussed previously, size-based separations of particles were achieved utilizing a spiral channel to separate beads suspended in a Newtonian fluid. However, since both of the focused peaks will occur within the inner half of the channel, there is a limit to the distance that peaks can be focused, and in our results, even in the case of the largest separation (3 μm from 20 μm beads at 0.2 mL/min), this separation is limited to less than one third of the channel width. However, by utilizing viscoelastic fluids we can selectively shift the peak of the smaller particle to drive it to the far outside of the channel. This allows us to create an even greater distance between the peaks of the two different particles. In Figure 9 we report the separation that we observe in the 9 tested configurations. The most dramatic improvement is in the case of 3 μm and 10 μm beads which have focused peaks within one tenth of the channel width in Newtonian fluid at all flow rates, but which can be separated by >40% of the channel width at 0.005% PEO at 0.2 mL/min.

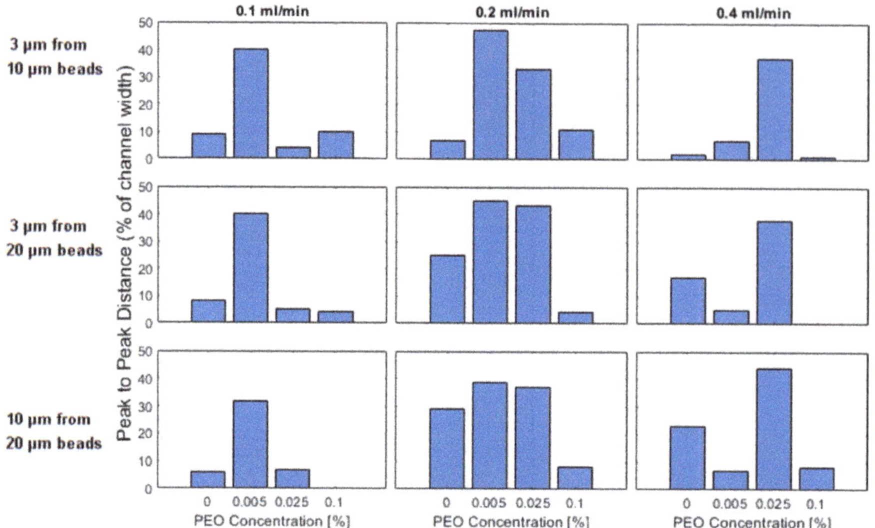

Figure 9. Peak to peak distance as a percentage of the channel width) between the peaks formed in the focusing of differently sized particles. In the first row, 3 and 10 µm beads are compared, in the second row 3 and 20 µm beads are compared and in the third row 10 and 20 µm beads are compared. The columns represent the flow rates of 0.1, 0.2, and 0.4 mL/min. The individual bars within a plot represent different PEO concentrations which increase from left to right.

For all combinations of bead sizes and flow rates, the best separation is identified in one of the two intermediate PEO concentrations. There is no separation that cannot be improved by adding a small concentration of PEO into the fluid. In preparing to apply these results, it is important to remember that the distances are reported without regard for the location, i.e., in the Newtonian regime 3 µm beads will be focused further inside while they will later be focused further to the outside of the channel.

The ability to increase the separation distance between focused peaks could prove helpful in separation and detection applications where the signal of similarly sized particles may otherwise overlap. More generally, especially in the transition region, since the focusing behavior represents such a strong function of flow rate, PEO concentration, and bead size, we imagine that the proposed spiral channel device could be employed as a sensor for any of the other variables if two of the values were known.

Increased separation between focused peaks is also an indicator that increased particle separation should be achievable; we demonstrate that with the aid of an additional performance metric, separation resolution, which attempts to quantify our ability to create separation between two focused peaks. Again, the performance metrics of all 9 operating conditions is presented in Figure 10 with many trends consistent with the plots of the distance between focused peaks. All separations can be improved with the introduction of low quantities of PEO, either at 0.005% or 0.025% PEO.

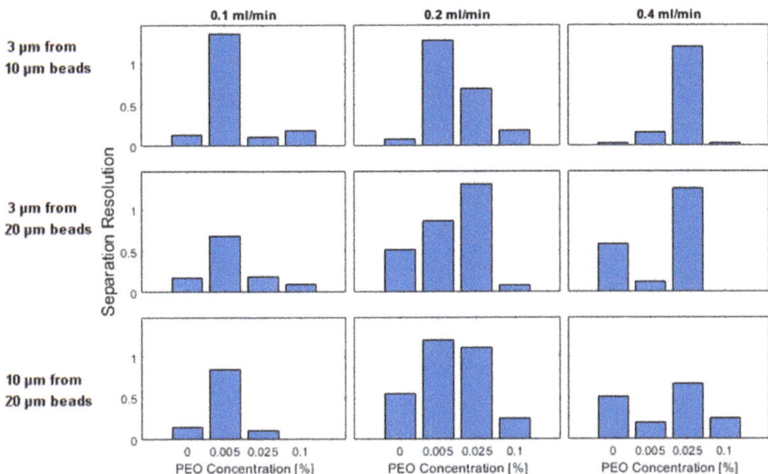

Figure 10. Separation resolution as a percentage of the channel width between the peaks formed in the focusing of differently sized particles. In the first row, 3 and 10 μm beads are compared, in the second row 3 and 20 μm beads are compared and in the third row 10 and 20 μm beads are compared. The columns represent the flow rates of 0.1, 0.2, and 0.4 mL/min. The individual bars within a plot represent different PEO concentrations which increase from left to right.

5. Conclusions

In this study, particle focusing position in PEO solutions with a wide range of concentrations was studied experimentally. The achieved results are used for the analysis of the particle focusing regime and improvement of separation resolution. PEO solution viscoelastic properties in high shear rate flow are characterized, and the variation of solution viscosity under different shear rates affects particle movement in the viscoelastic flow. Particles with different sizes bear different focusing positions with the increase of PEO concentration and flow rate, and the particle focusing process is determined by the balance of inertial flow, viscoelastic flow, and Dean flow. Therefore, by precisely manipulating the PEO concentration of the fluid, we can create a situation in which smaller particles are driven to the outer side of the channel while larger beads are driven to the inside of the channel. This condition allows for an increased separation distance between their focused peaks, which could assist with the detection of beads of different sizes, and an increased separation resolution which could help in separation protocols.

Author Contributions: Conceptualization, H.F.; formal analysis, A.R.J. and J.J.M.; funding acquisition, B.K.G.; Investigation, H.F., A.R.J., B.W. and H.B.; methodology, H.F. and B.K.G.; Supervision, B.K.G.; writing—original draft, H.F.; writing—review & editing, J.J.M. and B.K.G. All authors have read and agreed to the published version of the manuscript.

Funding: Research reported in this publication was supported by the Eunice Kennedy Shriver National Institute of Child Health & Human Development of the National Institutes of Health under Award Number 5R44HD095355-03; the Multi-Scale Fluid-Solid Interactions in Architected and Natural Materials (MUSE), an Energy Frontier Research Center funded by the U.S. Department of Energy, Office of Science, Basic Energy Sciences under Award # DE-SC0019285. The content is solely the responsibility of the authors and does not necessarily represent the official views of the National Institutes of Health.

Data Availability Statement: The data that support the findings of this study are available from the corresponding author, B.G, upon reasonable request.

Acknowledgments: Research reported in this publication was supported by the Eunice Kennedy Shriver National Institute of Child Health & Human Development of the National Institutes of Health under Award Number 5R44HD095355-03; the Multi-Scale Fluid-Solid Interactions in Architected and Natural Materials (MUSE), an Energy Frontier Research Center funded by the U.S. Department of Energy, Office of Science, Basic Energy Sciences under Award # DE-SC0019285. The content is solely the responsibility of the authors and does not necessarily represent the official views of the National Institutes of Health.

Conflicts of Interest: The authors declare no conflict of interest.

References

1. Pamme, N. Continuous flow separations in microfluidic devices. *Lab Chip* **2007**, *7*, 1644–1659. [CrossRef] [PubMed]
2. Sajeesh, P.; Sen, A.K. Particle separation and sorting in microfluidic devices: A review. *Microfluid. Nanofluid.* **2013**, *17*, 1–52. [CrossRef]
3. Samuel, R.; Feng, H.; Jafek, A.; Despain, D.; Jenkins, T.; Gale, B. Microfluidic—Based sperm sorting & analysis for treatment of male infertility. *Transl. Androl. Urol.* **2018**, *7*, S336–S347. [CrossRef] [PubMed]
4. Bhagat, A.A.S.; Bow, H.; Hou, H.W.; Tan, S.J.; Han, J.; Lim, C.T. Microfluidics for cell separation. *Med. Biol. Eng. Comput.* **2010**, *48*, 999–1014. [CrossRef]
5. Chung, A.J.; Gossett, D.R.; Di Carlo, D. Three Dimensional, Sheathless, and High-Throughput Microparticle Inertial Focusing Through Geometry-Induced Secondary Flows. *Small* **2012**, *9*, 685–690. [CrossRef]
6. Whitesides, G.M. The origins and the future of microfluidics. *Nature* **2006**, *442*, 368–373. [CrossRef]
7. Gascoyne, P.R.C.; Vykoukal, J. Particle separation by dielectrophoresis. *Electrophoresis* **2002**, *23*, 1973–1983. [CrossRef]
8. Petersen, K.E.; Shiri, F.; White, T.; Bardi, G.T.; Sant, H.; Gale, B.K.; Hood, J.L. Exosome Isolation: Cyclical Electrical Field Flow Fractionation in Low-Ionic-Strength Fluids. *Anal. Chem.* **2018**, *90*, 12783–12790. [CrossRef]
9. Jubery, T.Z.; Srivastava, S.K.; Dutta, P. Dielectrophoretic separation of bioparticles in microdevices: A review. *Electrophoresis* **2014**, *35*, 691–713. [CrossRef]
10. Ding, X.; Lin, S.-C.S.; Lapsley, M.I.; Li, S.; Guo, X.; Chan, C.Y.; Chiang, I.-K.; Wang, L.; McCoy, J.P.; Huang, T.J. Standing surface acoustic wave (SSAW) based multichannel cell sorting. *Lab Chip* **2012**, *12*, 4228–4231. [CrossRef]
11. Neuman, K.C.; Block, S.M. Optical trapping. *Rev. Sci. Instrum.* **2004**, *75*, 2787–2809. [CrossRef] [PubMed]
12. Zaman, M.A.; Padhy, P.; Ren, W.; Wu, M.; Hesselink, L. Microparticle transport along a planar electrode array using moving dielectrophoresis. *J. Appl. Phys.* **2021**, *130*, 034902. [CrossRef] [PubMed]
13. Zaman, M.A.; Wu, M.; Padhy, P.; Jensen, M.A.; Hesselink, L.; Davis, R.W. Modeling Brownian Microparticle Trajectories in Lab-on-a-Chip Devices with Time Varying Dielectrophoretic or Optical Forces. *Micromachines* **2021**, *12*, 1265. [CrossRef] [PubMed]
14. Pethig, R. Dielectrophoresis: Using Inhomogeneous AC Electrical Fields to Separate and Manipulate Cells. *Crit. Rev. Biotechnol.* **1996**, *16*, 331–348. [CrossRef]
15. Di Carlo, D.; Irimia, D.; Tompkins, R.G.; Toner, M. Continuous inertial focusing, ordering, and separation of particles in microchannels. *Proc. Natl. Acad. Sci. USA* **2007**, *104*, 18892–18897. [CrossRef]
16. Di Carlo, D. Inertial microfluidics. *Lab Chip* **2009**, *9*, 3038–3046. [CrossRef]
17. Martel, J.M.; Toner, M. Inertial focusing in microfluidics. *Annu. Rev. Biomed. Eng.* **2014**, *16*, 371–396. [CrossRef]
18. Hur, S.C.; Choi, S.-E.; Kwon, S.; Di Carlo, D. Inertial focusing of non-spherical microparticles. *Appl. Phys. Lett.* **2011**, *99*, 044101. [CrossRef]
19. Kuntaegowdanahalli, S.S.; Bhagat, A.A.S.; Kumar, G.; Papautsky, I. Inertial microfluidics for continuous particle separation in spiral microchannels. *Lab Chip* **2009**, *9*, 2973–2980. [CrossRef]
20. Martel, J.M.; Toner, M. Particle Focusing in Curved Microfluidic Channels. *Sci. Rep.* **2013**, *3*, 3340. [CrossRef]
21. Gossett, D.R.; Di Carlo, D. Particle Focusing Mechanisms in Curving Confined Flows. *Anal. Chem.* **2009**, *81*, 8459–8465. [CrossRef] [PubMed]
22. Son, J.; Jafek, A.R.; Carrell, D.T.; Hotaling, J.M.; Gale, B.K. Sperm-like-particle (SLP) behavior in curved microfluidic channels. *Microfluid. Nanofluid.* **2018**, *23*, 4. [CrossRef]
23. Volpe, A.; Paie, P.; Ancona, A.; Osellame, R. Polymeric fully inertial lab-on-a-chip with enhanced-throughput sorting capabilities. *Microfluid. Nanofluid.* **2019**, *23*, 37. [CrossRef]
24. Clark, A.S.; San-Miguel, A. A bioinspired, passive microfluidic lobe filtration system. *Lab Chip* **2021**, *21*, 3762–3774. [CrossRef] [PubMed]
25. Jiang, D.; Ni, C.; Tang, W.; Huang, D.; Xiang, N. Inertial microfluidics in contraction–expansion microchannels: A review. *Biomicrofluidics* **2021**, *15*, 041501. [CrossRef] [PubMed]
26. Kalyan, S.; Torabi, C.; Khoo, H.; Sung, H.; Choi, S.-E.; Wang, W.; Treutler, B.; Kim, D.; Hur, S. Inertial Microfluidics Enabling Clinical Research. *Micromachines* **2021**, *12*, 257. [CrossRef] [PubMed]
27. Volpe, A.; Gaudiuso, C.; Ancona, A. Sorting of Particles Using Inertial Focusing and Laminar Vortex Technology: A Review. *Micromachines* **2019**, *10*, 594. [CrossRef] [PubMed]
28. Jafek, A.; Feng, H.; Brady, H.; Petersen, K.; Chaharlang, M.; Aston, K.; Gale, B.; Jenkins, T.; Samuel, R. An automated instrument for intrauterine insemination sperm preparation. *Sci. Rep.* **2020**, *10*, 21385. [CrossRef]

29. Leshansky, A.M.; Bransky, A.; Korin, N.; Dinnar, U. Tunable nonlinear viscoelastic "focusing" in a microfluidic device. *Phys. Rev. Lett.* **2007**, *98*, 234501. [CrossRef]

30. Lim, E.J.; Ober, T.J.; Edd, F.J.; Desai, S.P.; Neal, D.; Bong, K.W.; Doyle, P.S.; McKinley, G.H.; Toner, M. Inertio-elastic focusing of bioparticles in microchannels at high throughput. *Nat. Commun.* **2014**, *5*, 4120. [CrossRef]

31. Zhou, J.; Papautsky, I. Viscoelastic microfluidics: Progress and challenges. *Microsyst. Nanoeng.* **2020**, *6*, 113. [CrossRef]

32. Li, D.; Lu, X.; Xuan, X. Viscoelastic Separation of Particles by Size in Straight Rectangular Microchannels: A Parametric Study for a Refined Understanding. *Anal. Chem.* **2016**, *88*, 12303–12309. [CrossRef] [PubMed]

33. Lee, D.; Brenner, H.; Youn, J.R.; Song, Y.S. Multiplex Particle Focusing via Hydrodynamic Force in Viscoelastic Fluids. *Sci. Rep.* **2013**, *3*, 3258. [CrossRef] [PubMed]

34. Kim, J.; Lee, J.; Wu, C.; Nam, S.; Di Carlo, D.; Lee, W. Inertial focusing in non-rectangular cross-section microchannels and manipulation of accessible focusing positions. *Lab Chip* **2016**, *16*, 992–1001. [CrossRef] [PubMed]

35. Amini, H.; Lee, W.; Di Carlo, D. Inertial microfluidic physics. *Lab Chip* **2014**, *14*, 2739–2761. [CrossRef] [PubMed]

36. Zhou, J.; Papautsky, I. Fundamentals of inertial focusing in microchannels. *Lab Chip* **2013**, *13*, 1121–1132. [CrossRef] [PubMed]

37. Bhagat, A.A.S.; Kuntaegowdanahalli, S.S.; Papautsky, I. Continuous particle separation in spiral microchannels using dean flows and differential migration. *Lab Chip* **2008**, *8*, 1906–1914. [CrossRef]

38. Norouzi, M.; Biglari, N. An analytical solution for Dean flow in curved ducts with rectangular cross section. *Phys. Fluids* **2013**, *25*, 053602. [CrossRef]

39. Feng, H.; Jafek, A.; Samuel, R.; Hotaling, J.; Jenkins, T.G.; Aston, K.I.; Gale, B.K. High efficiency rare sperm separation from biopsy samples in an inertial focusing device. *Analyst* **2021**, *146*, 3368–3377. [CrossRef]

40. Pipe, C.J.; Majmudar, T.S.; McKinley, G.H. High shear rate viscometry. *Rheol. Acta* **2008**, *47*, 621–642. [CrossRef]

41. Jain, A.; Posner, J.D. Particle Dispersion and Separation Resolution of Pinched Flow Fractionation. *Anal. Chem.* **2008**, *80*, 1641–1648. [CrossRef] [PubMed]

42. Macosko, C.W. *Rheology: Principles, Measurements, and Applications*; Wiley: Hoboken, NJ, USA, 1996.

43. Feng, H.; Magda, J.J.; Gale, B.K. Viscoelastic second normal stress difference dominated multiple-stream particle focusing in microfluidic channels. *Appl. Phys. Lett.* **2019**, *115*, 263702. [CrossRef] [PubMed]

micromachines

MDPI

Article

Electroosmotic Mixing of Non-Newtonian Fluid in a Microchannel with Obstacles and Zeta Potential Heterogeneity

Lanju Mei [1],*, Defu Cui [2], Jiayue Shen [3], Diganta Dutta [4], Willie Brown [1], Lei Zhang [1] and Ibibia K. Dabipi [1]

[1]   Department of Engineering and Aviation Sciences, University of Maryland Eastern Shore, Princess Anne, MD 21853, USA; wlbrown@umes.edu (W.B.); lzhang@umes.edu (L.Z.); ikdabipi@umes.edu (I.K.D.)

[2]   Department of Computational Modeling and Simulation Engineering, Old Dominion University, Norfolk, VA 23529, USA; dcui001@odu.edu

[3]   Department of Engineering Technology, SUNY Polytechnic Institute, Utica, NY 13502, USA; shenj@sunypoly.edu

[4]   Department of Physics and Astronomy, University of Nebraska at Kearney, Kearney, NE 68849, USA; duttad2@unk.edu

*   Correspondence: lmei@umes.edu; Tel.: +1-410-651-6813

Abstract: This paper investigates the electroosmotic micromixing of non-Newtonian fluid in a microchannel with wall-mounted obstacles and surface potential heterogeneity on the obstacle surface. In the numerical simulation, the full model consisting of the Navier–Stokes equations and the Poisson–Nernst–Plank equations are solved for the electroosmotic fluid field, ion transport, and electric field, and the power law model is used to characterize the rheological behavior of the aqueous solution. The mixing performance is investigated under different parameters, such as electric double layer thickness, flow behavior index, obstacle surface zeta potential, obstacle dimension. Due to the zeta potential heterogeneity at the obstacle surface, vortical flow is formed near the obstacle surface, which can significantly improve the mixing efficiency. The results show that, the mixing efficiency can be improved by increasing the obstacle surface zeta potential, the flow behavior index, the obstacle height, the EDL thickness.

Keywords: electroosmotic flow; micromixing performance; heterogeneous surface potential; wall obstacle; power-law fluid

Citation: Mei, L.; Cui, D.; Shen, J.; Dutta, D.; Brown, W.; Zhang, L.; Dabipi, I.K. Electroosmotic Mixing of Non-Newtonian Fluid in a Microchannel with Obstacles and Zeta Potential Heterogeneity. *Micromachines* **2021**, *12*, 431. https://doi.org/10.3390/mi12040431

Academic Editor: Chiara Galletti

Received: 2 March 2021
Accepted: 13 April 2021
Published: 14 April 2021

Publisher's Note: MDPI stays neutral with regard to jurisdictional claims in published maps and institutional affiliations.

1. Introduction

In recent decades, microfluidics has attracted significant attention with its increasing applications in chemical synthesis, biomedical analysis, drug delivery [1–3]. Mixing of species in a microfluidics plays an important role in many of these applications. However, the small scale of microfluidic system leads to a low Reynolds number and laminar flow behavior. The mixing under this situation becomes difficult and, thus, efficient mixing mechanism is a great demand in microfluidic devices [4–7].

Micromixers can be categorized into passive and active micromixers, depending on the actuation mechanism. Active micromixers require use of external energy source, such as pressure [8], acoustics [9], electric field [10], and magnetic field [11]. Passive micromixers, on the other hand, utilize surface structure modification, obstacles or grooves to enhance the mixing of the fluids. Compared to active micromixers, passive micromixers do not need active moving parts and are easier on fabrication and operation [4,12]. Among various passive mixing strategies, electroosmotic flow (EOF), with its flexibility of adjusting the flow patterns by manipulation of surface properties and geometry, is widely used to enhance the mixing performance in microfluidics [6,13,14]. A number of theoretical and experimental studies have been done to improve the mixing efficiency in microchannels by proper design of surface zeta potential [15], surface topology [16], and geometrical configuration [17], etc. Basati et al. [18] investigated the effect of zeta potential distribution and geometrical specifications on the mixing performance of EOF in converging-diverging microchannels. Bhattacharyya et al. [19] studied the vortex formation of combined pressure-driven EOF in a microchannel with a rectangular obstacle on the wall. Wang et al. [20] numerically investigated the vortex formation near a two-part cylinder under an external DC electric field. Chen et al. [21] presented a novel electroosmotic micromixer that consists of arrays of asymmetric electrodes and lateral which can enhance mixing efficiency with applied potential. Seo et al. [22] studied the mixing characteristics in straight microchannel with various obstacle configuration and concluded that the rectangular obstacle shows the most effective mixing enhancement. Many of these theoretical studies on the electrokinetic micromixing in the literature assume Newtonian fluid behavior. However, the biomedical and chemical applications often involve the use of complex solutions (e.g., polymer solution, blood) which exhibit non-Newtonian characteristics. Understanding the mixing performance of the EOF for non-Newtonian fluids is important for the experimental design of efficient micromixers. Various non-Newtonian models have been used to characterize the rheological behavior of the electrokinetically driven complex solution, such as the Carreau–Yasuda model [23], power-law model [24], Oldroyd-B model [25], and generalized Maxwell model [26].

In recent past, several numerical and analytical studies have been performed to investigate the electroosmotic mixing of non-Newtonian fluid in rectangular, cylindrical, and wavy microchannels. To describe the electric potential within the electric double layer (EDL) near the charged surface, the Boltzmann distribution [25,27,28] or the Smoluchowski slip velocity boundary condition [29,30] is commonly used in these studies. Compared to the general Nernst–Planck model, the use of Boltzmann distribution or Smoluchowski slip velocity boundary can reduce the computational effort, but has some limitations [31–33]. On the frame of Nernst–Planck theory, Banerjee et al. numerically investigated the electrokinetic micromixing of power-law fluid both in cylindrical microchannels with surface contraction/expansion [34] and in a wavy patterned microchannel with sinusoidal zeta potential distribution [35]. Mei et al. [36] investigated the EOF of Linear Phan–Thien–Tanner (LPTT) fluid in a nanoslit. To the best knowledge of the authors, on the electroosmotic mixing of a power-law fluid in straight microchannels with rectangular obstacle and surface potential heterogeneity, no study has been done using the Nernst–Planck theory. The mixing in rectangular microchannels is of importance as it can provide very useful information on the design of efficient T/Y-micromixers.

In this study, the full model consisting of Navier–Stokes and Poisson–Nernst–Plank equations is considered to analyze the mixing performance in the microchannel with rectangular obstacle and surface potential heterogeneity. The power-law model is used for non-Newtonian fluid due to its simplicity and the ability to characterize the rheological behavior of non-Newtonian fluids [37]. The paper is organized as follows. In Section 2, the mathematical model describing the electroosmotic mixing of power-law fluid in the microchannel is presented. Section 3 presents the numerical calculation details and validation of our numerical results. In Section 4, effects of the heterogeneous zeta potential, the flow behavior index, the obstacle dimension, and the EDL thickness on the mixing performance of the microchannel are examined in detail. Section 5 concludes the paper.

2. Mathematical Model

Figure 1 illustrates the schematic diagram of the 2D microchannel filled with incompressible KCl electrolyte solution that is driven by an external potential bias V_0 acting along the streamwise direction across the channel. The channel is of height $2H$ and length L, with two obstacles of height H_o and length L_o mounted on the lower and upper wall of the channel. The obstacles are located at a distance of L_1 and L_2 from the inlet, respectively. The channel wall is assumed to be distributed with constant negative zeta potential ζ_c, except on the obstacle surface, where oppositive zeta potential ζ_w is distributed to create surface potential heterogeneity. Two fluid streams containing uncharged sample species of different concentration are injected at the inlet of channel, represented by red and blue arrow/line, respectively. As the fluid flows downstream, the uncharged sample species within these two fluid streams are gradually mixed. Cartesian coordinate system O-*xy* is adopted with *x*-axis in the length direction, *y*-axis in the height direction, and the origin fixed on the bottom corner at the channel inlet.

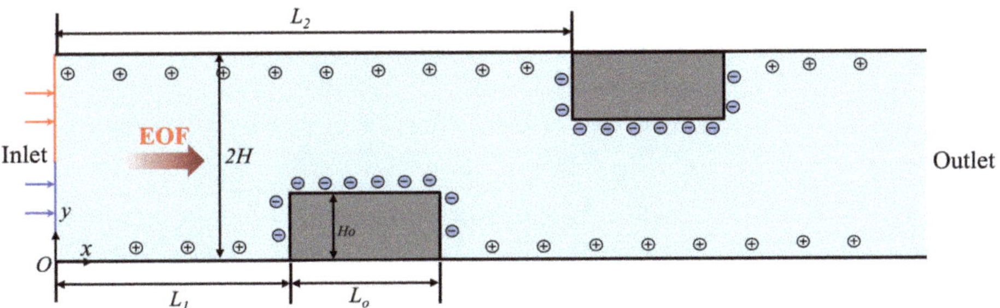

Figure 1. Schematic diagram of the EOF in the microchannel with wall-mounted rectangular obstacles and heterogeneous zeta potential. The EOF is induced by an external potential bias V_0 acting across the channel.

2.1. Governing Equations

The steady-state transport of the non-Newtonian electrolyte solution induced by the external electric field is governed by the mass and momentum conservation equation as:

$$\nabla \cdot \boldsymbol{u} = 0, \tag{1}$$

$$\rho \boldsymbol{u} \cdot \nabla \boldsymbol{u} = -\nabla p + \nabla \cdot (2\mu \boldsymbol{\Gamma}) - \rho_e \nabla \Phi. \tag{2}$$

where $\boldsymbol{u}$ is the velocity field; p denotes the pressure; Φ is the electric potential, and ρ_e is the volume charge density within the electrolyte solution; ρ represents the fluid density; $\boldsymbol{\Gamma} = \left[\nabla \boldsymbol{u} + (\nabla \boldsymbol{u})^T\right]/2$ is the strain rate tensor. The viscosity of the fluid is given by $\mu = m(\Gamma)^{n-1}$ for a power-law fluid, where m is the flow consistency index, n is the flow behavior index, and $\Gamma = \sqrt{\boldsymbol{\Gamma} : \boldsymbol{\Gamma}}$ is the magnitude of the shear rate tensor. It is noted that the shear thinning fluid, Newtonian fluid, and shear thickening fluid correspond to $n < 1$, $n = 1$, and $n > 1$, respectively.

The charged channel surface in contact with the electrolyte solution will develop an electric double layer (EDL) enriched with counterions in the vicinity of the charged surface. The electric potential distribution is determined by the superposition of the external electric potential ψ and induced electric potential ϕ (due to EDL). The electric potential and ion transport within the electrolyte solution are governed by the Laplace equation, Poisson equation, and the Nernst–Planck equation as:

$$-\varepsilon_f \nabla^2 \psi = 0 \tag{3}$$

$$-\varepsilon_f \nabla^2 \varnothing F(c_1 - c_2) \tag{4}$$

$$\nabla \cdot \left(\boldsymbol{u} c_i - D_i \nabla c_i - z_i \frac{D_i}{RT} F c_i \nabla \varnothing \right) = 0, \ i = 1, 2, \tag{5}$$

In the above, ε_f is the permittivity of the electrolyte solution; z_i, D_i, and c_i are the valence, diffusivity, and ionic concentration of ionic species K$^+$ ($i = 1$) and Cl$^-$ ($i = 2$), respectively; F, R, and T are the Faraday constant, gas constant, and absolute temperature, respectively.

The governing equation for the concentration of the uncharged sample species can be obtained from Equation (5), with the corresponding valance set to 0, which results in the convection-diffusion equation as:

$$(u \cdot \nabla)C - D\nabla^2 C = 0 \tag{6}$$

where C represents the concentration of the uncharged species, and D denotes its diffusivity.

2.2. Dimensionless Equations

The dimensionless form of the governing equations is derived in the following. Select the half of the channel height H as length scale, the EOF velocity under constant viscosity $u_0 = \varepsilon_f R^2 T^2 / (\mu_0 H F^2)$ as velocity scale, the constant viscosity μ_0 is the viscosity at $\Gamma = 1\ s^{-1}$ and has the same magnitude as m, ρu_0^2 as the pressure scale, the thermal potential RT/F as electric potential scale, the bulk concentration of the KCl electrolyte C_0 as the ionic concentration scale, the set of governing Equations (1)–(6) can be normalized as:

$$\nabla' \cdot u' = 0, \tag{7}$$

$$u' \cdot \nabla' u' = -\nabla' p' + \frac{1}{Re}\nabla' \cdot (2\mu' \Gamma') - \frac{\mu_0'(kH)^2}{2Re}(c_1' - c_2')\nabla'(\varnothing' + \psi'), \tag{8}$$

$$\nabla^2 \psi' = 0, \tag{9}$$

$$\nabla'^2 \varnothing' = \frac{1}{2}(kH)^2(c_1' - c_2') \tag{10}$$

$$\nabla' \cdot \left(u'c_i' - \frac{D_i}{Hu_0}\nabla' c_i' - \frac{z_i D_i}{Hu_0} c_i' \nabla' \varnothing' \right) = 0,\ i = 1,2. \tag{11}$$

$$(u' \cdot \nabla')C' - \frac{D}{Hu_0}\nabla'^2 C' = 0 \tag{12}$$

In the above, all variables with prime indicate their dimensionless form; the Reynolds number is $= \rho u_0^{2-n} H^n / m$; the dimensionless viscosity constant $\mu_0' = \frac{\mu_0}{m u_0^{n-1} H^{1-n}}$; the Debye length is $\lambda_D = \frac{1}{k} = \sqrt{\varepsilon_f RT / \sum_{i=1}^2 F^2 z_i^2 C_0}$; the dimensionless viscosity is $\mu' = (\Gamma')^{n-1}$.

2.3. Boundary Conditions

To solve for the coupled differential Equations (7)–(12), the boundary conditions are set as following.

On the channel wall, non-slip and no-ion penetration boundary condition is applied as:

$$u' = 0,\ \mathbf{n} \cdot \nabla' \psi' = 0,\ \varnothing' = \zeta,\ \mathbf{n} \cdot (-\nabla' c_i' - z_i c_i' \nabla' \varnothing') = 0,\ \mathbf{n} \cdot \nabla' C' = 0 \tag{13}$$

where $\zeta = \zeta_c$ on the channel wall, and $\zeta = \zeta_w$ on the obstacle surface, $\mathbf{n}$ represents the normal unit vector on the surface.

At the inlet, stress free boundary condition is applied, concentration of the KCl is set as the bulk concentration, and the concentration of the uncharged species follows a step-like concentration distribution, as:

$$\mathbf{n} \cdot \nabla' u' = 0,\ p' = 0,\ \psi' = V_0 \cdot \frac{F}{RT},\ \varnothing' = 0,\ c_1' = c_2' = 1,\ C' = \begin{cases} 1,\ y' > 1 \\ 0,\ y' \leq 1 \end{cases} \tag{14}$$

At the outlet, stress free boundary condition is applied, and the electric potential is set as 0:

$$\mathbf{n} \cdot \nabla' u' = 0,\ p' = 0,\ \psi' = 0,\ \varnothing' = 0,\ c_1' = c_2' = 1,\ \frac{\partial C'}{\partial x\prime} = 0 \tag{15}$$

3. Numerical Method and Validation

The coupled Equations (7)–(12) along with the boundary conditions (13)–(15) are numerical solved using the commercial finite element software COMSOL Multiphysics (version 5.1) with its AC/DC module, CFD module, chemical reaction engineering module, and MUMPS solver. As the flow field and electric field have larger variation within the EDL, finer mesh is distributed near the channel surface and the obstacle surface, and mesh independence study is carried out to ensure the accuracy of the simulation. To further validate the accuracy of the current simulation, we compare our simulation result with Choi et al. [38] who derived the analytical solution of EOF velocity of power-law fluids in a slit microchannel with asymmetric zeta potentials at top and bottom walls. The parameters are set as $V_0 = 1.5$ V, $kH = 15$ $D_1(D_2) = 1.96 \, (2.03) \times 10^{-9}$ m^{2}s^{-1}, $\varepsilon_f = 7.08 \times 10^{-10}$ CV^{-1}m^{-1}, $2H = 10$ µm, $L = 30H$, $m = 10^{-3}$ Pa·s^n, $F = 96,485$ C·mol^{-1}, $R = 8.314$ J·mol^{-1}K^{-1}, $T = 298$ K, zeta potential $\zeta = -10$ mV at the bottom surface and $\zeta = -15$ mV at the top surface. The profile of the dimensionless mainstream velocity component along the middle line $x' = 15$ for different fluid behavior index n is plotted in Figure 2. Here the velocity is scaled by $u_s = nk^{\frac{1}{n}-1}\left(\frac{\varepsilon_f V_0 \zeta_m}{mL}\right)^{\frac{1}{n}}$ with ζ_m being the average zeta potential of the top and bottom surface zeta potential. The results show that the velocity increases rapidly near the wall within the EDL, and the gradient of velocity is larger near the top surface due to the larger zeta potential. The dimensionless velocity decreases with increasing fluid behavior index n, due to the overall increased viscosity. It can be seen that the EOF velocity for power-law fluid under asymmetric zeta potential from the current simulation matches well with the analytical solution of Choi et al. [38]. In the following simulations, the parameters are set as $V_0 = 1$ V, $kH = 10$, $L = 20H$, $L_1 = 6H$, $L_2 = 10H$, $L_o = 2H$, $H_o = 0.2H$, $\zeta_c = -20$ mV, $\zeta_w = 20$ mV, other parameters are set as mentioned above unless otherwise specified.

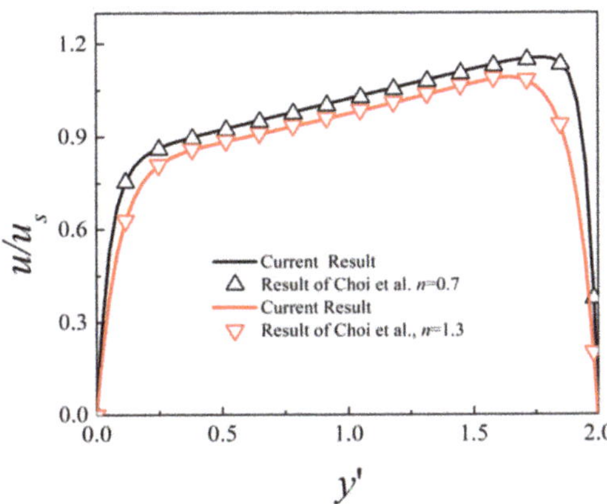

Figure 2. Dimensionless velocity distributions u/u_s within the microchannel of asymmetric zeta potentials on the walls for fluid behavior index $n = 0.7$ and $n = 1.3$: lines (current simulation result) and symbols (analytical result of Choi et al. [38].

To characterize the mixing performance within the microchannel, the mixing efficiency of the uncharged species is defined as:

$$\eta(x') = \left[1 - \frac{\int_{y'\_bottom}^{y'\_top}|C' - C_\infty|dy'}{\int_{y'\_bottom}^{y'\_top}|C_0 - C_\infty|dy'}\right] \times 100\%, \tag{16}$$

where $C_\infty = 0.5$ and $C_0 = 0$ (*or* 1) are the fully mixed concentration and totally unmixed concentration, respectively.

4. Results and Discussion

4.1. Effect of Obstacle Surface Zeta Potential

First of all, the effect of the obstacle surface zeta potential on the mixing performance for the fixed geometry is examined. Figure 3 presents the contour plot of the elute species concentration $C\prime$ in the microchannel and the concentration profile at the channel outlet for obstacle surface zeta potential $\zeta_w = 20$ mV, 40 mV, and 60 mV, respectively. For higher heterogeneous zeta potential ζ_w, significant improvement of mixing is achieved after the fluid flows past the obstacle. The distribution of concentration $C\prime$ at the outlet shows that the species approaches uniform distribution as ζ_w increases, revealing better mixing performance. The corresponding velocity contour and flow streamlines are plotted in Figure 4 to analyze the effect of the obstacle and zeta potential on the flow field. It can be observed that in the straight part away from the obstacles, streamlines are parallel to the channel surface. In the region where the obstacle is present, the streamlines are distorted and vortex is formed in the vicinity of the obstacle surface. The positive zeta potential at the obstacle surface induces the negative mainstream velocity near the surface, which in turn results in the vortex formation. The velocity profile along the cross-sectional line located at the center of the first obstacle (i.e., $x' = 7$) is shown in Figure 4b. It shows that as the magnitude of the zeta potential at the obstacle surface increases, the backward velocity near the surface becomes much larger, the vortex becomes stronger and the vortex center moves towards the centerline of the microchannel, which contributes to the better mixing performance of the elute species. Figure 5 further plots the variation of the mixing efficiency along the channel length direction and the dependency of the mixing efficiency at the outlet on the obstacle surface zeta potential ζ_w. It clearly shows that at the fixed zeta potential ζ_w, significant improvement of the mixing efficiency occurs right after the fluid flows past each obstacle. The mixing efficiency at the channel outlet monotonously increases with the heterogeneous zeta potential. The mixing efficiency at the outlet for $\zeta_w = 70$ mV is 2.7 times that of $\zeta_w = 0$ mV.

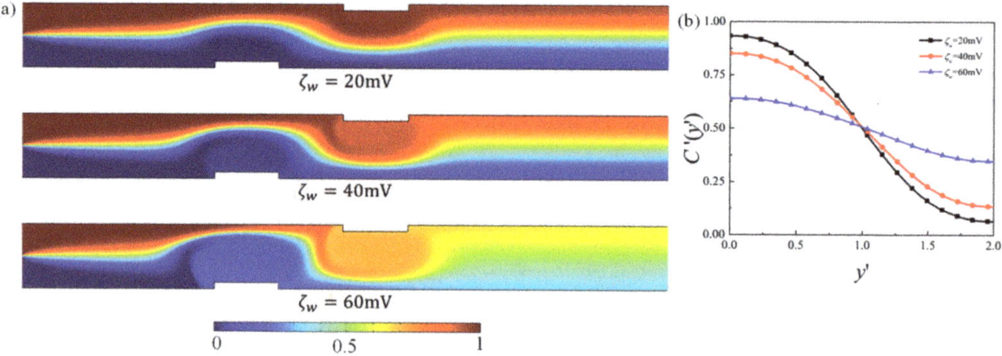

Figure 3. (a) Contour plot of elute species concentration C' in the microchannel, (b) distribution of species concentration at the outlet of the microchannel, for different obstacle surface zeta potential $\zeta_w = 20$ mV, 40 mV, and 60 mV.

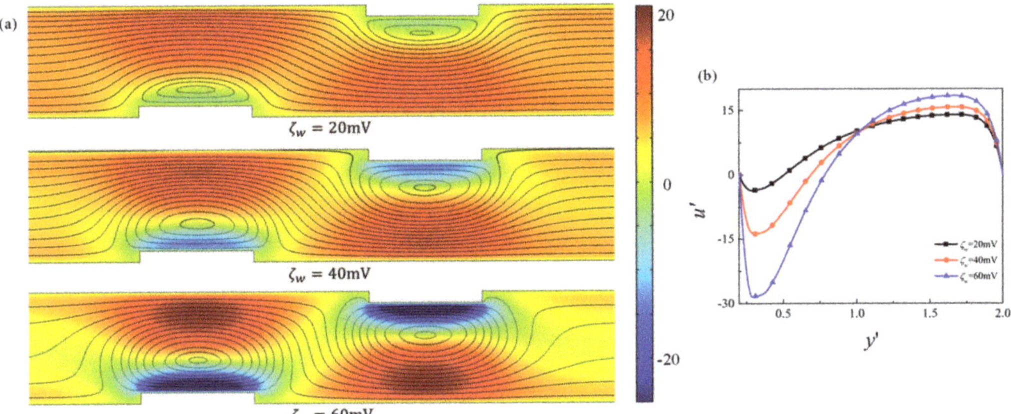

Figure 4. (**a**) Velocity contour and streamlines for different obstacle surface zeta potential, (**b**) mainstream velocity component u' along the cross section located at the center of the first obstacle $x' = 7$, for $\zeta_w = 20$ mV, 40 mV, and 60 mV.

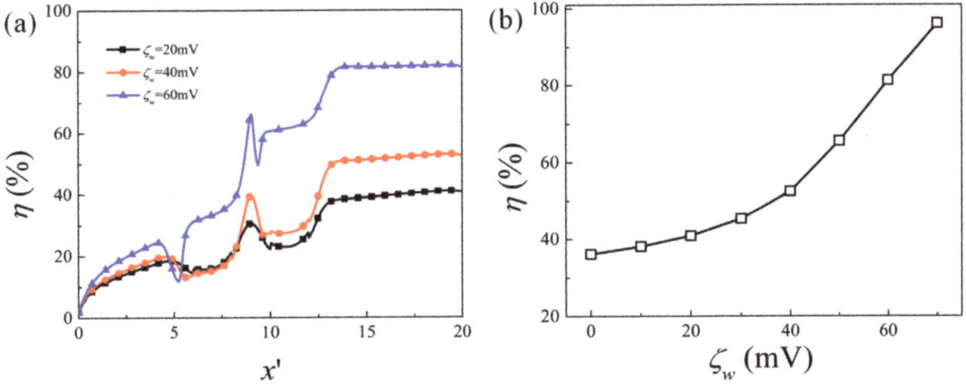

Figure 5. (**a**) Evolution of the mixing efficiency along the channel length direction for different obstacle surface zeta potential; (**b**) the variation of mixing efficiency at the outlet with obstacle surface zeta potential ζ_w.

4.2. Effect of Flow Behavior Index

Figure 6 presents the mixing efficiency along the channel length direction and mainstream velocity component u' along the cross-section located at the center of the first obstacle ($x' = 7$) for various flow behavior index n. It is obvious that the mixing efficiency becomes much higher for larger value of n. Under the same condition, the shear thickening fluid has better mixing performance than the Newtonian fluid ($n = 1$), and the shear thinning fluid has lower mixing efficiency. The mainstream velocity component is negative near the surface of the obstacle due to the positive zeta potential, and is positive within a large portion of the channel. The velocity varies more steeply near the wall and the overall velocity magnitude is much larger for a smaller value of n. The dimensionless flow rate $Q' = \int_{0.2}^{2} u' dy'$ is 14.4, 3.0, 1.8, and 0.8 for $n = 0.7, 0.9, 1.0$, and 1.2, respectively. The result is consistent with that from Banerjee et al. [34]. This is because, under the fixed shear rate, the viscosity for the power-law fluid is larger for higher n, which results in the lower velocity under the same electric condition. When the velocity is smaller, the solute species can get more diffusion flux to improve downstream mixing and, thus, better mixing performance is achieved for higher n.

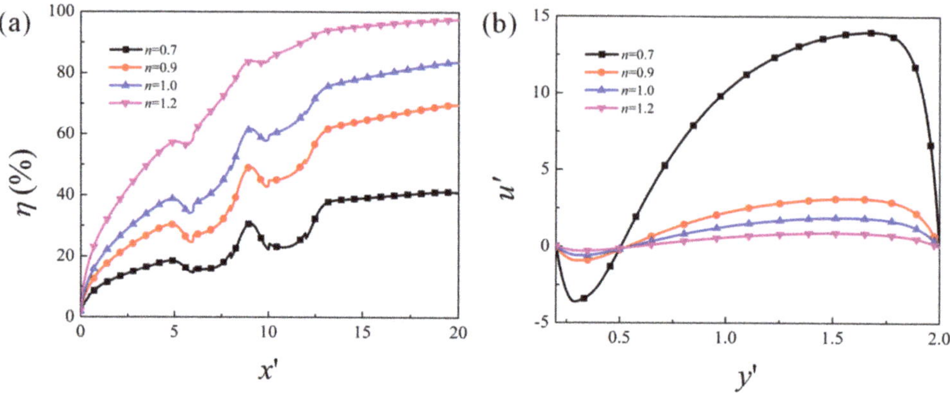

Figure 6. (**a**) Evolution of the mixing efficiency along the channel length direction for different flow behavior index n; (**b**) mainstream velocity component u' along the cross section located at the center of the first obstacle $x' = 7$.

4.3. Effect of Obstacle Height

The effect of the obstacle height on the mixing efficiency is presented in Figure 7, where the evolution of the mixing efficiency along the channel length direction and the variation of mixing efficiency at the outlet as a function of the obstacle height H_o are plotted. It can be seen that as the obstacle height becomes larger, the mixing performance is better. Compared to the microchannel without obstacle, the presence of the obstacle can improve the mixing performance very effectively. The mixing efficiency at the outlet for $H_o = 0.7H$ is 2.2 times of that without obstacle (i.e., $H_o = 0$).

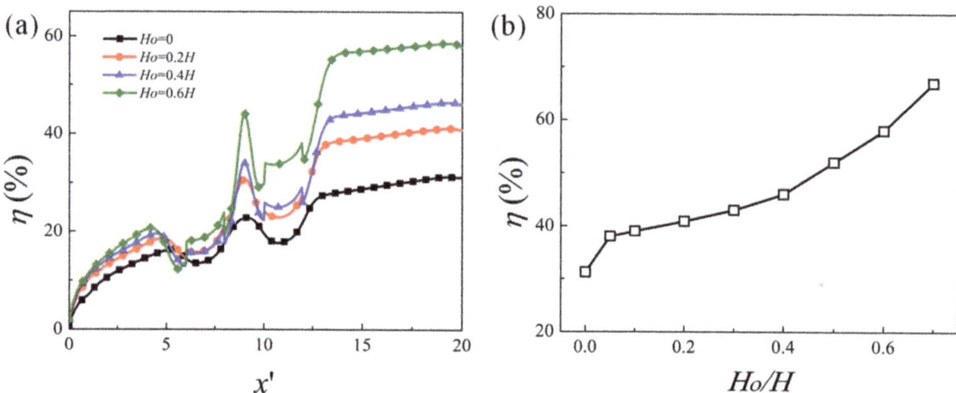

Figure 7. (**a**) Evolution of the mixing efficiency along the channel length direction for obstacle height $H_o = 0, 0.2H, 0.4H,$ and $0.6H$, (**b**) the dependence of the mixing efficiency at the outlet as a function of ratio H_o/H.

4.4. Effect of EDL Thickness

Finally, the effect of EDL thickness on the mixing performance is presented in Figure 8, where the variation of mixing efficiency along the channel length direction for EDL thickness $kH = 5, 20,$ and 50 is plotted. It can be seen that the mixing efficiency is slightly higher for larger EDL thickness (i.e., $kH = 5$) than that of thin EDL thickness (i.e., $kH = 20$ and 50).

As shown in the cross-sectional mainstream velocity profile in Figure 8b, when the EDL thickness is comparable to the channel height, which is true when the electrolyte concentration is low, the change of the velocity near the wall is small, and the overall velocity is relatively small. This means that the vortex near the obstacle surface is weaker. When the EDL is very thin (e.g., $kH = 50$), the gradient of the velocity is very large in the vicinity of the wall, and the velocity is much larger than that of the large EDL thickness. The effect of EDL thickness on the EOF velocity is consistent with that in the literature, and has been well explained [36,39]. When the velocity is smaller, the diffusion effect becomes stronger, which results in a slight increase in the mixing efficiency.

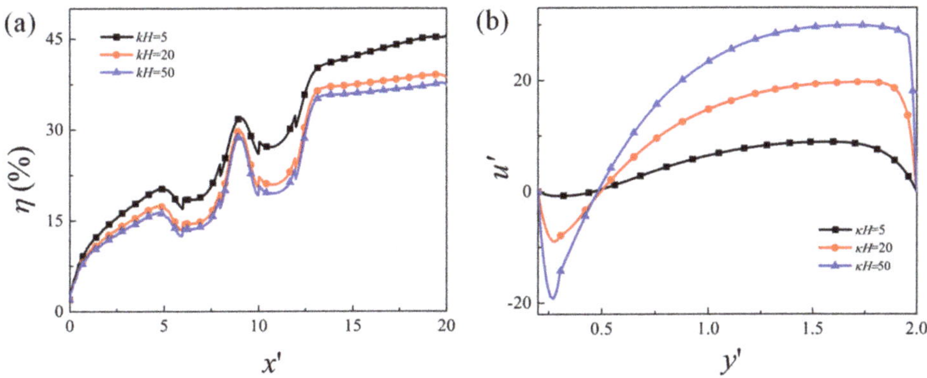

Figure 8. (**a**) Evolution of the mixing efficiency along the channel length direction, (**b**) mainstream velocity component u' along the cross-section located at the center of the first obstacle $x' = 7$ for EDL thickness: $kH = 5$, 20, and 50.

5. Conclusions

In this study, the steady-state mixing performance of electroosmotic flow of the power-law fluid is numerically investigated in a 2D microchannel with wall mounted obstacles and heterogeneous zeta potential. The numerical simulation is based on the full model consisting of the Poisson–Nernst–Planck and Navier–Stokes equations. Compared to the channel without obstacle, the presence of the obstacle can significantly increase the mixing efficiency. By increasing the obstacle height, the mixing efficiency can be further improved. The heterogeneous zeta potential on the obstacle surface induces vortical flow in the vicinity of the obstacle surface, and the vortex strength becomes stronger as the zeta potential increases, which results in the improvement of the mixing performance. Additionally, for larger behavior index of the power-law fluid, velocity becomes smaller, the transport of the uncharged species becomes diffusion dominant, resulting in better mixing performance. For relatively large EDL thickness, the variation of velocity near the surface is smaller, and the mixing efficiency is slightly higher than that of thin EDL thickness due to the overall lower velocity.

Author Contributions: L.M. developed the model, carried out the numerical simulation, and prepared the manuscript. D.C., J.S. and D.D. conducted the literature review and revised the paper. W.B., L.Z. and I.K.D. contributed to the review and revision. All authors have read and agreed to the published version of the manuscript.

Funding: This research received no external funding.

Conflicts of Interest: The authors declare no conflict of interest.

References

1. Karniadakis, G.; Beskok, A.; Aluru, N. *Microflows and Nanoflows: Fundamentals and Simulation*; Springer Science & Business Media: Berlin/Heidelberg, Germany, 2006; Volume 29.
2. Hsu, W.-L.; Daiguji, H. Manipulation of Protein Translocation through Nanopores by Flow Field Control and Application to Nanopore Sensors. *Anal. Chem.* **2016**, *88*, 9251–9258. [CrossRef] [PubMed]
3. Chen, X.; Zhang, S.; Zhang, L.; Yao, Z.; Chen, X.; Zheng, Y.; Liu, Y. Applications and theory of electrokinetic enrichment in micro-nanofluidic chips. *Biomed. Microdevices* **2017**, *19*, 19. [CrossRef] [PubMed]
4. Suh, Y.K.; Kang, S. A Review on Mixing in Microfluidics. *Micromachines* **2010**, *1*, 82–111. [CrossRef]
5. Chang, C.-C.; Yang, R.-J. Electrokinetic mixing in microfluidic systems. *Microfluid. Nanofluid.* **2007**, *3*, 501–525. [CrossRef]
6. Rashidi, S.; Bafekr, H.; Valipour, M.S.; Esfahani, J.A. A review on the application, simulation, and experiment of the electrokinetic mixers. *Chem. Eng. Process. Process. Intensif.* **2018**, *126*, 108–122. [CrossRef]
7. Johnson, T.J.; Ross, A.D.; Locascio, L.E. Rapid Microfluidic Mixing. *Anal. Chem.* **2002**, *74*, 45–51. [CrossRef] [PubMed]
8. Glasgow, I.; Aubry, N. Enhancement of microfluidic mixing using time pulsing. *Lab Chip* **2003**, *3*, 114–120. [CrossRef]
9. Shang, X.; Huang, X.; Yang, C. Vortex generation and control in a microfluidic chamber with actuations. *Phys. Fluids* **2016**, *28*, 122001. [CrossRef]
10. Daghighi, Y.; Li, D. Numerical study of a novel induced-charge electrokinetic micro-mixer. *Anal. Chim. Acta* **2013**, *763*, 28–37. [CrossRef]
11. Qian, S.; Bau, H.H. Magneto-hydrodynamics based microfluidics. *Mech. Res. Commun.* **2009**, *36*, 10–21. [CrossRef]
12. Cai, G.; Xue, L.; Zhang, H.; Lin, J. A Review on Micromixers. *Micromachines* **2017**, *8*, 274. [CrossRef]
13. Nayak, A. Analysis of mixing for electroosmotic flow in micro/nano channels with heterogeneous surface potential. *Int. J. Heat Mass Transf.* **2014**, *75*, 135–144. [CrossRef]
14. Wu, H.-Y.; Liu, C.-H. A novel electrokinetic micromixer. *Sens. Actuators A Phys.* **2005**, *118*, 107–115.
15. Kim, H.; Khan, A.I.; Dutta, P. Time-Periodic Electro-Osmotic Flow with Nonuniform Surface Charges. *J. Fluids Eng.* **2019**, *141*, 081201. [CrossRef]
16. Hu, Y.; Werner, C.; Li, D. Influence of the three-dimensional heterogeneous roughness on electrokinetic transport in microchannels. *J. Colloid Interface Sci.* **2004**, *280*, 527–536. [CrossRef]
17. Wu, Z.; Li, D. Mixing and flow regulating by induced-charge electrokinetic flow in a microchannel with a pair of conducting triangle hurdles. *Microfluid. Nanofluid.* **2007**, *5*, 65–76. [CrossRef]
18. Basati, Y.; Mohammadipour, O.R.; Niazmand, H. Numerical investigation and simultaneous optimization of geometry and zeta-potential in electroosmotic mixing flows. *Int. J. Heat Mass Transf.* **2019**, *133*, 786–799. [CrossRef]
19. Bhattacharyya, S.; Bera, S. Combined electroosmosis-pressure driven flow and mixing in a microchannel with surface heterogeneity. *Appl. Math. Model.* **2015**, *39*, 4337–4350. [CrossRef]
20. Wang, C.; Song, Y.; Pan, X. Electrokinetic-vortex formation near a two-part cylinder with same-sign zeta potentials in a straight microchannel. *Electrophoresis* **2020**, *41*, 793–801. [CrossRef]
21. Chen, L.; Deng, Y.; Zhou, T.; Pan, H.; Liu, Z. A Novel Electroosmotic Micromixer with Asymmetric Lateral Structures and DC Electrode Arrays. *Micromachines* **2017**, *8*, 105. [CrossRef]
22. Seo, H.-S.; Kim, Y.-J. A study on the mixing characteristics in a hybrid type microchannel with various obstacle configurations. *Mater. Res. Bull.* **2012**, *47*, 2948–2951. [CrossRef]
23. Ko, C.; Li, D.; Malekanfard, A.; Wang, Y.; Fu, L.; Xuan, X. Electroosmotic flow of non-Newtonian fluids in a constriction microchannel. *Electrophoresis* **2019**, *40*, 1387–1394. [CrossRef]
24. Cho, C.-C.; Chen, C.-L.; Chen, C.-K. Mixing enhancement of electrokinetically-driven non-Newtonian fluids in microchannel with patterned blocks. *Chem. Eng. J.* **2012**, *191*, 132–140. [CrossRef]
25. Mahapatra, B.; Bandopadhyay, A. Electroosmosis of a viscoelastic fluid over non-uniformly charged surfaces: Effect of fluid relaxation and retardation time. *Phys. Fluids* **2020**, *32*, 032005. [CrossRef]
26. Li, X.-X.; Yin, Z.; Jian, Y.-J.; Chang, L.; Su, J.; Liu, Q.-S. Transient electro-osmotic flow of generalized Maxwell fluids through a microchannel. *J. Non-Newtonian Fluid Mech.* **2012**, *187–188*, 43–47. [CrossRef]
27. Zhao, C.; Zholkovskij, E.; Masliyah, J.H.; Yang, C. Analysis of electroosmotic flow of power-law fluids in a slit microchannel. *J. Colloid Interface Sci.* **2008**, *326*, 503–510. [CrossRef]
28. Zhu, Q.; Deng, S.; Chen, Y. Periodical pressure-driven electrokinetic flow of power-law fluids through a rectangular microchannel. *J. Non-Newtonian Fluid Mech.* **2014**, *203*, 38–50. [CrossRef]
29. Ng, C.-O.; Qi, C. Electroosmotic flow of a power-law fluid in a non-uniform microchannel. *J. Non-Newtonian Fluid Mech.* **2014**, *208–209*, 118–125. [CrossRef]
30. Sailaja, A.; Srinivas, B.; Sreedhar, I. Electroviscous Effect of Power Law Fluids in a Slit Microchannel with Asymmetric Wall Zeta Potentials. *J. Mech.* **2019**, *35*, 537–547. [CrossRef]
31. Masliyah, J.H.; Bhattacharjee, S. *Electrokinetic and Colloid Transport Phenomena*; Wiley: New York, NY, USA, 2006.
32. Craven, T.J.; Rees, J.M.; Zimmerman, W.B. On slip velocity boundary conditions for electroosmotic flow near sharp corners. *Phys. Fluids* **2008**, *20*, 43603. [CrossRef]
33. Park, H.; Lee, J.; Kim, T. Comparison of the Nernst–Planck model and the Poisson–Boltzmann model for electroosmotic flows in microchannels. *J. Colloid Interface Sci.* **2007**, *315*, 731–739. [CrossRef] [PubMed]

34. Banerjee, A.; Nayak, A.K.; Weigand, B. A Comparative Analysis of Mixing Performance of Power-Law Fluid in Cylindrical Microchannels with Sudden Contraction/Expansion. *J. Fluids Eng.* **2020**, *142*. [CrossRef]
35. Banerjee, A.; Nayak, A. Influence of varying zeta potential on non-Newtonian flow mixing in a wavy patterned microchannel. *J. Non-Newtonian Fluid Mech.* **2019**, *269*, 17–27. [CrossRef]
36. Mei, L.; Zhang, H.; Meng, H.; Qian, S. Electroosmotic Flow of Viscoelastic Fluid in a Nanoslit. *Micromachines* **2018**, *9*, 155. [CrossRef]
37. Zhao, C.; Yang, C. Electrokinetics of non-Newtonian fluids: A review. *Adv. Colloid Interface Sci.* **2013**, *201–202*, 94–108. [CrossRef]
38. Choi, D.-S.; Yun, S.; Choi, W. An Exact Solution for Power-Law Fluids in a Slit Microchannel with Different Zeta Potentials under Electroosmotic Forces. *Micromachines* **2018**, *9*, 504. [CrossRef]
39. Mei, L.; Chou, T.H.; Cheng, Y.S.; Huang, M.J.; Yeh, L.H.; Qian, S. Electrophoresis of pH-regulated nanoparticles: Impact of the Stern layer. *Phys. Chem. Chem. Phys.* **2016**, *18*, 9927–9934. [CrossRef]

MDPI

St. Alban-Anlage 66

4052 Basel

Switzerland

Tel. +41 61 683 77 34

Fax +41 61 302 89 18

www.mdpi.com

Micromachines Editorial Office

E-mail: micromachines@mdpi.com

www.mdpi.com/journal/micromachines